D0699910

VEGETATION AND SOILS

A World Picture

TO
MY WIFE

Vegetation and Soils

A WORLD PICTURE

by

S. R. EYRE
B.Sc., Ph.D.

Second edition

EDWARD ARNOLD

© S. R. EYRE 1968

First published 1963
by Edward Arnold (Publishers) Ltd.,
41 Maddox Street, London W1R 0AN
Reprinted 1963
Reprinted 1964
Reprinted 1966
Second edition 1968
Reprinted 1968
Reprinted 1970
Reprinted 1971

ISBN : 0 7131 5129 3

Printed and bound in Great Britain by
The Camelot Press Ltd., London and Southampton

Preface

THIS book contains nothing which is beyond the understanding of anyone who has profited by the liberal education provided in a good secondary school. It assumes no basic scientific knowledge beyond the elementary chemistry and botany which can be gained from standard courses in general science and general biology; it demands no factual information about geology and climate beyond that which can be found in general text books on the physical aspects of geography. Provided the reader is conversant with such terms as 'molecule', 'transpiration' and 'evolution' no difficulties in terminology should be encounted. Although more specialized terms are introduced, every effort has been made to make their meaning clear from the context in which they occur; furthermore, a glossary has been appended containing all those technical terms which appear in italics in the text.

The requirements of students in upper sixth forms and in their first year at college and university have been borne particularly in mind. Essentially the book is intended as a contribution to the study of the physical aspects of geography; it should therefore be of particular interest to all who embark upon advanced, diploma or degree courses in geography. It is hoped however that students of botany and agriculture also will find here an ecological view of the world which will provide some intellectual stimulus and perspective.

Although this is a mere introduction to the study of vegetation and soil distribution, an effort has been made to avoid over-generalisation and scientific untruth. Too often in the past have the reasons for these distributions been much over-simplified and it is partly because of this that climatic correlations receive little mention here. Although climate is of such obvious importance in soil and vegetation development, its relationships with them are usually very complicated. Consequently, there is always too much room for misunderstanding when exact values of temperature and precipitation are mentioned in the same context as biotic distributions; the reader may be tempted to assume that the latter are determined by the former. Systematic treatment of climate is limited here to an appendix in which some mean monthly precipitation and temperature

figures are tabulated for two or three stations within the area occupied by
each main vegetation type. It will be found that these suggest a climatic
diversity over the terrain occupied by many plant formations rather than
the reverse. Some discussion of vegetation-climatic relationships will be
found in a number of places but the aim is to explore their complexity
rather than to arrive at facile explanations.

The main aim is to describe and outline the distributions of two of the
most important elements in the landscape; other elements are only
referred to insofar as they are relevant to an explanation of these distributions. The treatment adopted is a regional one; the land areas of the
earth are sub-divided according to the distribution of the main vegetation
types upon them and the vegetation and soil of each region are viewed
together. This approach seems reasonable and appropriate although,
again, soil-vegetation correlations must not be over-stressed. The nature
of the soil is not *determined by* vegetation any more than it is *determined
by* climate; the only point being made is that the nature of the vegetation
always has *some bearing on* the nature of the soil and *vice versa*. Vegetation
development and soil development are intimately connected.

Since the majority of readers will only have a first-hand knowledge of
countries outside the tropics, the vegetation and soils of middle and high
latitudes are examined before those of tropical regions. It is hoped that
this progression from the familiar to the unfamiliar will meet with
approval. In order to supply some corrective for any misconceptions
which might arise from the broad generalisations in Parts II and IV, three
chapters have been included on the vegetation and soils of a relatively
small area. No matter how realistic one tries to be, it is very difficult
indeed to present an accurate picture when dealing exclusively with vast
regions; there is always a tendency to give an impression of homogeneity
over large areas where, in fact, there is much variety and contrast. The
British Isles have been treated in some detail here in order to illustrate the
fact that vegetation and soil maps on a continental scale can be most
misleading. Some apology for this selection of area must be made to
readers in other countries; the British Isles have been selected to illustrate
this point merely because they are better known to me than any other
relatively small part of the world. Nevertheless, the point can be made
that the variety of the original vegetation here and the complexity caused
by protracted human occupance, make these islands an admirable subject
for more intensive study.

Section I has been devoted entirely to a study of the ways in which plant
communities and soil profiles develop. This is by far the most technical
part of the book, but it is hoped that it will not be neglected because of
this. For those who require regional information there will be the obvious
temptation to ignore this systematic introduction. It must be stressed
however that only an imperfect understanding of the regional sections

can be achieved if these earlier chapters have not been digested first. Not only are most of the technical terms explained here but also the nature of some salient ecological relationships.

In order to satisfy those readers who require a deeper insight into the nature of phenomena or more detailed regional information, a bibliography has been provided. Several major works of reference on soils, plant geography and plant ecology are included here, but the majority of the works cited have been selected because of their regional or topical value. The reader is referred to a selection of books and articles which, between them, cover most of the vegetation types of the earth and the majority of the more important topics which have been explored in the text.

Finally, a series of vegetation maps of the continents has been appended showing, as realistically as possible, the distribution of plant formations. Although constant reference is not made to these maps in the text, they have been devised so as to be of maximum help to the reader. The vegetation classification upon which the maps are based is exactly the same as the one which is outlined in the text, so that the formation boundaries described there can be traced with some precision.

S. R. EYRE.

University of Leeds.
February, 1962.

Preface to Second Edition

OVER the past five years reviewers and correspondents have made a number of helpful criticisms and suggestions for which the author has been most grateful. In the light of the advice received and the results of more recent research, some much needed amendments have now been made. In particular, the brief introduction to pedology in Chapter III has been subjected to close scrutiny and, although it has not been thought necessary to recast the material completely, much of it has been expressed in a different way. It is hoped that this will meet with the approval of specialists in this particular field and, at the same time, provide a clear introduction for the type of reader for whom the text as a whole was written. Substantial substitutions of material have also been made in a number of other chapters where newly published research, or misleading statements in the first edition, have rendered this desirable.

Only minor alterations have been made on the maps in Appendix I but, in response to insistent demands for wider documentation, the bibliography has been widened considerably. When this work was first devised, it was envisaged as an introductory text for students in sixth forms or in their first year at college and university, for whom teachers and tutors would provide the appropriate guidance for further reading. Experience has shown that many readers have regarded it more as a work of reference than a mere introduction, and an attempt must now be made to meet their requirements.

S. R. EYRE.

University of Leeds.
July, 1967.

Acknowledgements

I WISH to express my sincere thanks to all those colleagues and teachers, past and present, who have contributed so substantially to the more commendable material that may be found in this book. For all the intellectual stimulation that they have provided I am most grateful. My particular thanks are due to some of my present colleagues in the University of Leeds; although they must in no way be held responsible for any deficiencies, the advice and help that they have given so generously has added much to the value of what is written here. Dr. D. D. Bartley of the Department of Botany and Mr. J. Palmer of the Department of Geography have been so kind as to read through Chapters VII and VIII respectively, while Mr. G. R. J. Jones of the Department of Geography and Mr. W. N. Townsend of the Department of Agriculture have, on numerous occasions, given me the benefit of their knowledge and opinions.

I am also much indebted to Mr. A. F. Braithwaite, Mr. R. Holliday, Mr. J. Palmer, Professor R. F. Peel and Mr. J. Radley for their permission to use original photographs, and to Mr. G. Bryant for all his valuable suggestions regarding the maps and diagrams.

Many of the plates and figures used here are reproductions or adaptations from published sources and have been acknowledged individually.

Contents

CONCLUSION

List of Plates

Figures in the Text

PART ONE

VEGETATION AND SOIL
DEVELOPMENT

CHAPTER I

The Original Landscape and the Hand of Man

In the first chapter of the Book of Genesis one may read how the dry land emerged from the oceans and became populated with plants and animals and how, ultimately, man appeared on the scene to use the landscape for his own purposes. Although it has now been demonstrated that all this did not take place within the space of a mere seven days, it is clear that the ancients of Babylonia had a clear and almost inspired vision of the general order of world evolution. As the initially hot earth cooled down, one of the gases in its atmospheric envelope began to condense and liquid water accumulated on the lower portions of the surface of the globe; the higher portions remained above water-level. The oceans and continents thus came into existence. The remaining atmosphere continued to circulate around the earth, abstracting water from the oceans and depositing it on the continents as rain. The sun shone between the rains and warmed the land surfaces. In this warm, moist and relatively stable environment, terrestrial plants and animals evolved and multiplied. The plants rooted themselves firmly in the weathered surface of the land and their dead remains supplied it with organic material. True soils thus came into existence.

Because of mountain-building, erosion, sedimentation and other geomorphological processes, the land surfaces increased in complexity. A great variety of rocks made their appearance and the soils which developed in their weathering residues were very different. Mountain ranges formed climatic divides and precipitation and temperature varied greatly from one place to another. In consequence the vegetation of the earth, so dependent upon soil, rainfall and sunshine, evolved very differently in different regions. In turn, the various species of animal, dependent upon different plants for shelter and food, accommodated themselves to this varied vegetation pattern.

The original untamed landscape thus varied enormously from place to place. Everywhere however it was compounded of the same interdependent components (Fig. 1). Climate exercises a very strict and sometimes direct control on the development and distribution of plants, animals and soils; all species have a specific range of moisture and heat requirements and the rates of weathering, organic decay and leaching in soils are strongly influenced by the same climatic elements. On the other hand, climate itself is profoundly affected by the nature of the vegetation cover;

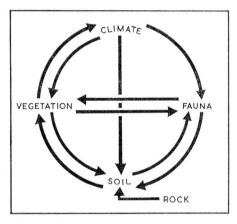

Fig. 1. The ecosystem or biotic complex.

wind speed, temperature régime and atmospheric humidity just above the surface of the ground beneath the shade of dense forest, are remarkably different from those at just the same height above soil level in near-by grassland. Vegetation controls the distribution of animals most rigorously; each animal must remain within reach of the plants upon which it feeds or, in the case of a carnivore, within reach of its prey which, in turn, is controlled in its distribution by vegetation. Conversely, the fauna exerts some effect on vegetation. Large herds of wild ungulates, because of their selective grazing, must favour some plants at the expense of others. Similarly there are reciprocal relationships between soil and vegetation and between soil and fauna.

This complex of interacting phenomena has frequently been referred to as the 'biotic complex' or 'ecosystem'. Any change affecting any single element within it must obviously have repercussions throughout the entire system. For hundreds of millions of years all such changes were due entirely to the operation of the 'blind forces of nature'; changes in climate occurred from time to time while, now and then, plant and animal evolution produced new species which fundamentally altered the structure and general appearance of the whole complex.

It was into this wild complex that modern man (*Homo sapiens*) intruded only a relatively short time ago. 'The Garden of Eden' awaiting him was thus a good deal wilder and more intransigent than the Scriptures imply. Man wandered in it for a long time with no thought of cultivation; he merely gathered fruits, leaves and roots, and preyed upon other animals. He thus modified his environment but little, probably affecting soil and vegetation no more than many other species of animal. Because of his potentialities for reasoning and intellectual development, however, *Homo sapiens* gradually changed his habits and increased his power over the other denizens of the primeval environment and over the forest habitat itself.

Ultimately, as his numbers increased and his technical skill became greater, man was able to sweep away much of the wild vegetation. In its place he planted orderly patches of those species which he had come to esteem most highly as food. More than that, he also changed the vegetation over even vaster areas by spreading fires and by pasturing those few species of grazing animal which he had domesticated. Only those plant species which could withstand firing and grazing were able to survive in these pastures. Directly and indirectly man also transformed the soil as well as the vegetation; he plied his spade and his plough in order to plant crops, thus destroying many of the natural features of the soil at one fell swoop. Furthermore, by gradually altering the vegetation on the grazing lands, he automatically, but more gradually, changed the soil over these wider areas.

During the very latest stage of his social evolution, man has wrought even more drastic changes in his environment. Over large areas, not only has the vegetation been almost completely eliminated, the soil also has disappeared beneath a veneer of asphalt, concrete, brick and tile. Here, the descendants of the hunters and gatherers of the primeval forest have continued to develop their skills as thinkers and toolmakers. In their millions they now pass their working days as mechanics, clerks and teachers, the majority of which never have occasion to think about the natural complex in which mankind evolved. On the other hand, many further millions are still engaged in using what remains of the original natural environment as the sole source of food and clothing for the whole of our species.

It is obvious that the evolution of human societies and the evolution of vegetation and soil over the past few thousand years must have been closely inter-related in many areas. The present landscape can be regarded as the product of two sets of forces; on the one hand there are the natural physical and biotic processes[1] and, on the other, the activities of human communities. One of the most rewarding approaches to geography is to view it as the study of the ways in which the present landscape has been

---

[1] The biotic complex with its geological foundations has often been referred to as 'the physical basis of geography'.

produced by the interaction between these two sets of forces. Unfortunately the nature of the interaction has frequently been gravely oversimplified; it has too often been assumed that the physical aspects of geography can be regarded as an almost static framework within which the human drama has been enacted. This has led to many misconceptions. The point has already been made that, quite apart from the effects of changes in the macroclimate, man's removal or modification of the original vegetation must have affected all the other elements in the biotic complex. Soil, microclimate and water-availability have all been modified enormously; even the relief of the surface has sometimes been significantly changed by accelerated soil erosion. One should therefore never assume that the natural attributes of the landscape are just the same today as they were before man's intervention; they may have changed out of all recognition. If one seeks to interpret the distribution of long-established human settlement in terms of the physical potentialities of the environment, the task is therefore far from easy. It is first necessary to reconstruct an image of the physical basis as it was at the time of the founding of the settlement. In subsequent chapters some attempt has been made to effect this reconstruction for those parts of the earth's surface where human interference has occurred.

CHAPTER II

Vegetation Development

It is probably some 400 or 500 million years since the ancestors of all the land plants which now inhabit the earth emerged from the oceans. They must have evolved from simple marine organisms similar in some ways to some of the present-day *Algae*. Before this time, for many hundreds of millions of years, plant-life must have been evolving, but it had probably been almost exclusively marine. The direct ancestors of all the most primitive classes of plants—the *Algae*, *Fungi* and *Bacteria*—probably evolved in the primeval oceans. These relatively simple plants are usually grouped together as one major *phylum* of the plant kingdom known as the *Thallophyta*. It seems likely that some primitive *Algae* invaded the land some time during the Cambrian or Ordovician Periods and that from them evolved the original ancestors of all the ferns, horsetails and club-mosses (Fig. 2). These more complex, spore-bearing plants, known collectively as the *Pteridophyta*, gradually gained ascendancy until they dominated the vegetation of all the land areas. By the end of the Carboniferous Period great forests of tree ferns, giant horsetails and tall club-mosses covered vast areas (Plate I*a*). The coal seams of Carboniferous age are composed almost entirely of their remains. Some time during the Palaeozoic Era the original ancestors of all the mosses and liverworts— the *Bryophyta*—must also have sprung independently from thallophytic stock but the time and mode of their doing so is still quite obscure. Fossil remains of the *Bryophyta* are rare but sufficient have been found to show that both mosses and liverworts were in existence before the end of Carboniferous times. Unlike the pteridophytes, however, these plants appear never to have been important in the gross structure of most of the vegetation communities of the earth.

Even during the period of pteridophyte supremacy seed-bearing plants had already come into existence [58]. Seed-bearing plants as a whole are of two distinct types, the more primitive of which are called the *Gymnospermae* and the more advanced, the *Angiospermae*. The only large surviving group of gymnosperms today are the conifers (*Coniferales*). Conifers had made their appearance before the end of Carboniferous times and, although they

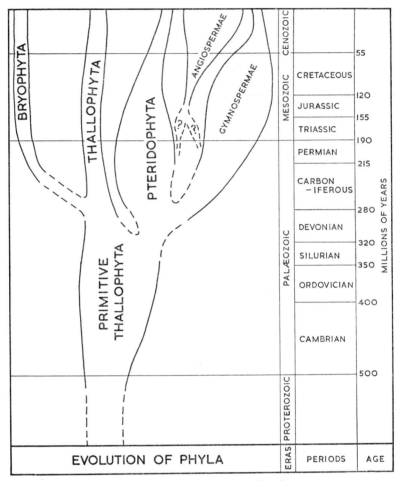

Fig. 2. Evolution of the plant kingdom.

must have originated in some way from pteridophyte stock, fossil evidence of their early evolution is incomplete. They remained quantitatively unimportant until the end of the Palaeozoic Era but, along with other primitive seed-bearing plants [58], they gained predominance over the woody pteridophytes with almost startling rapidity at the beginning of Mesozoic times, about 180 million years ago. They maintained this predominance throughout Triassic and Jurassic times (Plate I*b*). Some time during early Mesozoic or even late Palaeozoic times however, the *angiospermae* made their appearance. It is by no means a foregone conclusion that they evolved from gymnosperm stock; they may have arisen quite independently from one of the pteridophyte groups. During the Cretaceous

Period the angiosperms rapidly gained ground, mainly at the expense of the gymnosperms, until, by Eocene times, the former had become the predominant group of land plants. Some genera of conifers have remained important however.

Angiosperm and coniferous trees similar to or almost identical with those which now dominate the earth's vegetation have thus maintained their supremacy for 50 or 60 million years. Evolution, once started, has never stopped but it is clear that communities of plants can persist for millions of years with only slight changes in their floristic composition; as old individuals die they tend to be replaced by young ones of the same species so that changes in the composition of the community as a whole are negligible. From time to time species have become extinct and new ones have made their appearance but a significant percentage of the tree species which are common in our forests today are plentifully represented by almost identical ones in the fossil floras of the Cainozoic Era. Thus the thick beds of Miocene lignite near Cologne in Germany are found to be composed in very great part of the remains of cedars and redwoods (*Sequoia spp.*) which appear to have been identical with those still living in California.

Provided no catastrophic environmental change takes place, the vegetation of a particular area may therefore remain of almost constant composition over unimaginably long periods of time. Only occurrences like climatic changes and volcanic eruptions will cause wholesale changes in vegetation, some permanent and some only temporary. Nevertheless, every single square yard of the present land areas of the earth was, at one time, fresh and untenanted. All land areas began as upraised sea floors, lava flows, silted lake basins, abandoned ice-fields and the like. In some cases it is millions of years since the original invasion by vegetation took place, in others this occurred relatively recently. Not many thousands of years ago the whole of north-western Europe was covered by an ice cap which had advanced from the north and swept away almost every vestige of the former vegetation and soil. When the ice again retreated it left a completely inorganic surface—partly bare rock, partly smooth boulder clay plains and partly uneven surfaces of moraine or fluvio-glacial sand and gravel. The whole natural drainage system had also been disrupted by glaciation so that much of the area was ill-drained and studded with lakes. The whole of the present vegetation of north-western Europe is therefore of recent provenance having only just established itself on this relatively new surface. On the other hand vast areas of tropical rain forest in the Congo and Amazon basins have occupied their present positions for millions of years without interruption.

When a new surface is exposed it has no soil upon it and yet, when we examine the vegetation existing on the earth today, we discover that most types—forest, grassland and scrub—are rooted in quite a deep and complex soil. Furthermore, particular classes of vegetation seem to be associated with certain types of soil and vice versa. It is difficult to see how a dense forest could establish itself on a bare rocky surface and, on the other hand, how a brown forest soil could come into existence if the forest did not precede it. The dilemma is similar to the familiar one concerning the chicken and the egg and the answer even more complicated. In fact, within an area where the climate is quite suitable for the existence of a particular type of forest with its characteristic associated soils, neither the forest nor the soil will appear immediately upon a fresh surface, whether this is a raised sea floor, a lava flow or a plain of glacial till. Many complicated changes have to occur, frequently extending over a period of centuries, before a vegetation of fairly constant species-composition, associated with a mature soil, comes into existence.

Climatic climax vegetation

This stable type of vegetation, in complete equilibrium with climatic and soil conditions, has been referred to traditionally as the 'natural vegetation' of an area. This term is not without its ambiguities however, and ecologists now prefer the term 'climatic climax vegetation'. The basic premise is that if a naturally well-drained surface is left completely undisturbed for a protracted period, with no human activity, climatic change or other natural cataclysm, a whole series of plant communities, one after another, will occupy it but, ultimately, a community will establish itself and persist, unchanged, quite indefinitely. This climatic climax community will be dominated by plants which, of all those available, can compete most successfully in the existing physical conditions.

Quite apart from those areas where man grows his crops, a large proportion of the surface of the earth is not covered with climatic climax vegetation. New land of one kind and another is constantly appearing and, at a given time, various new areas will be at a variety of stages of development towards climatic climax. New mud banks, lava fields, landslip scars and sand dunes as well as areas burned over by forest fires, are soon invaded by vegetation, but this pioneer vegetation cannot be regarded as climatic climax. Any world classification, if it is going to be realistic and comprehensive, however, must take cognizance of the vegetation of disturbed and 'undeveloped' areas as well as of the climatic climax communities.

Vegetation classification

Any attempt to classify and regionalise the vegetation of the earth must have limitations since in no two areas is vegetation identical. Plant communities of very similar general appearance have evolved thousands of miles apart on different continents but, on closer examination, they are always found to be composed of almost entirely different species. When these communities are placed in the same class therefore, important differences are consciously ignored. Even in a relatively small area of only a few acres, distinct differences in the vegetation cover can usually be detected. Any vegetation map of the world must therefore be viewed with an educated and critical eye. Not only must there have been a great deal of selection and generalisation in order to distinguish between the 'vegetation regions', but the placing of the boundaries between these regions must also have been somewhat arbitrary. Recognisable plant communities do not occupy sharply-defined areas; they merge imperceptibly one into another through zones of competition in which typical species of the one community intermingle or interdigitate with those of the other. These zones of transition between communities are often referred to as *ecotones*, and in most cases, on maps on a continental or world scale, the boundary between two vegetation regions is drawn, rather arbitrarily, through the ecotone zone.

The distribution of plant communities, even under undisturbed conditions, is determined by a great number of inter-related factors. The chemical composition and physical characteristics of the soil-parent-material are just as important as temperature, precipitation, insolation, wind speed and a number of other purely climatic factors. An area of particularly well-drained soil in a region of generally heavy rainfall may carry rich forest while areas of less pervious materials are covered with thick peat and moorland grasses. On the other hand a pervious limestone or sandstone outcrop may carry only thin forest in a region where, on other types of rock, the rainfall is sufficiently heavy to support luxuriant rain forest. Quite obviously a varied geological pattern can affect the range of species and communities very considerably, but, in spite of this, the concept of climatic climax vegetation has proved to be a most useful one as a basis for classification on a world scale. In spite of lithologically-determined aberrations (and, perhaps, historical relics), much of the world pattern is very obviously controlled by climatic factors.

Nevertheless, it is very easy to adopt an over-simplified view of the way in which climate affects the distribution of plant communities. Ever since the classic work of nineteenth-century biologists it has been realised that competition or the 'struggle for existence' between different species has an important bearing on their ultimate distributions. Thus, if in an undisturbed area, physical conditions permit the growth of both tall, shady trees and light-loving, herbaceous plants, one outcome is inevitable; the herbaceous plants will be 'shaded out' and eliminated. The tall trees will become the *dominants* and the climatic climax vegetation will be forest. Climate is only a permissive factor, since competition may eliminate whole communities which could otherwise flourish. An observer looking across parts of the lowlands of England from a convenient hill top two or three thousand years ago, might have concluded that the climate precluded the growth of meadow and pasture grasses. Where the land was naturally well-drained, trees formed a complete cover over the landscape and plants like ryegrass (*Lolium spp.*) and cocksfoot grass (*Dactylis glomerata*) must have been rare. Today we know that the climate in this country is as favourable as any in the world for the growth of meadow and pasture grasses; they were precluded, before the intervention of man, not by climatic conditions but by competition; they cannot tolerate the low light intensities experienced beneath forest cover.

Low-growing shrubs and herbaceous plants can only be the most prominent elements in the climatic climax vegetation in regions where they are not shaded out by a continuous cover of trees. This occurs in the arctic fringes of Canada and Eurasia, around the fringes of the Sahara and other deserts and, more locally, on windswept hills and cliff tops in generally forested areas. Trees have certain ecological characteristics which make them more susceptible than low-growing plants to unfavourable climatic conditions. In the first place they project further into the atmosphere and are, in consequence, much more at the mercy of desiccating winds. Secondly, since they are perennial woody plants, they produce soft green shoots each year from pre-existing, woody growth. In any area subject to seasonal cold or drought these young green shoots must have a sufficiently long growing season in which to ripen and become woody. Without this they would be killed back each year and the tree could not continue to grow. Thirdly, the tree, because of its height, is faced with the problem of transporting mineral nutrients a relatively great distance from its roots to the place where they are required in the leaves. Since this transport can only take place in dilute solution, very large amounts of water are required by the tree throughout the growing season. Different species of tree certainly differ considerably in the degree to which they can withstand climatic rigours but, because of their very physical structure, they are all subject in some degree to these three types of limitation.

The hierarchy of climatic climax communities

Because of the above points, it is convenient to divide the climatic climax vegetation of the earth into two classes—forest communities and non-forest communities. Each of these can then be sub-divided according to differences in detail in the form and functions of its dominant plants. One type of forest is the *tropical rain forest*[1] which still covers vast areas in the Congo Basin, Amazonia and Indonesia. Many thousands of species of tree are dominant in this type of forest but they are nearly all of one

[1] This is often referred to as 'equatorial forest' but since structurally similar communities occur in areas like coastal Queensland, south-eastern Brazil and Assam, all of which are far from the Equator, the term is obviously undesirable.

morphological type: they are broad-leaved, evergreen trees. Such a set of morphological characteristics, possessed by the vast majority of dominant species in a plant community, is usually referred to as a *life form*. In the case of the tropical rain forest, however, it is clear that, in spite of morphological similarities, the species in the Congo Basin will be very different taxonomically from those in Amazonia; indeed large families of plants which form an important element in the flora of one are absent from the other. This is inevitable since plant life in the two areas has been evolving quite separately for a long period. Nevertheless, because all areas of tropical rain forest have dominants characterised by the same life-form, they are all classed together as a single *plant formation-type*. This formation-type is then sub-divided into three distinct *plant formations*—the American tropical rain forest, the African tropical rain forest and the Indo-Malaysian tropical rain forest, the latter stretching discontinuously from India to eastern Queensland.

Similarly the original vegetation of western Europe must be placed in the same formation-type as that of the eastern U.S.A. In both cases the dominant plants are broad-leaved, deciduous trees which lose their leaves in the cold season of the year. This plant formation-type is often referred to as the *deciduous summer forest* of which the western European and North American sub-divisions are representative formations, each with its own distinctive assemblage of species. The climatic climax vegetation of the whole earth, both the forest and the non-forest communities, can be divided and sub-divided in this way.

Even further sub-division is desirable however since plant formations are by no means homogeneous. In England, for instance, quite distinct forest communities are found, though all of them belong to the west European deciduous summer forest formation. It is quite certain that, after 500 B.C., the forests in the clay vales of the Midlands were oak forests dominated by the pedunculate oak (*Quercus robur*) while those on the slopes of the Chalk uplands and some other well-drained areas in the south were dominated by beech (*Fagus sylvatica*). In northern England the forests on the Millstone Grit slopes of the Pennines were dominated by sessile oak (*Q. petraea*) and those on the Carboniferous Limestone of the Derbyshire Peak District and Craven by ash (*Fraxinus excelsior*) or elm (*Ulmus spp.*). All these are distinct sub-divisions of the one formation and each one is referred to in British literature as a separate *plant association*. All associations have their own characteristic dominant species or assemblages of species and are given titles based upon the generic names of these species.

All the climatic climax vegetation of the earth can be divided and sub-divided in this way into distinct vegetation regions and sub-regions.

It is useful to attempt to summarise the underlying reasons why climatic climax vegetation can be regionalised in this way. Firstly, it seems that each formation-type occupies a region which possesses certain climatic characteristics to which a particular life-form is most suitably adapted. Quite regardless of family relationships therefore, species possessing this life-form have a competitive advantage and come to dominance. It is quite obvious, for instance, that the deciduous habit is a very favourable adaptation in an area where there is a moderate or short cold season when delicate green foliage would be destroyed by frost. In any case, leaves are superfluous in such a season since food synthesis

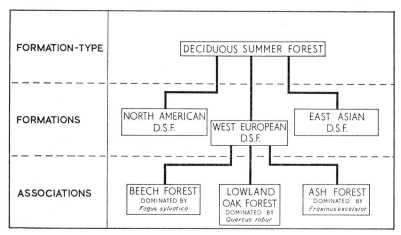

Fig. 3. The hierarchy of climatic climax communities.

cannot be carried on. Secondly, formations of the same formation-type occupy areas which are separated by oceans, deserts or mountain ranges; as a consequence of this geographical separation over a long period of time, they have very different floras. Finally, the different associations of a formation occupy territories which are often characterised by distinct conditions of soil, slope or drainage. It must be realised, however, that a gradual transition in climatic conditions is normally found as one passes across the wide area occupied by a plant formation. Climate therefore must play some part in determining the distribution of associations within a plant formation. Thus, in eastern North America, although the deciduous summer forest occupies an area of roughly similar climate, there is a general increase in dryness from east to west. Species of beech and maple dominate in the associations of the east whereas, in lowlands west of the Mississippi, oak and hickory associations are almost universal up to the edge of the prairie (Chapter VI).

Priseres

It has already been shown that, for various reasons, very large parts of the land masses now carry a vegetation which is not climatic climax.

Indeed, at any one point of time, even without human interference, many places will have a vegetation which is not in equilibrium with physical conditions. In Sunda Strait, between Java and Sumatra, lies the island of Krakatau. A visitor during the last decade of the nineteenth century might have received the impression that here was an island which, for some peculiar reason, would not support forest. The steep slopes of the island up to a height of about 1,200 feet were covered with savanna grassland and, above that height, the grasses gradually gave way to ferns. Only a thin sprinkling of woody plants was to be seen. Since near-by parts of the islands of Java and Sumatra supported tropical rain forest, this herbaceous vegetation was obviously anomalous. Those who visited Krakatau at this period were, in fact, well aware of the reason for the anomaly. In 1883 a gigantic volcanic explosion had completely removed part of the island and buried all the rest beneath many feet of white-hot volcanic ash. All life on the island was completely destroyed with the doubtful exception of a few deep-seated roots. Until this occurrence, throughout historical times, the island had been clothed up to and beyond the 1,000-foot contour with tropical rain forest of the Indo-Malaysian formation and with sub-montane forest above that height. The dominant trees of these two formations could not re-invade immediately however. The seeds of rain forest trees, even if agents of dispersal are available to re-introduce them, are not able to germinate on a bare, inorganic surface. A complicated development of both vegetation and soil had to take place before the climatic climax could be re-established [24].

During the first few years only herbaceous plants were able to establish themselves in the fresh, weathering volcanic ash. These rapidly invaded the ground so that a cover of savanna, already referred to, had come into existence before 1900. Careful observation of this savanna would have indicated, quite conclusively, that it was not climatic climax vegetation; it was not in a state of equilibrium. As month followed month and year followed year, changes constantly occurred. Shrubby plants gradually ousted the grasses, and the shrubs were followed by trees of more drought-resistant species than those which are the dominants in tropical rain forest. Ultimately the whole of the developing soil was shaded from the direct rays of the sun by a dense cover of forest. Beneath this, seeds of the tropical rain forest dominants which were brought in by birds or by wind, were able to germinate, grow up and, ultimately to over-top and shade out the lower-growing species. This series of changes took place through the decades of the first half of the twentieth century so that today Krakatau is clothed again in rain forest which is very similar in appearance to the original cover. The flora is certainly much poorer than it was before 1883 because large numbers of species which are typical of the area have not had sufficient time to find their way back across the waters of the Sunda Strait. If this volcanic eruption had occurred within a larger land area, however, there is little doubt that, by now, after a period of eighty years, the flora would probably have been much richer.

A whole series of communities had to flourish and, successively, to displace each other in order to effect the re-establishment of the climatic climax on Krakatau. These transient communities, beneath which a new soil gradually developed, are normally spoken of as *seral communities* and the whole series of communities leading from the virgin surface back to tropical rain forest are referred to as a *prisere.* The essential difference between a climatic climax community and a seral community is that, whereas the former is in equilibrium with its environment, the latter modifies it continuously. By deepening the soil, adding more organic material to it and by casting deeper shade upon it, it makes the site more and more favourable for the invasion of a more advanced community.

In the case of Krakatau the first community to re-invade—the *pioneer community*—was composed almost entirely of blue-green algae. Only a few months after the volcanic surface had cooled down it was seen to be almost covered by a film of algal slime. These primitive plants are able to live on a completely inorganic surface, with no true soil, provided the climate is humid. Within a short time however a *second stage community* of herbaceous plants, particularly grasses, began to invade and rapidly formed a closed-cover over the ground. After between ten and twenty years this, in turn, gradually gave way to a *third stage community* of shrubs and trees, primarily those with very light, wind-borne seeds. During all this time weathering, along with the addition of more and more organic material, brought into existence a true soil with ever-increasing powers of water retention. In consequence more demanding trees of the climatic climax type were now able to invade; their seedlings could germinate beneath the shade of the third stage community which protected them from the drying effect of direct sunlight and wind as well as providing conditions of low light intensity to which such species are adapted in the early stages of their development. The whole course of the prisere was run within relatively few decades mainly because the material on which development took place was very pervious, easily penetrable and very rich in plant nutrients. Nevertheless, even on such material, priseral development was necessary before tropical rain forest could return.

The story of Krakatau is rather an unusual one; very rarely in recent times has a whole island been completely sterilised in this way. Nevertheless, on a greater or lesser scale, fresh mineral surfaces are constantly being exposed within the general areas of all types of climatic climax vegetation so that many types of priseral vegetation, at various stages of development, are to be found. In Britain, for instance, a landslip or flood may strip away vegetation, soil and subsoil from a sloping area of granite on the side of Dartmoor or of Millstone Grit on the slopes of the Pennines. The climatic climax vegetation here is oak forest but, quite obviously, oak trees cannot grow on a bare rock surface. Such a surface presents conditions which are just as dry as the surface of the Sahara Desert for a considerable part of the year; even in an area of humid climate a sloping surface of bare rock can become completely dry only a few minutes after heavy rain has ceased. The first plants to invade such a surface must therefore be adapted to withstand the frequent occurrence of complete drought. These pioneer community plants must also be able to adhere to a surface completely devoid of soil, a normal rooting system being quite useless. The crustose lichens which form the familiar orange or grey patches on the coping stones of old stone walls are obviously adapted to this type of environment and frequently compose the pioneer community in this kind of prisere. They anchor themselves to the rock by short outgrowths from their under-surfaces; they are able to withstand desiccation for protracted periods, to take up water as soon as it becomes available and to resume growth immediately. These plants gradually etch away particles of rock and also provide small amounts of organic material. A thin, primitive soil thus begins to form over the rock and, in this, a second stage community of mosses often establishes itself. Soil-forming processes are accelerated so that a water-retaining layer gradually develops over the rock and the habitat becomes less susceptible to drought. The third stage community thus consists frequently of turf-forming grasses like sheep's fescue (*Festuca ovina*) along with annual and ephemeral plants. The fourth stage community often comprises low-growing shrubs such as the bramble (*Rubus fruticosus*) and the dog rose (*Rosa canina*) and, with an ever thickening soil, subsequent stages include trees like the birch (*Betula spp.*) and the ash (*Fraxinus excelsior*) which have lightweight, wind-borne seeds. Ultimately, provided there is no interruption by natural or human agencies, the deep-rooting trees of the climatic climax formation will be able to establish themselves. Since hard, consolidated rock weathers only slowly, however, the development of this prisere must inevitably take many years.

The seral community, by its very existence, creates conditions which are more and more favourable for more complex and demanding communities. Provided there is no interference it is ultimately eliminated by competition. In the example given, the sere gradually overcomes conditions of periodic drought by creating a water-retaining soil so that each stage needs to be less drought-resistant than the preceding one. Because of the nature of the original conditions this type of prisere is called a *xerosere*. There are several types of xerosere however (Fig. 4). The one already described which begins on bare, consolidated rock is called a *lithosere* whereas another type which begins its development on bare sand dunes is referred to as a *psammosere*. Because of the unstable nature of this latter type of environment, quite different species from those in the lithosere are involved, particularly in the earlier stages. In Britain the

Cvs

pioneer community nearly always consists almost entirely of marram grass (*Ammophila arenaria*). This plant not only survives protracted drought; its *rhyzomes* can stabilise and accommodate themselves to the shifting sand so that the plant is not overwhelmed. By this means the sandy habitat is rendered suitable for the less xerophytic, less resilient plants of the second stage community.

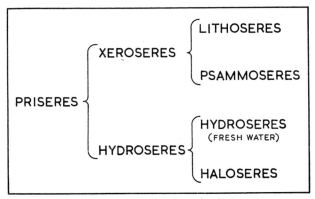

Fig. 4. A classification of priseres.

Other types of prisere develop in face of exactly the opposite type of conditions unfavourable to plant growth. These are the *hydroseres*. When a lake is reduced to a depth of a few feet by silting, a pioneer community of water lilies (*Nymphaea alba*) often establishes itself. This has the effect of accelerating silting and when the depth has been further reduced a second stage community often dominated by reeds (*Phragmites spp.*) and reed-mace (*Typha spp.*) normally follows. Silting continues so that the lake becomes a swamp, then a marsh and, ultimately, may develop into quite stable ground (Chapter XII). Each stage is characterised by a community with specific tolerances to a degree of wetness. Another kind of prisere develops on an emerging estuarine flat but, in this case, the pioneer community, which is often of glasswort (*Salicornia herbacea*), has to be resistant to a high concentration of sodium chloride in the rooting medium. When once the mud flat has achieved a certain height above high tide mark, rainfall will begin to leach out the salt so that subsequent stages in the sere will consist of plants with quite different tolerances. This prisere is known as a *halosere*.

Subseres and subclimaxes

In any area of varied relief one must expect to find a patchwork of climatic climax and priseral communities. Indeed pronounced steepness or extreme flatness of land may act as long-term *arresting factors* which prevent the prisere from proceeding to the climatic climax. A hillside may be so steep as to prevent the accumulation of soil upon it; consequently the sere is arrested at an early stage. At the other extreme, a flat plain, particularly if near sea-level, can cause the persistence of marshy conditions on to which large trees cannot encroach. Under undisturbed conditions the Fens of eastern England were an example of this, often carrying *carr* vegetation dominated by alder (*Alnus glutinosa*) and willows (*Salix spp.*). The climate here, as in the rest of lowland Britain, was quite suitable for oak forest but hydrological conditions arrested the prisere at a fairly late stage. An almost permanently arrested community such as this is referred to as a *subclimax*. If the natural arresting factors are removed, either gradually or quickly, by a lowering of sea-level or by artificial drainage, then seral development can be resumed and further progress towards the climatic climax can take place. This delayed sere is known as a *subsere*.

Subseres of another type have been initiated because of man's interference with climatic climax vegetation. Quite often this has been swept away by fire or in forestry operations. A mature soil is thus laid bare upon which vegetation re-invasion can begin and, provided there is no further interference, development back to the climatic climax will take place. All stages in this type of subsere can be observed in Britain where stands of timber have been removed without subsequent replanting. The original floor of the woodland may be almost bare of flowering plants though usually there will be a sparse ground flora of soft grass (*Holcus mollis*) and bluebell (*Endymion non-scripta*). With the removal of the tree canopy, strong sunlight penetrates to the ground and, within a year or two, a dense cover of plants such as bracken (*Pteridium aquilinum*) and rose-bay willow herb (*Epilobium angustifolium*) along with low-growing shrubs like dog rose (*Rosa canina*) and bramble (*Rubus fruticosus*) will have formed. Most of us have had the experience of revisiting a well-known wood only to find that not only have the trees been felled, but that an impenetrable thicket has replaced them. Subsequently, through a number of stages, climatic climax forest could establish itself though, in Britain, continuing disturbance usually prevents this from taking place. In a very similar way, if arable or pasture land is abandoned, subsere development begins immediately though the early stages comprise plants which are quite different from those in the early stages of the woodland subsere. Arable weeds like chickweed (*Stellaria media*), groundsel (*Senecio vulgaris*) and persicaria (*Polygonum spp.*) compose the pioneer community on arable land while plants like the thistles (*Cnicus spp.*) and ox-eye daisy (*Chrysanthemum leucanthemum*) are pioneer invaders in meadow and pasture. These invading plants are constantly opposing man's efforts and, should he relinquish his efforts for only a short time, the subsequent stages dominated by shrubs and trees soon follow. It is obvious in fact that subsere development is much more rapid than prisere development because the former has the advantage of a ready-made, water-retaining soil.

Plagioseres and plagioclimaxes

A final class of communities, now widespread on all the continents, has been brought into existence by the protracted action of man and his domesticated animals. The development of the English landscape from mediaeval times onwards again helps one to understand what has happened over much wider areas, particularly in recent centuries. William the Conqueror's Domesday Book provides evidence that, in A.D. 1086, a great deal more of the English landscape was forested than at the present day. In a very large percentage of the townships or 'vills' there was a large extent of 'woodland for pannage' or 'woodland for grazing'; indeed, it is quite clear that one of the mainstays of the Anglo-Saxon farming economy was a large area of forest where the animals predominantly pigs, goats and cattle, were turned out to forage. Because of this the annual 'crop' of acorns and beech-mast would be nearly all devoured and the few seeds which did survive would rarely escape being grazed off after germination. Consequently, the forest would almost cease to regenerate and, as old trees died, there would be few saplings to replace them. The forest canopy therefore gradually opened, allowing sunlight to stream in. In most places sporadic firing and felling would accelerate this process. The forest, by these means, was gradually reduced to scrub in which spiny shrubs like hawthorn (*Crataegus oxyacantha*) and blackthorn (*Prunus spinosa*) would be dominant. As heavy grazing continued amongst these shrubs, even they receded, this process being particularly rapid if large numbers of goats were present. Indeed, it is known that the Cistercian monks had large numbers of goats in the thirteenth and fourteenth centuries, probably for this very purpose. Presumably they were reducing the scrub as a prelude to increasing the numbers of the animal which, to them, was of much greater value—the sheep. Ultimately, through the agency of grazing animals, the forest and scrub were removed and replaced by a continuous turf of pasture grasses. These thrived with the high light intensity even in face of heavy, continuous grazing. This was the way in which the bent-fescue (*Agrostis-Festuca*) grasslands of our acidic heaths and the almost pure, fine fescue turf of our chalk downlands came into existence. It is as though the natural regeneration of the vegetation was gradually 'deflected' from trees to grasses; the series of communities that resulted is therefore referred to as a *plagiosere* and the final community, in equilibrium with both climate and grazing animals, as a *plagioclimax*.[1]

These plagioclimax grasslands existed for many centuries on much of upland Britain; indeed many of them still persist where grazing has been maintained continuously. Since the last decades of the nineteenth century, however, interesting changes have taken place on many of these rough

[1] '*Disclimax*' in American literature.

grazings. With the importing of large quantities of meat and wool, agricultural depression occurred and the intensity of grazing was much reduced; in many places it almost ceased. Grazing animals were also much reduced in numbers on vast areas purchased by water boards. With the removal of the chief deflecting factor, subsere development was permitted though sporadic firing has usually prevented any rapid progress towards the re-establishment of climatic climax. The bracken (*Pteridium aquilinum*) has been the most important pioneer invader of the long-established turf and, in many places, this has been followed or accompanied by shrubs and trees. The bracken has the advantage of very tiny spores which can penetrate into a close turf and thus germinate on the shaded soil beneath, whereas even small seeds are often too large to do this. The bracken, when once established, spreads rapidly by means of underground stems or rhyzomes and casts quite a dense shade over the turf it has invaded. This may open the way ultimately for tree regeneration (Chapter XIII). Today, in many places, quite thick birchwoods and ashwoods are established where, less than a century ago, there was quite valuable grazing land (Plate XX*a*). Locally, even young oak trees have followed though this is rather rare. This is an example of another type of subsere, this time leading from the plagioclimax community back to climatic climax vegetation.

Ecological status as a basis for vegetation classification

In Britain the effect of man on the vegetation is most profound even where he has not actually ploughed the ground and sown his crops. In other parts of the world, where human populations are not so great or for which there is no written history, the effect of man is more obscure. Nevertheless, wherever man has been present as a cultivator, herdsman or fire-carrier for even a few centuries, one cannot expect to find true climatic climax vegetation. A realistic vegetation classification on a world scale must take cognisance of the fact that very large areas are covered with plagioclimax, subclimax or purely seral vegetation. Over much of the earth indeed, climatic climax vegetation is the exception and not the rule. The exact mode of origin of much vegetation is frequently not obvious however. Even in Britain where so much ecological research has been done, there is much doubt about the exact nature of the climatic climax vegetation and the *ecological status* of the present vegetation. How much more profound is our ignorance of the ecological status of the vegetation in the vast areas of forest, grass and scrubland in Africa, Asia, Australia and South America!

It is clear that two quite distinct types of world vegetation map can be attempted. The present vegetation can be mapped with strict objectivity, showing evergreen forests, deciduous forests, wild grasslands and so on where they occur and rice lands, wheat fields, rubber plantations and other cultivated vegetation wherever it has been planted. On the other hand, an attempt can be made to construct a theoretical map of climatic climax vegetation, ignoring all the effects of natural arresting factors and human interference. Most world vegetation maps that have been made were intended to belong to one or the other of these categories; unfortunately, in nearly all cases, the compilers found it difficult, if not impossible, to be consistent. Most published vegetation maps on a continental or world scale fall between two stools. Because of the great difficulty of representing crop-growing areas in sufficient detail on small-scale maps, the tendency has been to attempt to adopt the second alternative and to show cultivated areas with the vegetation they are supposed to have carried before human interference. Unfortunately this task of ecological reconstruction has been beyond the competence of those who have attempted it. Even today far too little is known about the original vegetation over huge parts of the earth for anyone to map it with any degree of confidence. Reconstructions have usually been based on the assumption that the original vegetation of areas which are now cultivated was the same as the *present* wild vegetation on near-by uncultivated land. The hazards of such an assumption are obviously very great. One has no grounds for assuming that the present cultivated lands of central Nigeria, for instance, were originally covered with elephant-grass savanna merely because this type of vegetation is almost universal on uncultivated land there today. This savanna is burned regularly and grazed extensively by domesticated animals; it has been treated in this way for millennia. The effects of this treatment on both vegetation and soil must have been very great but their magnitude cannot yet be gauged with any degree of precision.

Because of these difficulties, the maps presented here (Appendix I) show the theoretical climatic climax vegetation only where its nature has been established with a reasonable measure of agreement. In all those areas about which there is still much speculation and disagreement, the commonest type of wild vegetation which prevails at the present day has been indicated. A clear distinction between these two categories of information has been made in the keys to the maps so that serious misconceptions should not arise.

All small-scale vegetation maps, no matter how carefully devised, can be very misleading in another way unless the user possesses a great deal of additional information. No matter how homogeneous vegetation may appear to be on the map, one can be fairly confident that, on the ground, it consists of an intricate patchwork of communities. The smallest unit that can possibly be shown is usually the formation; component associations could only be shown realistically on maps of much larger scale. Similarly, no attempt can be made to show the distribution of seral, subclimax and local plagioclimax communities on a map of continental

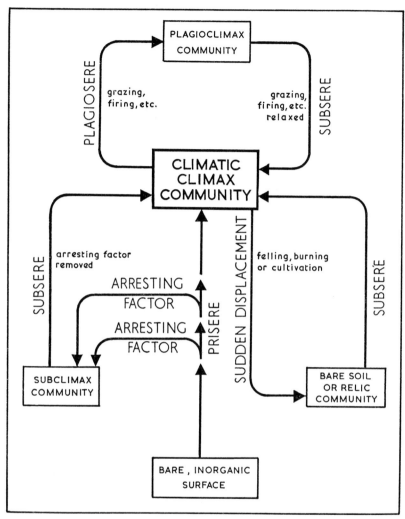

Fig. 5. The ecological status of plant communities.

scale. In order to appreciate fully the vegetation complexity that one would be likely to encounter within an area which is shown to be occupied by a particular formàtion, one must know what type of seral, subclimax and plagioclimax communities are normally associated with that formation. Every plant community can be conceived as having a specific *ecological status* or, in other words, of occupying a specific position in a hierarchy of plant communities all of which are related to the climatic climax community in terms of ecological succession (Fig. 5). Every association of every formation thus has its own hierarchy of related plant communities.

The concept of the climatic climax community as a basis for the classification and mapping of vegetation is thus a very useful one; it enables one to think of the vegetation of the earth, not merely as a static patchwork, but as a dynamic system which accommodates itself to environmental changes and to human interference. Nevertheless, like other systems of classification, the climatic climax concept is subject to certain limitations as will become apparent in later chapters.

CHAPTER III

Soil Development

Soil-parent-material

IN the late eighteenth and early nineteenth centuries many farmers and amateur geologists became interested in the reasons for the distribution of the different types of soil. They came to the general conclusion that there was a direct and fairly simple relationship between the character of the soil and the nature of the rock beneath it. They found chalky soils on chalk, sandy soils on sandstone, clayey soils on shales and organic soils on peat and so they devised soil classifications which were based on the nature of the underlying soil-parent-material. Nearly all these workers had experience of only small areas however; it was not until the latter part of the nineteenth century that individual pedologists gained a knowledge of soils on a continental scale. When this was achieved there ensued a change in attitude towards soil classification. Soils of very similar character were found in great belts extending across large parts of European Russia and the U.S.A. and individual soil belts frequently crossed rock outcrops of very diverse lithology. It became apparent that other factors, particularly climate and vegetation, were just as important as parent-material in determining the ultimate nature of the soil.

Weathering

The fact remains, however, that much of the material composing most soils has been derived from the underlying rock; no matter how long a soil has been in process of formation, the original 'raw material' must therefore have some effect on its ultimate nature. The processes by which the parent-rock-material is broken into small fragments are known collectively as '*weathering*'. Two distinct types of weathering occur. Firstly there are those by which rock is broken down into progressively smaller pieces without any change in the chemical composition of the component minerals. The shattering of quite hard rocks by frost action and the splitting up of fissile, argillaceous rocks like shales, by alternate wetting and drying, are good examples of this and they are known collectively as *mechanical weathering*. Secondly there are the processes of *chemical weathering*.

These not only cause the original rock to be split into smaller fragments; they also bring about chemical alterations in some of the minerals composing the rock. Many rock minerals are attacked by such substances as dissolved carbon dioxide (CO_2) derived from the atmosphere and by organic acids from decaying vegetation.

Both types of weathering usually operate together on the same site but usually one of them exceeds the other one in importance. Thus, in cold and dry regions, or in places where the overlying soil is thin or non-existent, mechanical weathering is the more important. On the other hand, in hot and wet regions, or in places where a thick soil overlies the rock, chemical weathering predominates.

The products of weathering

Rocks vary enormously in their physical and chemical properties; similarly the weathered materials derived from different rocks are themselves very different. The physical and chemical properties of individual *weathering complexes* are so complex that simplified statements can be very misleading; nevertheless a number of valid generalisations can be made, to show how soils develop and function. From the point of view of the soil scientist it is probably most useful to consider weathering products in two categories; on the one hand there are those which are readily soluble and on the other those which are almost insoluble.[1] The more soluble ones are predominantly carbonates since carbonic acid (H_2CO_3) occurs almost universally in soils. The atmosphere contains small amounts of carbon dioxide (CO_2) and any raindrop inevitably dissolves a little of this as it falls to earth:

$$CO_2 + H_2O = H_2CO_3$$

This acid, though chemically weak, is able to detach substances such as potassium, magnesium and calcium from rock minerals and to bring them into the soil solution as carbonates (K_2CO_3, $MgCO_3$, $Ca(HCO_3)_2$). These bases contain some of the most important mineral foods required by plants. The almost insoluble products of chemical weathering comprise a vast number of chemical substances amongst which, however, only a few types are common. These poorly-soluble substances go to form what is usually called the *inorganic fraction* of the soil.

As an illustration one may select the weathering complex which develops

_____

[1] Under natural conditions probably no substance is *absolutely* insoluble; the most resistant materials dissolve very slowly. It is particularly important to remember this point when studying soils since, although some processes operate quickly, many developments can only take place over thousands of years. During such long periods of time the slowest rates of solution may have measurable effects.

on a granite surface. The simplest type of granite consists[1] almost entirely of:

(a) Orthoclase felspar (potassium aluminium silicate: $K_2O.Al_2O_3.6SiO_2$)

(b) Biotite mica (a mixed silicate of potassium, magnesium, iron and aluminium: $K_2Mg_6O.2Fe_2O_3.Al_2O_3.6SiO_2.2H_2O$)

and(c) Quartz (silica: SiO_2)

On being weathered chemically, the felspar loses its potassium in the form of potassium carbonate (K_2CO_3) and a residue of a fairly simple clay mineral called kaolinite ($Al_2O_3.2SiO_2.2H_2O$) is left. The biotite mica breaks down to form potassium and magnesium carbonates (K_2CO_3, $MgCO_3$), the sesquioxides of iron and aluminium (Fe_2O_3, Al_2O_3) along with some clay (*vide infra*). The quartz, which is very inert chemically, is merely released from the rock as sand grains when the other minerals disintegrate. The clay, sand and the sesquioxides released from the minerals are added to the inorganic fraction of the soil while the soluble carbonates enter the soil solution.

Similar kinds of change take place when other types of rock are weathered although often the nature and variety of the substances produced is even more complex. For instance, the clay minerals produced by the weathering of many rock minerals have a much more complicated structure than kaolinite; most clay minerals are mixed silicates of aluminium and other metals (*vide infra*). Many of the minerals of igneous and metamorphic rocks also contain a greater variety of bases to be liberated as carbonates. Furthermore, whereas acid igneous rocks like granite contain a great deal of quartz and thus weather to a predominantly sandy material, basic igneous rocks such as basalt produce a residue which is predominantly clay. Generally speaking, a sedimentary rock produces a weathered residue which is much more homogeneous than that from an igneous rock. This is because sedimentary rocks are composed of material which has, to a greater or lesser extent, been sorted during a previous cycle of erosion and deposition. Shales thus weather to a predominantly clayey or *argillaceous* residue while sandstones and grits produce a sandy or *arenaceous* one. Even then, however, different types of sandstone give rise to weathering residues of different fertility; some sandstones consist of grains cemented together with calcium carbonate ($CaCO_3$) while others are cemented by yet more quartz. The former will obviously release a great deal of lime during weathering while the latter, as well as being much harder and more resistant to weathering, will produce remarkably little soluble material during the process. The above are merely a few examples

[1] The formulae here are presented in such a way as to indicate the components into which the minerals are liable to break down on weathering and not to show true molecular structure.

of the lithological contrasts found in different rocks; any good textbook of physical geology will provide some account of the mineral contents of all the main types of igneous, metamorphic and sedimentary rocks [49] from which one can obtain a clear indication of their inherent potentialities as soil-parent-materials.

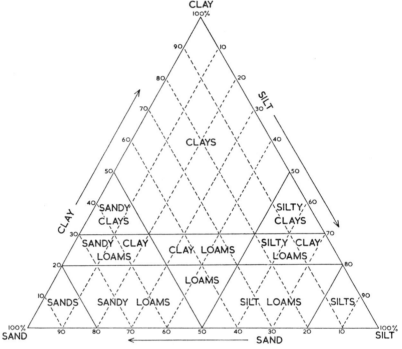

Fig. 6. One American system of soil nomenclature according to the texture of the inorganic fraction. (From S. G. Brade-Birks, 1944.)

From the point of view of the pedologist, differences in the lithological characteristics of the parent rock are of salient importance for two main reasons. Firstly, rocks differ enormously in the amount of soluble plant nutrients that they liberate on weathering. Secondly, the residue of weathering can vary between almost 100% clay and almost 100% quartz although most rocks yield a residue which contains a certain amount of both. This is of considerable importance even from the purely physical standpoint. Individual clay particles are exceedingly small, all having a diameter of less than 0·002 mm. Quartz grains on the other hand are nearly all much larger in size, those with a diameter between 0·002 mm. and 0·02 mm. being referred to as *silt* and those with above 0·02 mm. as *sand*. It is clear that the larger the percentage of clay particles in weathered material, the slower will be the rate at which water will percolate through it. On the other hand, very sandy materials will allow water to flow through them almost unimpeded. Soil textures are classified as sandy, silty, clayey or loamy according to the proportions of the different types of particle present (Fig. 6).

Apart from the purely physical property of permeability, however, the properties of soil chemistry and fertility are also very closely bound up with the relative amounts of clay and quartz in the inorganic fraction of the soil (*vide infra*).

Organic matter in the soil

It has already been pointed out that soil development is normally accompanied by the development of vegetation in the form of some recognisable prisere (Chapter II). Even as rock waste begins to be produced by weathering, some organic material also accumulates as the roots wither and aerial organs of the plants die down. The effect of organic material must therefore be taken into account from the very first stages of *pedogenesis*. In mature soils, because of differences in rates of growth and decay, the organic fraction may vary between almost 100% (as in peat) and much less than 1% (as in many desert soils).

Organic matter which has been partially decomposed and has become part of the soil is usually spoken of as *humus*. When plant organs die they initially suffer a certain amount of decay and disintegration; generally speaking this takes place very quickly in the tropics and may also be quite rapid in higher latitudes in summer. Everywhere, however, the rate of decay appears to slow down markedly when a particular chemical state has been achieved. It seems as though some of the lignin or woody material of the original vegetable matter combines with certain derivatives of the plant protein to form substances which are much more resistant to the attacks of decay organisms [15]. Although the details of their molecular formulae have not yet been ascertained, these decay-resistant substances probably comprise a series of distinct chemical compounds; they are not merely undifferentiated masses of organic material. Nevertheless there are numerous grades of humus, the type varying according to the nature of the associated plant community. Some communities are composed of plants which demand relatively large quantities of mineral nutrients from the soil; they abstract substances like potassium, calcium and ammonium (K, Ca, NH_4) and incorporate them in their leaves, stems and roots. Consequently these organs are rich in mineral matter and nitrogen. When the plant organs die the nutrients are returned to the soil where they tend to be retained by the humus. Such humus has, in consequence, a neutral or only mildly acid reaction and is known as 'mild humus' or 'mull'. On the other hand many plant communities are composed of plants which are adapted to an environment which is very poor in nutrients. They require little in this respect and therefore they have little to return. The humus they yield is thus poor in minerals and strongly acid in reaction and is referred to as '*raw humus*' or '*mor*'. Although it is customary to divide humus into these two main categories, it must be realised that all gradations between very alkaline mull and the most acid mor can be found.

Humus produced by the disintegration of the fibrous rooting systems of herbaceous plants is already in a finely-divided state and is disseminated throughout the soil. On the other hand the dead stems and leaves of plants fall on the surface of the soil where they humify to form a distinct layer of peat, turf or leaf-mould. This organic layer frequently becomes quite thick, particularly when it is composed of mor or in places where the ground is persistently waterlogged. Acidity and wetness tend to preserve a surface layer of peat or mor for two reasons; firstly, decay bacteria are much reduced, both in numbers and activity, by both these conditions; secondly, the larger members of the soil fauna such as earthworms and beetles are either few in number or absent altogether in both acid and waterlogged soils. In more base-rich and well-drained soils the opposite is usually the case; decay bacteria operate quickly in the destruction of humus while earthworms and other organisms mix it with the underlying inorganic material, often with astonishing rapidity. In his treatise on the earthworm [21] Darwin demonstrated that these creatures can deposit the equivalent of ten inches of top soil in the form of worm-casts on the surface of a pasture in the course of only 50 years. These casts are composed in part of inorganic material derived from below. This becomes mixed with dead organic material which would otherwise accumulate on the surface.

Soil structure

The incorporation of humus into the inorganic, weathering material of the soil does not result merely in a simple mixture of organic and inorganic particles. There is abundant evidence that the humus compounds enter into quite intimate associations with the tiny clay particles [15] to form what is usually referred to as the '*clay-humus complex*'. In fact, from the point of view of soil structure and soil fertility, clay resembles humus much more than it resembles the coarser fragments of silt and sand. The latter are usually referred to as the '*soil skeleton*' and are quite inert chemically. On the other hand, the clay-humus complex is chemically active; not only do the particles or *micelles* of clay and humus effect linkages between themselves, they are also capable of retaining mineral elements in the soil in exchangeable form (Fig. 7). If this active complex were not present in the soil, one substantial shower of rain would be sufficient to wash out such highly soluble salts as the carbonates of potassium, magnesium and ammonium. Inorganic salts, when dissolved in water, *ionize* to a considerable extent. Dissolved molecules of potassium carbonate in the soil solution are thus partly dissociated into separate potassium and carbonate *ions*:

$$K_2CO_3 \rightleftharpoons K^+ + K^+ + CO_3^{--}.$$

Each of the free potassium ions carries a positive charge or valence with which it is able to re-link with the carbonate ion or with any other similar ion in the soil. In fact the micelles of the clay-humus complex are chemically analagous to the carbonate ion so that the potassium ions can effect loose, temporary linkages with these soil compounds. Magnesium, calcium, ammonium and hydrogen ions, along with many others, can be retained similarly by the clay-humus. The actual molecular structures so formed are very complex but it is quite permissible to represent them diagrammatically.

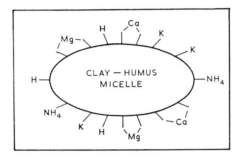

Fig. 7. Diagrammatic representation of a clay-humus micelle with appurtenant ions of hydrogen and nutrient minerals.

It is clear, therefore, why a mere mixture of silt and sand, completely devoid of clay and humus, could never be made into a fertile soil. Vast amounts of soluble mineral fertiliser could be applied to such a coarse, skeletal mixture but would be lost after only a few hours' rain. On the other hand, soils with a large clay-humus fraction have the potential for holding plant nutrients in chemical bonds for considerable periods. Such nutrients can be detached by the root hairs of plants, however, as these push their way between the soil particles.

In order to understand how soil development takes place under natural conditions and thus to appreciate how soils like podzols and chernozems come into existence, it is necessary to explore the chemistry of soils a little further. It would be easy to obtain a misleading impression from the points that have just been made. In spite of the fact that the clay-humus particles can retain mineral ions temporarily, the latter are constantly being detached by the purely inorganic action of rain-water. Because of this, soils in humid regions soon lose most types of exchangeable mineral ion unless the latter are constantly being replaced.

When the clay-humus particles lose their nutrient ions because of percolating rain-water, increased acidity must occur because hydrogen ions are substituted. The sulphate radicle can be used to illustrate this. When this radicle has both its valencies satisfied with hydrogen it forms a molecule of sulphuric acid (Fig. 8). In a similar way the more complex clay-humus radicles can have almost all their free valencies satisfied with hydrogen. Such hydrogen-saturated material is, of course, almost devoid of exchangeable plant nutrients.

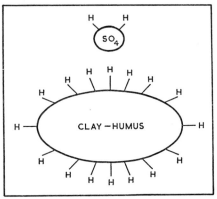

Fig. 8. Diagrammatic representation of sulphuric acid and hydrogen-saturated clay-humus.

Leaching

It is important to recollect that rain-water is, in effect, a dilute solution of carbonic acid (H_2CO_3). Consequently, in a region of humid climate where downward percolation of rain-water is persistent, the nutrient ions which are detached are carried away into the drainage water in the form of carbonates (Fig. 9). The carbonic acid molecules exchange their hydrogen for mineral ions as they pass through the clay-humus complex and then continue to move downwards as carbonate salts. It is clear, therefore, that unless replacement takes place, either through the further weathering of rock minerals or by decay of organic matter, ultimately the clay-humus complex could become completely hydrogen-saturated by this _leaching_ process. In fact this rarely happens because, when a high degree of acidification has been achieved, the clay itself begins to decompose (Chapter IV) and the aluminium ions released therefrom displace some of the exchangeable hydrogen.

Base status and pH value

The actual content of exchangeable mineral ions is referred to as the _base status_ of a soil. In theory the concept of base status is easy to appreciate but, in practice, its determination is difficult. One method of obtaining

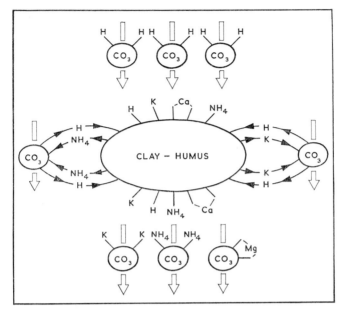

Fig. 9. Diagrammatic representation of the leaching of nutrient ions by percolating carbonic acid.

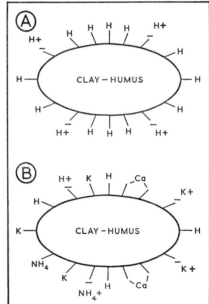

Fig. 10. Diagrammatic representation of the ionization of clay-humus micelles (A) in a complex which is hydrogen saturated and (B) in one which is partly hydrogenised and partly satisfied with basic ions.

Dvs

some index of the state of the clay-humus complex is by measuring the acidity of the soil solution. The assumption is that, the more exchangeable hydrogen there is attached to the clay-humus radicles, the less 'room' there will be for mineral ions. The usual way of stating the acidity of a solution is in terms of its *pH value*. The pH value of a soil is a statement of the number of free hydrogen ions in the soil solution. As in the case of potassium carbonate (*vide supra*), the clay-humus radicles and their attachments also ionise to a certain extent. At any point of time an ascertainable percentage of the appurtenant ions will actually be dissociated in the soil solution. These dissociated ions will be a representative sample of all the exchangeable ions present. It follows from this that the greater the total quantity of hydrogen ion, the more ionised hydrogen will be present (Fig. 10). In the case of a clay-humus complex which is completely hydrogen-saturated therefore (A), all the dissociated ions will be hydrogen; in the case of one which is partially satisfied with hydrogen and partly with bases (B), there will be less ionised hydrogen while, in the case of a completely base-saturated complex, all the freed ions will be bases.[1]

A solution which, by weight, is exactly 1/10,000th part hydrogen ion is said to have pH 1; one which is 1/100,000th part hydrogen ion—pH 2; 1/1,000,000th part—pH 3 and so on. It will be seen that the pH progression is a logarithmic one but that the lower the acidity happens to be, the higher the pH value. In fact the pH value of a solution is *the logarithm of the reciprocal of the hydrogen ion concentration.*

It can be demonstrated that a soil whose clay-humus complex is almost saturated with hydrogen, with no other acids present, will have approximately pH 4. On the other hand, one which is base-saturated may have a pH of anything between 7 and 10, depending upon the proportions of the different bases present. All gradations between pH 4 and pH 7 (neutral point) can be found, corresponding with the gradation between almost complete hydrogen-saturation and near base-saturation of the clay-humus complex. There are also some soils which, apart from having their clay-humus complex hydrogen-saturated, also contain certain amounts of free organic acids derived from humus. These soils may have pH values of 3·3, 3·2 or even less. At the opposite extreme, most soils in sub-humid and semi-arid regions, apart from having a completely base-saturated clay-humus complex, also contain an excess of bases like calcium carbonate ($CaCO_3$) or sodium carbonate (Na_2CO_3). They therefore have

---

[1] Even in this case, however, the soil solution will not be entirely devoid of hydrogen ions since water itself (H_2O) ionises into $H+$ and $OH-$. In fact no chemist could regard the above explanation of pH as entirely satisfactory; fundamentally the reaction of an aqueous solution is determined by the balance between the $H+$ and $OH-$ ions present, neutral point being achieved when they are numerically equal [15].

alkaline reactions and pH values of about 8 (in the case of $CaCO_3$) or as high as 10 (with Na_2CO_3).

A statement of the pH value of a soil thus gives some indication of its fertility and one can be fairly confident that, if the pH of *a particular* soil falls, this is due to an increase in exchangeable hydrogen with a concurrent decrease in exchangeable bases. Unfortunately, however, the same kind of inference cannot be made when comparing the pH values of *different* soils. Two different soils may have similar pH values and yet contain very different amounts of exchangeable bases; conversely, two other soils may have very different pH values and yet contain very similar amounts of exchangeable bases. This can be understood by a consideration of two hypothetical cases. The first of these is a soil whose clay-humus complex is exactly base-saturated and which has a pH of 7. The second is a soil of quite different constitution whose clay-humus complex is approximately half-satisfied with bases and half with hydrogen and which, in consequence, has a pH of 5·5 (i.e. half-way between 4 and 7). At first glance one might infer from this information that the first of the two soils must inevitably contain a larger quantity of exchangeable bases than the second one but this is not necessarily so. If it is further postulated that the first soil consists of about 95% skeletal material and only about 5% clay and humus while the exact opposite is true of the second soil, then the two are seen in an entirely different light. Even though the first soil is base-saturated, it contains such small amounts of chemically active clay and humus that the actual *quantity* of bases being held cannot be anything but small. On the other hand, the second soil is so rich in clay and humus that it retains much larger quantities of bases even though it is partly satisfied with hydrogen. In other words the *base exchange capacity* of the first soil is small while that of the second is large. It follows from this, that a simple statement of pH value does not, of necessity, give an accurate impression of the base status of a soil. Consequently modern soil analyses include quite separate assessments of all the main exchangeable bases such as potassium, calcium and ammonium.

Colloidal properties

The two types of substance which combine to form the clay-humus complex have a further property in common which is important in soil development. Both clay and humus are colloidal. A *colloid is a* substance which, without actually changing its chemical composition, can exist in quite different physical states: it may be in either a flocculated or a deflocculated condition dependent upon the physical and chemical nature of its environment. These soil colloids remain flocculated provided the pH value does not fall far below 7, but deflocculate very readily with strong acidity (low pH). Consequently, in acid podzols, whole particles of clay and humus are moved rapidly downwards by percolating water. The upper mineral horizons thus tend to become poorer in clay and relatively richer in skeletal material. Any humus in these uppermost mineral horizons also tends to be transient; it is in process of being moved downwards from the peaty humus of the uppermost layer.

Other changes also take place in a hydrogen-saturated clay-humus complex. Not only do whole molecules become mobile and move downwards, the clay molecules also become chemically unstable and begin to dissociate. As has been noted already, clays are complex silicates and their generalised formula can be written as follows:

$$xAl_2O_3 . yFe_2O_3 . zSiO_2.$$

For practical purposes therefore they can be assumed to consist of different proportions of molecules of iron sesquioxide, aluminium sesquioxide[1] and silica though in many cases (e.g. kaolinite) the iron sesquioxide is lacking and, in most, metals such as potassium and magnesium are incorporated. Generally speaking, however, when clay dissociates, the two sesquioxides and silica are the substances which are most likely to be released. It is important to note, however, that whole molecules of clay do not disintegrate instantaneously. In some environmental conditions they shed sesquioxide molecules faster than silica molecules; this happens in the cases of podzolisation. In other conditions, silica is lost faster than sesquioxides; this is the laterisation process. The clay in the upper horizons thus changes its general composition and may ultimately disintegrate altogether. Much more will be said on this subject in subsequent chapters.

The separate molecules of sesquioxide and silica may themselves be colloidal so that they too can become mobile in a soil which is periodically soaked by rain. As they move downwards along with particles of undissociated clay and humus, they usually encounter different chemical conditions in the lower, less weathered layers. Because of this they are frequently re-deposited. Thus, in most soils which are subject to leaching, it is possible to distinguish two quite distinct sets of horizons beneath the uppermost mat of organic material. The upper set are referred to as '*eluviated horizons*' and the lower as '*illuviated*' ones. There is an almost infinite variation in the nature and arrangement of materials within this general scheme of things however. Indeed there are so many major and minor variables operating in soil development that it is quite difficult to find two soils from different sites which are alike in all their main characteristics.

Pedalfers and pedocals

Leached and acid soils whose clay minerals have suffered some dissociation into their component parts are referred to as '*pedalfers*'. Soils of this class occur, generally speaking, in areas where the amount of water

[1] A sesquioxide is an oxide which has one and a half oxygen atoms to every one metallic atom.

⊠	Tundra soils	⊞(+++)	Chestnut-brown soils
≡	Podzols	⊞(''')	Light-brown semi-desert soils
≣	Brown podzolic and brown forest soils	▦	Grey and grey-brown semi-desert soils
⠿	Degraded chernozem (grey forest soils) and leached chernozem of the .meadow steppe	⊞(ᴐ)	Blown sand
⠄	Chernozems	⊞(∧)	Mountain complexes

Fig. 11. Generalised zonal soil map of western U.S.S.R.

reaching the surface of the soil as precipitation exceeds the amount which is evaporated from it. The excess water percolates down through the soil removing any excess bases which might otherwise accumulate. In drier areas there is no such excess of water and here some of the less soluble bases, particularly calcium carbonate ($CaCO_3$), are able to accumulate. This ensures that the clay-humus complex remains base-saturated and furthermore, the excess of lime gives rise to a pH in excess of 7. Because of this alkalinity, the clay remains quite stable with no tendency to dissociation. These lime-accumulating soils are classed together as *pedocals*.

The zonal concept

Several main factors are operative in determining the course of soil development under natural conditions. The importance of the lithological characteristics of the parent-material has already been demonstrated. If this is coarse and siliceous it weathers to a material which permits rapid leaching; it will also probably be poor in plant nutrients from the outset and incapable of retaining any considerable amount of nutrient which might be supplied. On the other hand, parent-material such as calcareous marl will weather to a much less pervious clay material which is rich in calcium and other nutrients from the start. From these two examples it is clear that rates of leaching and innate fertility must be influenced by the physical and chemical properties of the parent-rock.

Climate also is of obvious importance since it influences the rate at which leaching, weathering and organic decay take place. Other things being equal, leaching of nutrients and the podzolisation process will proceed much more quickly with heavy rainfall than with light rainfall. Similarly the weathering of rock minerals and the decay of organic materials go on much more quickly in a hot, moist climate than in a cool, dry one. Generally speaking, the average rate of decomposition processes, both mineral and organic, doubles with every increase of 18° F. (10° C.) above freezing-point.

The pioneers of modern soil science were so impressed with the apparent correlation between climatic and soil distributions over the continents that they devised a *zonal soil classification* which has persisted to the present day. The basic premise underlying this classification is that, regardless of the original nature of the parent-materials, given a certain set of climatic conditions, a specific type of soil will ultimately come into existence. Russian pedologists such as Glinka [31] were the chief protagonists of this classification. They noted a belt or 'zone' of podzols across northern European Russia (Fig. 11) in a region characterised by cool, moist climatic conditions and a vegetation of coniferous forest. Further south was a zone with warmer summers and less intense leaching where brown forest soils were found beneath a cover of deciduous forest. South again, in the Ukraine, was a zone of hotter, drier conditions in the growing season where grasslands were found prior to clearance and cultivation. Here black earths or *chernozems* were the commonest soil type. The zonal school were further confirmed in their views by the discovery that soils from analagous climatic regions in North America (Fig. 12) appeared to be very similar to those of Russia. It seemed to them that climate, rather than parent-material, was the more important factor accounting for the distribution of soils on a continental scale.

Limitations of the zonal concept

Even at its inception, however, the zonal classification had to be hedged around with provisos. Just as all wild vegetation cannot be regarded as climatic climax vegetation, so many natural soils are obviously not 'climatic climax' or 'zonal' soils. As in the case of vegetation, there are numerous factors which arrest, deflect or in some way inhibit soil development.

Firstly, it is clearly necessary for weathered material to remain quite

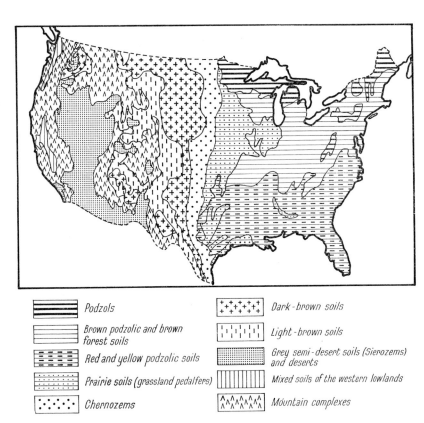

≡≡≡	Podzols	+++++	Dark-brown soils
	Brown podzolic and brown forest soils	¦¦¦¦¦	Light-brown soils
≡-≡-≡	Red and yellow podzolic soils	▓▓▓	Grey semi-desert soils (Sierozems) and deserts
	Prairie soils (grassland pedalfers)	‖‖‖‖	Mixed soils of the western lowlands
•·•·•	Chernozems	ΛΛΛΛΛ	Mountain complexes

Fig. 12. Generalised zonal soil map of the U.S.A.

undisturbed *in situ* for a long period of time in order for weathering and *pedogenic processes* to produce a mature profile (i.e. one which does not alter in its essential features with further lapse of time). It is this mature soil, in equilibrium with climate, which is implicit in the concept of 'the zonal soil'. But in many places on the earth, relief and erosional forces do not permit the same material to be undisturbed, with no additions or sub-tractions, for long periods of time. In any area of considerable slope the process of soil creep, even beneath dense vegetation, will be sufficiently intense to remove or mix the developing horizons long before maturity has been achieved. At the opposite extreme, on flat alluvial plains or in volcanic areas, frequent deposition of mineral-rich alluvium and volcanic ash respectively, will effectively offset the pedogenic process of leaching. Addition or removal of inorganic material thus effectively prevents the attainment of maturity; soils subject to persistent truncation or to frequent deposition are therefore referred to as *permanently immature soils*. A special category had to be made for them in the world zonal classification: they were classed together as *azonal soils*.

Apart from the latter there are other soils which are anomalous within the general zonation of soils. Firstly, there are distinct types in naturally ill-drained areas. Soils with natural free drainage experience persistent percolation of water in areas of heavy rainfall but this can be prevented by a high water-table. Even in areas of low rainfall, impeded drainage often prevents the leaching of the most soluble substances such as common salt ($NaCl$) so that they are liable to accumulate in the soil. Impeded drainage thus causes anomalous soils to develop in both humid and dry regions. Anomalous soils also develop on certain types of rock outcrop, those which form on limestones in humid areas being particularly note-worthy. Because limestones are composed almost entirely of calcium carbonate, the weathering complex derived from them can be particularly lime-rich. This, in many cases, is sufficient to offset losses by leaching so that areas of limestone within regions of generally podzolised soils may carry soils which are actually alkaline in reaction. An island of soil thus arises which has characteristics very similar to those possessed by soils in dry regions. It is because of this that soils which, for all kinds of sporadic reasons, are anomalous within the areas where they are found, have been classed together as *intrazonal soils*.

Providing its limitations are clearly appreciated, the soil classification devised by the zonal school (Fig. 13) is a useful framework in which to view pedogenesis on the world scale. Its main limitation is analogous to that which restricts the concept of climatic climax vegetation. In both cases one factor—that of climate—is taken as the basis for classification; all other factors are regarded as subsidiary. Thus, for an area like England and Wales where the climate is everywhere relatively humid and cool, a small-scale zonal soil map shows a homogeneous area of brown forest soils. Because England and Wales have a very varied lithology, however, the true picture of soil distribution, even under natural conditions, was very much more complicated than this. Indeed, over large areas, the zonal soil-type was the exception rather than the rule; limestone soils, truncated soils, alluvial soils and soils with impeded drainage predominated. Quite clearly, climate is not everywhere the predominant factor determining soil distribution; for areas such as this, the zonal soil map may give a very misleading picture.

The zonal soil concept has other limitations. In most kinds of parent-material and in all types of climatic conditions, many centuries are required for the evolution of a mature soil profile. The zonal concept requires a gradual development of soil characteristics in response to climatic conditions so that, ultimately, equilibrium between soil structure and climate is reached. From then onwards, although continued weathering may cause further increase in depth, no change in the salient features

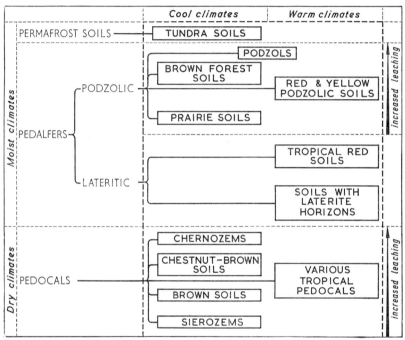

Fig. 13. A classification of zonal soils.

of the profile will occur. For such equilibrium to be reached it is quite clear that climatic conditions must remain constant throughout the whole period of profile evolution; if a major change in climate occurs, the whole trend of soil development will also be altered. It is well known, however, that during the past 10,000 years there have been profound climatic changes throughout much of middle and high latitudes. Many places here are experiencing very different soil-forming processes today from those which were operating in 1000 B.C. Knowledge of the climatic history of inter-tropical areas is still very scanty but here again evidence of considerable climatic change during the past 10,000 years is accumulating rapidly.

This being the case, it seems likely that many profile features which can be observed today may not be the product of present climatic conditions; they may have been formed in the past under quite different conditions and thus be mere survivals. Relict horizons of this nature may be particularly plentiful in the tropics (Chapter XIX). Some of those which may traditionally have been regarded as typical features of present-day 'tropical zonal soils' may be fossil horizons of this kind.

The zonal soil concept is useful as long as these limitations are fully recognised. It cannot be denied that beneath grassland in a dry sub-humid region in mid-latitudes a zonal chernozem can confidently be expected, provided the slope is not too great and the drainage is quite free. It will have been noted, however, that many types of zonal soil are associated with specific types of vegetation as well as particular types of climate. Indeed, it seems probable that associated vegetation may be of even greater significance than climate. Much more will be said on this subject in subsequent chapters but some preliminary explanation of the role of vegetation in pedogenesis is a necessity.

The nutrient cycle

Many pedologists in the past, while recognising the importance of vegetation as a supplier of humus, have had an over-simplified view of the relationship between vegetation and other factors of the environment. Many protagonists of the zonal concept, while postulating that the sub-humid climate was the reason for the belts of chernozem in Eurasia and North America (Chapter IX), cannot fail to have noticed that this chernozem has only come into existence beneath grassy vegetation; it has never formed beneath forest. They seem to have assumed, however, that the grass communities were the *inevitable* vegetation in a sub-humid climate and, this being the case, they saw no point in treating climate and vegetation as independent soil-forming factors. Many observations in the past half-century have shown this to be fallacious however. In Texas for instance, woodland and grassland are often found side by side in areas of identical climate, both types of community inhabiting soils which are freely drained. Beneath the grassland, however, a typical chernozem is found while beneath the woodland a quite distinct forest soil is maintained. A similar situation is found in the Ukraine; here the soil beneath the trees is often distinctly podzolised while that beneath the grass is an unleached chernozem (Chapter IX).

In the light of this kind of evidence, quite clearly, vegetation must be treated as a completely independent variable; completely different types of vegetation can be found in areas of identical macro-climate. Indeed, so specific are the effects of vegetation types on the soil profile, that many soil scientists now seem to view vegetation almost as the 'machine' in the soil-making 'factory'. It is as though the parent-rock provides much of the

raw material, the climate lubricates and determines the speed of manufacture but the vegetation ultimately determines the nature of the
finished product. This view probably overstates the importance of
vegetation to some extent just as the earliest pedologists over-emphasised

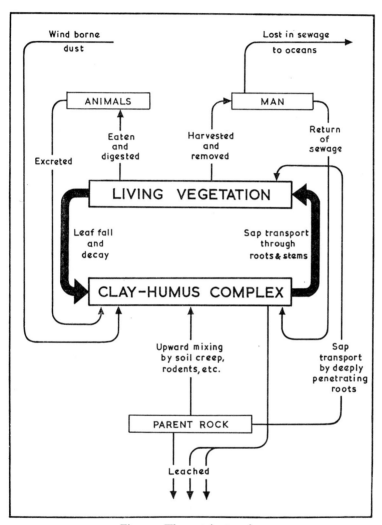

Fig. 14. The nutrient cycle.

lithology and the pioneers of the zonal concept overstated the case for
climate. Nevertheless, it cannot be denied that, without a cover of
nutrient-demanding vegetation, no soil could maintain indefinitely a high
base status in a humid climate, no matter how rich in bases the underlying

parent-material. Without such a vegetation cover to abstract bases from the soil and to return them in base-rich humus, leaching can rapidly remove all the bases as they are released from the rock minerals by chemical weathering. There is, clearly, a *nutrient cycle* in which vegetation is an indispensable link (Fig. 14). If a less demanding vegetation type supersedes a richer one the effect may be profound; from then onwards, even though more mineral nutrient is made available by further rock weathering, soluble plant foods will be permitted to slip away rapidly and to be lost to the ground water. On the other hand, if a more demanding vegetation takes the place of a poorer one, a richer soil will gradually develop as more and more bases are incorporated into the cycle. This must have happened over large areas of lowland Britain where the original forests have given place to more demanding grass communities. Because of human interference the original brown forest soils were thus changed into soils which, in many ways, resemble true prairie soils (Chapter XIII).

Since the relative amounts of exchangeable bases and exchangeable hydrogen associated with the clay-humus complex have such an influence on the rate of processes like podzolisation and laterisation, quite clearly the richness (or poverty) of the nutrient cycle is of salient importance. The extent to which vegetation, quite independently, can affect the cycle is still open to much speculation. Some demanding plant communities may be capable of invading a poor environment and of raising the level of the nutrient cycle; other communities may be less resilient and flexible in their reactions. There can be no simple answer to the general problem; all communities and all individual environments must be studied quite independently. Indeed the complexity of the soil-vegetation relationship is so great that it does not seem inappropriate to devote all the following chapters to a world survey in which the two are viewed together, region by region.

PART TWO

VEGETATION AND SOILS
OUTSIDE THE TROPICS

CHAPTER IV

Coniferous Forests

Of the two classes of flowering plants the Gymnospermae are the more primitive and ancient. Nevertheless, they contain a group of plants which still dominate the vegetation over considerable portions of the land masses of middle and high latitudes, particularly in the northern hemisphere. The coniferous tree, with its tapering, symmetrical form and needle-shaped leaves can still compete successfully with the broad-leaved trees of the Angiospermae in quite a variety of environments. Species of pine (*Pinus spp.*), spruce (*Picea spp.*) and fir (*Abies spp.*) are particularly common and widespread. The vast majority of conifers are evergreen, (the deciduous larches (*Larix spp.*) being exceptional in this respect), and there can be little doubt that the evergreen habit, linked with other common characteristics of the conifers, account for their widespread dominance.

I. THE BOREAL FOREST

The coniferous evergreen tree has proved itself to be particularly well adapted to those areas with cold winters and short growing season in Eurasia and North America. The reduced area of the needle-shaped leaves enables these trees to reduce water-loss by *transpiration* to a very low level during the winter when water is unobtainable from the frozen soil. On the other hand, as soon as air and soil temperatures are sufficiently high to permit *photosynthesis* and other physiological processes to commence in the spring, these trees are able to profit by it. They have an advantage over deciduous trees which lose their leaves in the cold season in that the latter must grow a new set of leaves before photosynthesis can begin. A deciduous tree thus loses a significant proportion of the short growing season and, in consequence, will tend to be over-topped and eliminated by more luxuriantly-growing evergreens. This point must not be overstated however. Under even more rigorous conditions, the evergreen conifer may be unable to survive the winter merely because it is an evergreen. In some of the northernmost forest areas the winter climate is so bitter that a tree which carries any leaves at all finds it difficult to survive. In such areas therefore, very slow-growing deciduous trees may be the dominant plants. This is the reason why deciduous birches (*Betula spp.*), aspens (*Populus spp.*) and larches (*Larix spp.*) often survive on the edge of the tundra beyond the northern limit of the evergreen coniferous forests (*vide infra*).

It is, therefore, because of their ability to survive a fairly rigorous winter and their adaptation to a relatively short summer, that the evergreen conifers have come to dominance in the great northern belts of forest stretching from east to west across Eurasia and North America. These forests extend almost to the Arctic shore in relatively sheltered areas such as the lower Mackenzie Valley (Map 4). The factors determining the position of this northern forest limit will be discussed later (Chapter VIII). The northern coniferous forests are usually regarded as comprising a single formation-type known as 'the boreal forest' and consisting of two separate formations—one in Eurasia and one in North America.

Dominant species

Even within one formation, however, the species-composition and floristic richness varies very much. In the Eurasian formation the European section is much poorer in species than the Asiatic one. From western Norway to the Urals the pine (*Pinus sylvestris*) and the spruce (*Picea excelsa*) are absolutely predominant but as one passes eastwards across European Russia there is a gradual increase in the frequency of occurrence of other species. The Siberian fir (*Abies sibirica*), the Siberian larch (*Larix sibirica*), the dwarf Siberian pine (*Pinus pumila*) and the Siberian spruce (*Picea obovata*) are the first ones to become frequent and the tree flora continues to become richer eastwards. Ultimately, in Sakhalin and Hokkaido, although the boreal forest is actually dominated by only a few species like the firs *Abies veitchii* and *A. sachalinensis* [19] and the Yezo spruce (*Picea ajanensis*), a great variety of subsidiary species are present. This is the outstanding difference between the boreal forest of Europe and that of eastern Asia. Another significant feature of the Asiatic boreal forest is the abundance of larches. Several species of these deciduous conifers are common in northern Japan, Manchuria and south-eastern Siberia but it is in eastern Siberia that the most interesting situation is found. From the valley of the Yenisey to the Sea of Okhotsk and extending as far as 73° N., is an area of coniferous forest which survives the coldest winters experienced in the northern hemisphere. The forests here also have to contend with a very shallow layer of soil above the permafrost. The dominant tree almost everywhere in these forests is the shallow-rooted dahurian larch (*Larix dahurica*) (Plate II*a*). Although the dwarf Siberian pine (*Pinus pumila*) is a frequently associated species, it appears that no tall evergreen conifer can survive here, so that the dahurian larch has no serious competitor. It is thus able to maintain dominance even though the average annual growth is very small indeed. Though this species is capable of

B

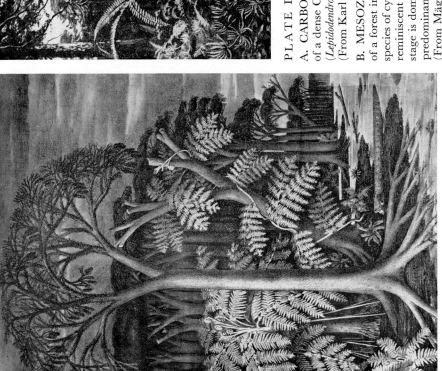

A

PLATE I

A. CARBONIFEROUS FOREST. An artist's reconstruction of the edge of a dense Carboniferous forest dominated by tree club-mosses (*Lepidodendron*) with subsidiary tree ferns (*Dactylotheca*) and climbing ferns. (From Karl Mägdefrau, *Paläobiologie der Pflanzen*, 1953; after Potenie, 1899).

B. MESOZOIC FOREST. An artist's reconstruction of the riverine edge of a forest in Rhaetic-Liassic times. In the foreground there are various species of cycad and cycadeoid (primitive gymnosperms with many features reminiscent of the pteridophytes); in the middle distance the hydroseral stage is dominated by tall horsetails; in the background is the edge of a predominantly coniferous forest dominated by a species of araucaria pine. (From Mägdefrau, after Gothan, 1926.)

A

B

PLATE II

A. BOREAL FOREST. A stand of well-developed Dahurian larch (*Larix dahurica*) in Sakhalin.
(From a photograph by Krasnov in Schimper, 1903.)

B. COAST FOREST. A stand of coastal redwoods with thick herbaceous and shrub undergrowth dominated by ferns and evergreen huckleberries.
(U.S. Forest Service photograph, by permission of the U.S. Department of Agriculture.)

developing into quite well-formed trees, over large areas it grows to no more than a few feet in height and individuals which appear superficially to be no more than saplings may be of considerable age. It is doubtful if these areas would be 'forested' at all were it not for the great climatic and edaphic tolerance of this tree. Although it has become customary to include this larch forest in the boreal forest belt, strictly speaking it should be regarded as distinct since its main dominant is a deciduous and not an evergreen tree.

In the general richness of its flora the North American boreal forest formation resembles that of Asia more than that of Europe. Once again, however, there is a gradual transition between the western and eastern parts of the formation. Thus, on the southern part of the Laurentian Shield, in the provinces of Quebec and Ontario, the white spruce (*Picea glauca*) and balsam fir (*Abies balsamea*) dominate on the better-drained soils, with the black spruce (*P. mariana*) and tamarack (*Larix laricina*) in areas of impeded drainage. The jack pine (*Pinus banksiana*) is also common on poorer or wetter soils which are unfavourable for the more demanding species. On the other hand, as one passes westwards across the northern Prairie Provinces, these eastern dominants become less frequent until ultimately, in northern British Columbia and central Alaska, they have given way to the lodgepole pine (*Pinus contorta* var. *murrayana*), the alpine fir (*Abies lasiocarpa*) and a number of other western species.

Distinct associations

A journey through certain portions of the virgin boreal forest can leave one with the impression of extreme monotony. For mile upon mile the dense cover of evergreen conifers casts an unbroken shadow on the ground beneath. In consequence undergrowth is scanty or non-existent, the ground being covered with a litter of needle-shaped leaves and dead, decaying wood. The sombre, shadowy tones of trunks, branches and foliage are rarely relieved by the brighter hues of grass and flowers. However, a more analytical view of the boreal forest as a whole, reveals an interesting patchwork of distinct communities. Firstly, over wide areas, the forest itself is far from continuous; flat areas are frequently too ill-drained to support tree growth and are occupied by bogs or 'muskegs' in which plants like bog moss (*Sphagnum spp.*) and cotton grass (*Eriophorum spp.*) are dominant. Also exposed slopes or hill tops are frequently almost 'bald' since the strong, cold winds of winter only permit the growth of low-growing, heathy shrubs and herbaceous plants, identical with those which dominate in the continuous tundra to the north (Chapter VIII).

Quite apart from this, however, the forest itself shows a clear division into distinct associations, often within quite small areas. In northern Europe, for instance, the pine (*Pinus sylvestris*) is normally dominant on sandy soils, quite often forming almost pure stands. On the other hand, the

Ev5

spruce (*Picea excelsa*) is frequently completely predominant on loams and clays [44]. The spruce, quite clearly, is the more demanding species and thus cannot compete with the pine where the soil-parent-material is poor in nutrients. On the other hand, on richer parent-material, the spruce is able to maintain a relatively rapid rate of growth and to cast a deep shade. Though the pine could flourish here as it does on the sandy areas, it is prevented from doing so by competition. Similar differentiation into distinct associations is also found in Asia and North America. No one species of coniferous tree seems to have exactly the same demands and tolerances as any other species so that, in the process of competition for nutrients at root level and for light above soil level, the species gradually segregate themselves into communities with only one or two common dominants. Climate and soil would permit most of the species to have wider ranges than they actually possess; competition limits them to their more restricted territories.

Seral communities

The existence of distinct associations in the boreal forest with the frequent occurrence of almost pure stands (*consociations*) has facilitated their exploitation on a large scale. Maximum efficiency can be achieved where the most economically desirable species grow in pure stands and there is no necessity to 'extract' them from the forest. During the last half-century the boreal forest has been the main source of wood pulp for the world's newspapers so that today many thousands of square miles of former virgin boreal forest have been very much modified. Part of this area has been felled; part of it has been inadvertantly cleared of trees. With the advent of human populations one of the persistent problems in the forests is that huge numbers of trees are lost each year because of accidental fires and burned-over areas are now just as extensive as cut-over areas.

The economic and conservation problems created by this rapid forest depletion arise mainly from the fact that very large clearings do not revert quickly to their former state. Experiments in felling in Scandinavia have shown that if the forest is cleared in relatively narrow strips, leaving untouched belts of forest in between, natural regeneration of the original spruce or pine will take place within one or two seasons. On larger clearings this does not occur; a quite protracted subsere is experienced on all large burned-over and cut-over areas and, in most cases, the first stage in this subsere is dominated by deciduous trees. In both Eurasia and North America species of birch (*Betula spp.*) and aspen (*Populus spp.*) are the first to re-invade. In most years these trees produce vast quantities of light-weight, wind-borne seeds which are capable of being carried many miles from their parents. Furthermore, these seeds appear to germinate much more successfully than the seeds of the conifers on areas which are open to the drying agencies of wind and sun. Consequently, at the present time, large areas of these deciduous trees are to be seen within the general area of the boreal forest where formerly they only occurred in small, infrequent clumps or as isolated individuals. Frequently all the canopy in these large stands of birch and aspen is formed by trees of exactly the same age, each one having germinated in the year immediately following that in which the burning or felling of the climax forest took place.

These stands of deciduous trees are gradually invaded by coniferous species. On the Laurentian Shield it is often the jack pine (*Pinus banksiana*) which first intrudes but this is subsequently followed by the true climatic climax dominants of the area such as the white spruce (*Picea glauca*) and the balsam fir (*Abies balsamea*). As these conifers grow to maturity, they gradually over-top and shade out the deciduous trees which again become an unimportant element in the forest. This common type of subsere illustrates very clearly that the deciduous, broad-leaved tree, as a life-form, is not precluded from these areas by climatic conditions. Birch and aspen can flourish here if not subjected to competition by taller conifers. If the evergreen conifers are kept out or removed by climatic or human agencies, then the deciduous trees demonstrate how well they are able to withstand the climatic rigours of these northern latitudes.

The nutrient cycle and the soil

Nearly all the boreal forest experiences a mean annual precipitation of less than forty inches, indeed most of this formation-type is found between the fifteen- and twenty-inch isohyets. Because evaporation rates, even in summer, are never very high and because of the concentration of the precipitation into the growing season, however, the climate here must be regarded as a moist one. Very rarely, if ever, will the surface soil beneath the forest dry out and cause a prolonged upward or *capillary* movement of the soil solution. During the months of winter the ground is frozen continuously and during the summer the melt-water and the rain-water seep downwards through the soil. Since the forest casts a deep shade and maintains a calm atmosphere near the ground, the surface of the soil is protected from rapid evaporation. Because of this combination of circumstances movement of water in the soil, whenever it takes place, is nearly always downwards. In pervious soils leaching is therefore rapid and in all types of material the development of a pedalfer type of soil inevitably takes place.

Coniferous trees vary a great deal from species to species in the demands they make upon the soil for mineral nutrients. Generally speaking, however, they are much less demanding in this respect than are most broad-leaved trees. In particular most species of pine require very little mineral material and are thus able to thrive on very poor, sandy soils. Consequently pines normally dominate on sandy glacial outwash, which is of such frequent occurrence in Sweden and Finland, as well as on sandstone, quartzite and acid igneous rocks on the Laurentian and Scandinavian Shields. Since the pine trees abstract little mineral material from the soil, the vegetable debris which they return to the soil must also be very deficient in this respect and the humus resulting from the partial decomposition of it will be a very acid *mor* or *raw humus*. It follows from this that, on sandy parent-material beneath stands of pine in this type of humid climate, conditions are very favourable for *base-desaturation* and *podzolisation* (Plate IV*a*). Whatever mineral nutrients are released by weathering from the parent-rock-material are not required by the undemanding vegetation and are thus leached away rapidly.

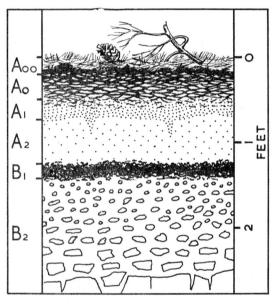

Fig. 15. A podzol profile beneath pine forest showing eluviated horizons (A) and illuviated horizons (B).

Because of the rapid leaching of minerals and the presence of raw humus, the whole soil becomes very acid and any clay minerals which are initially present in the weathering parent-material become very mobile and unstable (Chapter III). They are consequently either leached downwards in a chemically unaltered state or they are decomposed. In this process of decomposition the sesquioxides of aluminium and iron (Al_2O_3, Fe_2O_3) are released most rapidly and, since they too are very soluble in an acid medium, they are leached rapidly from the upper parts of the soil profile. So thorough is this *podzolisation process* under these conditions that normally the whole of the clay fraction is completely removed from the upper part of the profile leaving *eluviated horizons* which are almost completely composed of sandy skeletal material (Fig. 15).

The clay and sesquioxides which are removed from the upper horizons are, in part, re-deposited lower in the profile thus forming a lower set of *illuviated horizons*. The exact reasons for this are not fully understood but the basic reason is almost certainly that mineral substances which are encountered in the less weathered materials in this lower half of the profile cause the clay and sesquioxides to lose mobility and to become stable again [15]. Normally two distinct illuviated horizons are formed. The upper one is very often dark brown or blackish in colour due to the presence of re-deposited humic material in complex chemical combination with the sesquioxides of both iron and aluminium. This horizon is normally compact but friable in nearly all podzols found beneath pine forest. Beneath some plagioclimax heaths however (Chapter XIII) and beneath some cultivated soils which have been persistently ploughed to only shallow depth, this uppermost horizon may have a much paler, reddish-brown colour and be markedly indurated with hard concretions of ferruginous material. These are referred to in Germany as "ortstein" and by various terms such as "hard pan", "plough pan" and "fox bench" in different parts of Britain. The lower horizon may be any colour from a dark, reddish brown to a pale yellow-brown depending upon the relative proportions of the two sesquioxides present and on the exact chemical nature of the iron sesquioxide. It is impossible to make any meaningful generalisations regarding the detailed characteristics of either of these horizons however since the variations in their appearance and composition from one site to another are almost infinite.

It is customary to label the horizons of soils from the uppermost downwards referring to the eluviated ones as the 'A' horizons and the illuviated ones as 'B' horizons. Normally the uppermost 'A' horizon consists of a loose mat of undecayed pine needles and other organic debris whose original structure is still quite obvious. This is the A_{00} horizon. It normally merges into an underlying layer of humified organic material where decay organisms, predominantly fungi, have partially decomposed the material, destroying its original structure. This horizon is still almost entirely organic however and is referred to as the A_0 horizon. Beneath this there is normally quite a sharp discontinuity where one passes from predominantly organic to predominantly inorganic material. This underlying A_1 horizon is composed of mineral matter from which the clay minerals have been almost completely removed; it is almost entirely sand but is stained with organic material which has either percolated from above or has been derived from decaying roots. The A_1 horizon is thus dark grey above, becoming progressively paler downwards. Ultimately this horizon gives way to the A_2 horizon which is almost entirely composed of light grey or whitish, bleached sand. The junction between the A_2 and

B_1 horizons is usually very sharp indeed. The latter is often the most shallow horizon in the whole profile and gives way with varying degrees of sharpness to the underlying B_2 horizon. Finally this grades into the C horizon which is composed of only partially weathered parent-material. The 'typical' podzol profile just described is of very common occurrence beneath pine forest but it must be appreciated that many variants are encountered. Frequently the B_1 horizon is completely lacking while, in other profiles, the A_2 horizon may be indistinct. Furthermore the depth of the whole profile can vary enormously from a few inches to several feet. Normally, however, a soil would not be referred to as a podzol unless some kind of bleached A_2 horizon were present and unless there was a clear accumulation of acid humus in an A_0 horizon. It seems likely that complete removal of the clay minerals from the eluviated horizons would be impossible without the presence of colloidal organic acids derived from this overlying raw humus. Given a vegetation cover of *acidophyllous* pine trees, a cold climate and slow, fungal decay, however, it is probable that the formation of such an A_0 horizon is inevitable.

The sharpness of the junction between some of the podzol horizons is significant. Quite clearly there can be very few mixing agents present in the podzol otherwise this marked stratification could not come into existence. Podzols are almost devoid of earthworms and most of the other types of organism which so effectively prevent the development of marked stratification in other soils which contain a more base-rich type of humus. In the evolution of the normal forest podzol the only structures which break through the developing horizons are the anchoring roots of the trees themselves.

The above description of podzol evolution relates to those soils beneath associations of pine on poor, pervious materials. With other species and other materials within the boreal forest the resultant soil is often very different. Species of spruce for instance are much more demanding; they usually grow on heavier, richer soil-parent-material from which they are able to abstract plentiful supplies of mineral nutrients. In consequence, the organic debris which they return to the soil is more base-rich and less acid. Furthermore the heavier parent-material permits only a slower rate of percolation. The nutrient cycle associated with an association dominated by spruce is therefore richer than that associated with one of pine. Although the podzolisation process goes on beneath spruce forest, it proceeds at a much slower rate. Quite often a really well-bleached A_2 horizon is not obvious in these soils, the whole of the A_1 and A_2 horizons being much browner than in the podzol because of the presence of unleached sesquioxides and undecomposed clay. Indeed these soils are much more akin to brown forest soils (Chapter V) than to true podzols. They are consequently referred to as *brown podzolic soils* by many authorities [11].

Settlement in the boreal forest

Vast areas of boreal forest in Canada and the U.S.S.R. remain almost unpopulated. Even in Sweden and Finland the boreal forest areas remain the most sparsely populated sections of their respective countries. Some writers have attempted to explain this sparseness of population merely in terms of climatic inhospitability and shortness of growing season but there is strong evidence to suggest that this is by no means the whole story. For hundreds of years, from mediaeval times onwards, the country which is now Sweden consisted of two entirely different types of landscape. The northern two-thirds of the country were almost unpopulated while the southern third, generally speaking, was a closely-settled territory. The boundary between these two was quite sharply-defined and ran from the north of Lake Vaner east-north-eastwards to the coast of Gävle Bay. It corresponded very closely in position with the boundary between the boreal forest and the mixed forest (Chapter V). To the north of this boundary the soils were either podzols or very heavily podzolised loams and clays; to the south there were frequent expanses of brown forest soils from which the original stands of deciduous trees had been removed. It was only in the nineteenth century, with the great expansion in wood-pulp manufacture and mineral exploitation that people settled north of this line in considerable numbers. A similarly close correlation is found in the U.S.S.R. where a line running roughly between the Leningrad and Moscow areas separates an anciently-settled landscape to the south from a recently-settled pioneer area to the north. No simple climatic or geological correlation can be found to explain these sharp discontinuities in land-use. Modern techniques and heavy fertilisation now permit successful agriculture in the former area of the boreal forest; originally the impoverished nutrient cycle maintained by the coniferous trees had held the soil in such a state of infertility as to discourage settlement by populations whose livelihood depended entirely upon a fairly primitive agriculture. It was necessary to transform the natural podzols but, when once this had been effected, agriculture was climatically possible.

2. THE SUB-ALPINE FOREST

In spite of their obvious adaptations to high latitude areas, forests dominated by conifers extend far south of the boreal forest belt. In part this is due to the fact that air temperature decreases with increase in altitude so that, on mountain ranges in middle latitudes, climatic conditions are found which, in some ways, resemble those of the boreal forest belt. Because of this, evergreen conifers can survive on these mountains as far south as the Tropic of Cancer, and even beyond it, in both the Old World and the New [81]. Just as the boreal forest gives way northwards to treeless tundra, so this sub-alpine forest gives way upwards to low-growing alpine vegetation. Generally speaking, the altitude of both the alpine and sub-alpine zones increases with decrease in latitude. Some of the species of conifer in the sub-alpine forest are identical with ones in the boreal forest of the same continent but most species, though they have closely-related counterparts in the boreal forest, are quite distinct. It is quite clear, however, in almost all cases, that the boreal and sub-alpine counterparts have sprung from common ancestors in the recent past.

Although there are obvious similarities between the mountain climates and those in the boreal forest region, there are also many differences. Increase in altitude certainly entails decrease in mean daily temperature throughout the year, so that late spring and early autumn frosts can occur on mountains when the same seasons are frost-free in near-by lowlands; the frost-free season in sub-alpine forest may therefore be no longer than in parts of the boreal forest. It is here that temperature similarities end however. Although the spring and autumn nights on mountains may be frosty, the day temperatures at these seasons are higher than in the boreal forest.[1] Even bigger contrasts are discovered when sun temperatures and light intensities in the two environments are compared. The light and heat intensity at midday will be much greater all the year round in the mountain environment because of the higher sun and the decreased thickness of the overlying layer of atmosphere. On the other hand, the length of day in the boreal forest is much longer in summer though much shorter in winter. The trees in the two environments thus have to contend with very different environmental régimes. It is small wonder that distinct species and varieties, with different tolerances and resistances, have evolved in each of the two environments.

Central and western Europe

Varieties of the pine (*Pinus sylvestris*) and spruce (*Picea excelsa*) as well as distinct species like the European larch (*Larix decidua*) and the silver fir (*Abies excelsa*) compose the sub-alpine forests of western and central Europe [45]. Remnants of the original pine and spruce forests are still to be found on the higher slopes of the Black Forest, Bohemian Forest and other Hercynian blocks in central Europe while the remains of distinct belts of coniferous forest can still be discerned along the flanks of the higher alpine fold mountains. Much forest of pine and silver fir still remains between 5,000 and 7,000 feet on the Pyrenees. Throughout the Alps, spruce forest persists in many places between 3,000 and 6,000 feet,

---

[1] *Mean daily maxima and means of extreme monthly minima (in parentheses) for a station in the boreal forest area and for one at sub-alpine heights in the Rocky Mountains (shown in °F.).*

	M.	J.	J.	A.	S.	O.
Kapuskasing, Ontario (752 ft.)	58 (*21*)	70 (*30*)	75 (*37*)	72 (*36*)	62 (*25*)	47 (*16*)
Cheyenne, Wyoming (6,139 ft.)	62 (*25*)	74 (*36*)	80 (*43*)	79 (*42*)	71 (*29*)	58 (*16*)

while the larch and cembran pine (*Pinus cembra*) form closed stands up to 7,000 feet in the central Alps. In the central Carpathians the surviving spruce forests begin just above 3,000 feet; they gradually give way upwards to larch which, in turn, gives way to cembran pine. The latter often extends upwards to almost 5,000 feet. A similar pattern of associations is found at sub-alpine heights on the mountains of central Asia but here close relatives of the Siberian boreal forest dominants compose the forests. In the western Caucasus, forests dominated by oriental spruce (*Picea orientalis*) and Caucasian fir (*Abies nordmanniana*) are extensive between 5,000 and 6,500 feet. Further east, in the Tien Shan, almost untouched forests dominated by Schrenk's spruce (*Picea schrenkiana*) extend between 4,000 and 7,500 feet while, on the Altai Range further to the north-east, larch (*Larix sibirica*) and fir (*Abies excelsa* and *A. sibirica*) dominate between 2,500 and 4,000 feet. Above this, *Pinus cembra* gradually asserts itself and persists up to the tree limit at between 4,500 and 5,500 feet.[1] Similar belts of coniferous forest are found on the other mountain ranges of central and eastern Asia, extending southwards to the Himalayas in Kashmir and Nepal and almost as far as the tropic in the mountains of northern Burma and southern China. Vast quantities of timber of many types and qualities exist but are almost unexploited because of their remoteness from areas of dense population and the ruggedness of the terrain that they occupy.

The Mediterranean area

The sub-alpine forests of Europe extend southwards into the Mediterranean peninsulas and even as far as the Atlas Mountains in North Africa and the mountains of southern Turkey and Syria in Asia. In parts of the European peninsulas some of the dominant species are identical with those on the mountains of central Europe. Thus the coniferous forests between 5,000 and 6,500 feet on the Sierra Nevada in Spain are dominated by *Pinus sylvestris*, and the silver fir (*Abies excelsa*), frequent north of the Alps, also dominates some of the forests between 3,000 and 6,000 feet on the Apennines. Since forests in the mountains of the Mediterranean area have to resist a certain degree of summer drought as well as the reduced temperatures of the mountain climates, however, distinct species have evolved and survived as the dominants here. Although the summer drought on the mountains is generally not so intense as in the Mediterranean lowlands, it does persist for several consecutive months in quite a high proportion of years. Species of pine and fir seem to have been best adapted to withstand these conditions. Before the intervention of man, species of these genera probably formed almost unbroken belts of forest between about 4,000 and 6,000 feet in the Mediterranean peninsulas. In

[1] It must be appreciated that the upward limit of trees is very varied on nearly all mid-latitude ranges. Normally, in the northern hemisphere, trees extend much higher on the south-facing than on the north-facing slopes.

Spain the firs are represented by *Abies pinsapo*, in Greece by *A. cephalonica* and in the Taurus by *A. cilicica* [19]. The pines are more important further south, the Austrian, Corsican or black pine (*Pinus nigra*) being found in the southern Apennines, Crete, Greece, Asia Minor, Corsica and Sicily. Most interesting of all, however, are the coniferous forests of North Africa. In Algeria and Morocco, forests of Aleppo pine (*P. halepensis*) and fir (*A. numidica*) still survive between 2,500 and 4,000 feet. Above this level, extending upwards to about 6,000 feet, are the famous cedar forests (*Cedrus atlantica*). Here, on the middle slopes of the Tell Atlas and the Maritime Atlas (Plate III*b*), forests have survived which must be very similar in appearance to those which biblical references have made famous. 'The cedars of Lebanon' (*Cedrus libani*) must once have clothed the mountains at the eastern end of the Mediterranean; they have now almost disappeared from the Lebanon Mountains though are still plentiful in some more remote parts of the Taurus [19]. Although the North African cedar forests and other sub-alpine forests around the Mediterranean can still be seen, there is no doubt that they too are no more than scanty fragments of formerly more extensive forests. The shipbuilding Phoenicians and Venetians and many generations of transhumant sheep and goats have reduced former forests to poor mountain pastures and rocky slopes.

North America

The general north-south alignment of mountain ranges in North America has provided a much closer connection between the sub-alpine and the boreal forests than in Eurasia. There is an almost continuous belt of sub-alpine forest extending the whole length of the Sierras from Canada as far as southern California in the west, and another one, along the whole length of the Rockies as far south as New Mexico, in the interior of the continent. Both these belts of sub-alpine forest merge with the boreal forest in the north so that free migration of species from the one to the other has been possible. The Appalachians in the east are of lower elevation so that, although almost pure coniferous forests dominated by species like *Picea nigra* and *Abies fraseri* are found above about 5,000 feet in the southern Appalachians, these were cut off from the northern coniferous forests by an area dominated by broad-leaved deciduous trees.

The general height of the sub-alpine forest in the western part of the continent rises as one moves to lower latitudes. In southern Alaska it lies between about 2,000 and 4,000 feet but it rises to between 3,000 and 7,000 feet in British Columbia and to between 8,000 and 12,000 feet in New Mexico [81]. It clothes the crests of the highest mountains (over 10,000 feet) in southern California and Lower California.

Although the sub-alpine forests of these two great mountain ranges do possess their own distinct species and assemblages of species, they have sufficient in common to be regarded as the same formation. The Engel-mann spruce (*Picea engelmanni*) and the alpine fir (*Abies lasiocarpa*) occur throughout, except in California, and the lodgepole pine (*Pinus contorta*) is present almost everywhere from the mountains of the Yukon to the San Pedro Martir Range in California and to the Front Ranges in Colorado. The limber pine (*Pinus flexilis*) is also widely distributed. Apart from these few species which are almost universal in the formation, however, there are many more local ones indigenous within the sub-alpine forest, some of which are mere varieties of species found in the boreal forest and some of which are full species.

The sub-alpine forest of the whole of the northern hemisphere can be regarded as a single formation-type. Not only are its dominants predomin-antly of the same life-form, they are drawn from remarkably few coniferous genera. More than 90% of the species which are dominant in both Eurasia and North America belong to three genera—*Pinus*, *Abies* and *Picea*. All are characterised by the needle-shaped leaves and pronounced *monopodial* development which gives the northern coniferous forests their unmistakable appearance.

3. THE LOWLAND AND MONTANE CONIFEROUS FORESTS OF NORTH AMERICA

Apart from its boreal forest, North America originally possessed vast areas of lowland coniferous forests belonging to quite different formations. In this it was entirely different from Eurasia where all the lowland forests outside the boreal forest were either mixed or almost entirely dominated by broad-leaved, deciduous trees. Although much of this North American coniferous forest has been removed, large areas of it still remain unexploited.

The lake forest

Extending from Minnesota across Michigan, northern Pennsylvania and southern Ontario to northern New England, there lies an area which was originally covered, in great part, by luxuriant coniferous forest (Map 5). This forest, usually referred to as 'the lake forest', was quite different from the boreal forest to the north. The dominant species of the lake forest were quite distinct and, at maturity, they achieved much greater heights, often rising to 200 feet. The white pine (*Pinus strobus*), red pine (*P. resinosa*) and eastern hemlock (*Tsuga canadensis*) were predomin-ant. Most ecologists regard this lake forest as a distinct plant formation. Unfortunately, because its component species were so attractive as constructional timber, nearly all those areas of the forest which had survived earlier phases of settlement were swept away during the last two decades of the nineteenth century. So rapid and complete was this deforestation that hardly any coniferous trees survived over vast areas of Minnesota, Wisconsin and Michigan and natural regeneration was effectively prevented. To this day thousands of square miles of these areas of sandy glacial outwash remain as derelict 'cut-over' and poor scrub. The reasons for the existence of this isolated and unusual plant formation will be discussed later (Chapter VII).

The western forests

Equally anomalous are the extensive coniferous forests of western North America. Stretching southwards from southern Alaska down the coastal lowlands as far as central California was a continuous belt of forest completely dominated by tall evergreen conifers. It extended inland across the inter-montane basins and plateaux of British Columbia, Washington, Oregon and Idaho as far as the Rocky Mountains and covered the lower flanks of much of the Cascades, Rockies and Sierra Nevada. Much of this forest still remains. It is dominated by quite different species in different places and is often regarded as belonging to two quite distinct formations each composed of a number of associations.

One of these formations is often referred to as the 'coast forest', but this title is somewhat misleading since the territory occupied extends as far inland as the eastern part of the Snake Basin in Idaho [81]. As might be expected in so vast an area, there is a considerable range of climatic conditions and a patchwork of associations adapted to these conditions has evolved. In the coastal lowlands of Alaska and northern British Columbia, where the climate is cool throughout the year and where the mean annual rainfall commonly exceeds eighty inches, the forest is dominated by the Sitka spruce (*Picea sitchensis*). This tree gradually becomes less important as one passes southwards in southern British Columbia and Washington and the western cedar (*Thuja plicata*) and western hemlock (*Tsuga heterophylla*) gradually assume dominance. Both these species commonly attain a height of between 150 and 200 feet at maturity and it is not remarkable that, when once the lake forests in the east had been swept away, this coastal forest became the chief source of constructional timber in North America. Also frequent in the coastal forest of British Columbia and Washington are extensive and almost pure stands of Douglas fir (*Pseudotsuga taxifolia*). There is strong evidence, however, that the latter are not true climatic climax vegetation but have established themselves as a subclimax on burned-over areas [81]. It is thought that in this area, with its cool summers and relatively moist, mild winters, the Douglas fir cannot compete with the cedar and hemlock; given sufficient time it would be over-topped and crowded out in spite of the fact that it is capable of achieving a height of 300 feet—well in excess of the maximum capabilities of its competitors. Further east, in the interior, the Douglas fir is undoubtedly an element in the climatic climax communities however. It is apparently a species adapted to relatively dry conditions and probably it can only become common in the moister, western area where competition has been reduced by fire and the forest floor has become drier because the wind and sun have been allowed to penetrate. Without fire and felling, the Douglas fir would probably be an infrequent species here flourishing only in abnormally exposed places or on exceptionally dry soils. From Oregon southwards the dominance of the cedar and hemlock is challenged by another species which attains an even greater stature. The gigantic coastal redwood (*Sequoia sempervirens*) (Plate II*b*) is a common forest dominant in this area of marked summer drought. It seems likely, however, that it has been able to survive at the southern end of its range in California only because of the relatively high humidity and the frequent fogs in the dry season. The coastal redwood, though quite closely challenged by some species of Australian eucalypt, is the tallest known species of tree. It frequently attains a height exceeding 300 feet and the tallest recorded living specimen is probably over 360 feet in height [58]. Many of the largest surviving individuals are probably several thousand years old. Because of their attractive and valuable timber the coastal redwood forests have suffered serious inroads and only by careful conservation can they be saved from annihilation.

Eastwards from the Cascades the 'coast forest' consists of different associations. The cedar, hemlock and Sitka spruce gradually decrease in importance and ultimately disappear altogether. The coastal redwood also disappears completely. White pine (*Pinus monticola*) and the western larch (*Larix occidentalis*) are almost universally dominant here but giant fir (*Abies grandis*), ponderosa pine (*Pinus ponderosa*) and lodgepole pine (*P. contorta*) are also of frequent occurrence. On the eastern flanks of the Cascades in Oregon and Washington this type of forest extends upwards to a height of between 4,000 and 5,000 feet where it merges gradually with the sub-alpine forest, but elsewhere another type of forest (*vide infra*) generally intervenes between the two. Throughout the whole of its range the interior forest experiences colder winters and warmer summers than do the associations on the seaward side of the ranges. It also receives an average of only twenty to thirty-five inches rainfall per annum although this is much more evenly distributed throughout the year than is the rainfall in the coastal regions.

The other type of western coniferous forest is usually referred to as the 'montane forest'. On most of the middle slopes of mountains from the Rockies westwards, it is found interposed between the sub-alpine forest and the 'coast forest'. In the drier parts of south-western U.S.A. it again occurs beneath the sub-alpine formation but gives way down-slope to woodland and scrub communities (Chapter X). The elevation of the montane forest varies enormously from region to region. On the Coastal

Range in Washington it occurs between 1,000 and 3,000 feet while on the western side of the Cascades it lies between 2,000 and 6,000 feet [81]. On the flanks of the Rockies in norther U.S.A. it extends from 4,000 feet up to about 7,000 feet whereas in New Mexico it is between 6,000 and 9,000 feet. The unity of the formation is demonstrated by the fact that several dominants are found throughout. The three most important ones are the ponderosa pine (*Pinus ponderosa*), the Douglas fir (*Pseudotsuga taxifolia*) and the white fir (*Abies concolor*). Several other species of pine (*P. contorta, P. flexilis* and *P. albicaulis*) are also common dominants on both the Rockies and the Sierras.

Although the montane forests and the interior associations of the 'coast forest' are not so valuable for timber as the forests in the actual coastal strip, nevertheless many mature trees reach a height of 150 feet or more. Indeed, on the western slopes of the Sierra Nevada, the montane forests contain another species of redwood (*Sequoiadendron giganteum*) as well as giant trees such as Jeffrey's pine (*Pinus ponderosa* var. *jeffreyi*), sugar pine (*P. lambertiana*) and incense cedar (*libocedrus decurrens*) [19]. It is only because of greater difficulty of access that it has been less exploited.

Connections between the lake forest and the western forests

The origin of the tall coniferous forests of North America will be discussed later (Chapter VII), but it is appropriate at this point to note the close relationship between the lake forest and the luxuriant western forests. Every single common dominant in the lake forest has a very close relative in the western forests. The eastern white pine (*Pinus strobus*) is represented by the western white pine (*P. monticola*) in the western forests, the eastern red pine (*P. resinosa*) by the ponderosa pine (*P. ponderosa*), the jack pine (*P. banksiana*) by the lodgepole pine (*P. contorta*), the eastern hemlock (*Tsuga canadensis*) by the western hemlock (*T. heterophylla*), the Atlantic cedar (*Thuja occidentalis*) by the western cedar (*T. plicata*) and the tamarack or eastern larch (*Larix laricina*) by the western larch (*L. occidentalis*). These pairs of species are very closely related and it is clear that, although the eastern and western coniferous forests are now partly separated by the dry heart of the continent, it is not so very long ago since they were connected. The pairs of species have evolved, in the recent past, from common ancestors.

The pine forests of south-eastern U.S.A.

There is only one further large area of coniferous forest to be found in middle latitudes. This includes much of the coastal plain of south-eastern U.S.A. from New Jersey southwards to northern Florida and Alabama and then westwards to Texas. Locally it extends upwards on to the edge of the Piedmont Plateau. There are several species of conifer, particularly of pine, which are found only in this region (Plate IIIa) and in some tropical areas further south. Almost pure stands of a single species are of frequent occurrence though the actual species varies from one place to another. The loblolly pine (*Pinus taeda*), the shortleaf pine (*P. echinata*), the pitch pine (*P. rigida*), the longleaf pine (*P. palustris*) and the slash pine (*P. caribaea*) are the most widespread. Careful analysis of the sites occupied by these pine communities reveals that they are almost entirely areas of either immature, sandy soil or low-lying, marshy soil. This has led many authorities to the conclusion that these pine forests, extensive though they may be, are not climatic climax communities. This view is almost inescapable in view of the fact that the areas to the north, west and south were originally occupied by broad-leaved forests; indeed enclaves of this type of forest, dominated by various species of oak (*Quercus*), hickory (*Carya*) and chestnut (*Castanea dentata*), are to be found on better drained, richer soils actually within the pine forest region [81]. The climate here is obviously quite suitable for the typical broad-leaved trees of the North American deciduous summer forest (Chapter VI), but the true climatic climax is not achieved in the areas of pine. On the sandy areas some of the pine forests can thus be regarded as a late stage in the *xerosere* though in many places they may well be a *plagioclimax* maintained by frequent burning of the forest [57]. In connection with the latter point it will be recollected that the Indian tribes in this area were cultivators even before the arrival of Europeans and that quite extensive pine forests also occur as far south as the coastal plain and uplands of Nicaragua where there is a long history of human settlement [23]. These forests must therefore have been disturbed extensively and persistently for many centuries. Given sufficient time without disturbance, there can be little doubt that the pine forests on these well-drained lands would ultimately give way to broad-leaved species. On the marshy and swampy lands there can be little doubt that the pine and cypress (*Taxodium distichum*) communities are a late stage or subclimax in the *hydrosere*. A persistently high water-table here prevents the invasion of the true climatic climax dominants. Only a rise in the level of the land relative to sea-level or a great deal more deposition of silt could permit this invasion.

The south-eastern coniferous forests have played a significant though often negative role in the history of the United States. The great swamps, such as the Dismal Swamp in North Carolina, dominated by cypress and pine, were a great deterrent to settlement during the early years of European expansion on the North American continent. Furthermore the barren sandy areas beneath pine forest offered little incentive to agricultural development. Many parts of the South Atlantic and Gulf Coasts of the U.S.A. thus served mainly as supply areas for the navies of Europe during the seventeenth and eighteenth centuries. Here they could obtain ample supplies of materials like turpentine, pitch and resinous timber. It is only very recently, however, that these pine forests have suffered large-scale exploitation. The wood of these southern species of pine is so resinous that it was unsuitable for wood-pulp manufacture when this industry became important in the early twentieth century. It is only during the last two or three decades that techniques have been devised for using this wood for paper manufacture.

4. THE NATURE AND ADAPTABILITY OF CONIFEROUS FOREST

It will have been noted that purely coniferous forests on an extensive scale are confined entirely to the northern hemisphere. Furthermore, although 'mixed forests' of conifers and broad-leaved trees are frequently encountered in the southern hemisphere, the conifers here are different from the pines, spruces and firs already described; they belong to different genera and their morphological and physiological characteristics are often distinct (Chapter VII). In spite of their confinement to the northern hemisphere, however, coniferous forests occupy areas with a remarkably wide range of climatic conditions. Cedar and hemlock forests flourish in British Columbia in places where the mean annual precipitation far exceeds 100 inches; on the other hand, stands of ponderosa pine (*Pinus ponderosa*) and piñon (*P. cembroides*) form the climatic climax on the flanks of the Rockies and on mountains in the south-western U.S.A. where the mean annual precipitation is not more than eighteen inches. Spruce and pine in the boreal forest can survive temperatures as low as − 60° F. or − 70° F. and the larches of north-east Siberia (*Larix dahurica*) sometimes experience even greater cold; the cedar forests on the middle slopes of the Atlas Mountains experience blistering temperatures day after day during the north African summer. Quite clearly the evergreen coniferous life-form has been remarkably adaptable. It seems to compete most successfully in areas subject to seasonal adversity and, nearly everywhere where forests are found with a growing season of less than half the year, they are dominated completely by conifers—usually evergreen conifers. On the other hand this life-form has remained dominant in some other areas where growing conditions are much more favourable.

Because these forests extend through regions of such diverse climate and terrain, a great variety of soils are found beneath them. Different conifers do have different requirements but, generally speaking, they demand less of the soil than do most of the broad-leaved angiosperms. The nutrient cycle in the mid-latitude coniferous forests, though often not so impoverished as that in the boreal forest, is normally poor; careful husbandry and fertilisation over a protracted period are often necessary before these forest soils can be transformed into good agricultural soils.

A

B

PLATE III

A. SOUTHERN PINE FOREST. A virgin stand of longleaf pine (*Pinus palustris*)
in Mississippi.
(U.S. Forest Service photograph, by permission of the U.S. Department of Agriculture.)
B. ALGERIAN CEDAR FOREST. Algerian cedars (*Cedrus atlantica*)
at Feniet el Haad.
(From Schimper, 1903.)

A

B

PLATE IV

A. A PODZOL FORMED BENEATH CONIFEROUS FOREST. The darker-coloured B_1 horizon, separating the lighter-coloured A_2 and B_2 horizons, is clearly visible at the level of the top of the ferrule of the spade.
(From G. V. Jacks, *Soil*, 1954.)

B. A BROWN FOREST SOIL. This almost homogeneous profile in Connecticut is about a yard in depth. Because of the presence of a continuous grassy ground cover the A_{00} and A_0 horizons do not show typical development.

CHAPTER V

The Mixed Forests
of Middle Latitudes

I. ECOTONE MIXED FORESTS

WITH the exception of the important anomaly in western North America, the high latitude and high altitude coniferous forests of the northern hemisphere give way, in humid regions, to broad-leaved deciduous communities. Ultimately, in most cases, the coniferous life-form almost disappears. This change is usually achieved through a transition zone where the two life-forms exist side by side. The areas of competition are usually spoken of as *zones of ecotone*. The change in species-content as one passes through the ecotone is often imperceptible so that it is impossible to define its limits realistically and to show them on a map. In some areas, however, the ecotone is very wide and consists of coniferous and broad-leaved species in almost equal quantities. Obviously such forests cannot be regarded as either coniferous or broad-leaved and it is customary to put them in a separate category known as 'mixed forest'. It seems likely that broad and distinct ecotones of this nature were the original vegetation in parts of north-central Europe, eastern Asia and north-eastern North America.

A certain amount of confusion has arisen in the past because forests of an entirely different kind have often been placed in the same category as the mixed forests just referred to. It has already been pointed out (Chapter IV) that several species of deciduous tree such as birch (*Betula spp*), aspen (*Populus spp.*) and willow (*Salix spp.*) occur in seral communities within the boreal forest proper. Large areas of boreal forest in Alberta, Saskatchewan and central Siberia are particularly rich in such species. In a certain sense therefore, much of the boreal forest itself can be spoken of as 'mixed forest'. To adopt such terminology, however, is to abandon the concept of climatic climax as the main basis for the classifying and mapping of vegetation; consequently all such areas are regarded here as parts of the true boreal forest (Maps 2 and 4). All the mixed forests described in the first part of this chapter are true ecotone forests in which

Fvs

the dominants of the contiguous boreal and broad-leaved deciduous forests are both present in more or less equal quantities.

European Russia

The term 'mixed forest' can be misinterpreted for another reason. It must not be imagined that, in any single acre, coniferous and broad-leaved trees will be found intermingled as individuals. Indeed the type of forest where this occurs is exceptional. Mixed forests are normally mosaics of diverse associations nearly all of which are either mainly coniferous or mainly of the broad-leaved deciduous type. The mixed forest of European Russia furnishes a clear example of this type of mosaic. It will be remembered that the boreal forest of European Russia is composed basically of distinct associations of pine and spruce, occupying the lighter and heavier soils respectively. As one passes southwards from a line running eastwards from north-eastern Estonia (approximately 59° N.), one notices that stands of spruce gradually become fewer, their place being taken by stands of oak (*Quercus robur*). The pine, however, remains quite important; indeed it remains as an element in the vegetation far south into the wooded steppe of the Ukraine (Chapter IX). Though other species are important locally, the oak association and the pine association are thus the dominant elements in the mixed forest of European Russia [44]. This stretched in an almost continuous zone as far east as the Urals but decreased in width eastwards. In White Russia it extended from 53° N. to 59° N. while on the western flank of the Urals it was little more than one degree in width.

Unlike the boreal forest, this Russian mixed forest has, in great part, been swept away. It was here that the Russian nation had its origin. Here the land proved suitable for the development of a mixed farming economy which ultimately proved to be a sufficiently sound basis for the urban, industrial society centred on Moscow. In particular the brown forest soils (Plate IV*b*) beneath the deciduous associations were readily converted into responsive agricultural land. Trees like the oak, root deeply in the soil; indeed their tap roots normally penetrate downwards through the weathering sub-soil into the joints of the rock itself. Here they are able to secure not only a reliable supply of water but also the dissolved mineral nutrients as these are released by rock weathering or are leached down-wards from decaying organic material in the soil. The oak thus tends not only to retain nutrients already in the cycle but also to bring in new supplies from below. A rich supply of nutrients is returned to the surface of the soil each autumn when the leaves fall.

This leaf litter forms the A_{00} horizon of the brown forest soil which, in undisturbed virgin forest, may be many inches in thickness (Plate V*b*). Normally it merges imperceptibly into an A_0 horizon of dark-brown *mild humus* or *mull* (Chapter III) which, though still acid, is much richer in plant foods than the raw humus in the A_0 horizon of the podzol. The pH value of the mild humus is usually between 4·5 and 6 but varies very much with different parent-materials and beneath different species of deciduous tree.

Beneath the A_0 horizon are found the predominantly inorganic horizons of the brown forest soil. Though some contrasts can usually be discerned in the profile of these mineral horizons, the marked contrasts of the podzol profile are lacking. Between the A_0 horizon and the weathering material in the C horizon (Fig. 16), there is usually a fairly homogeneous, brown

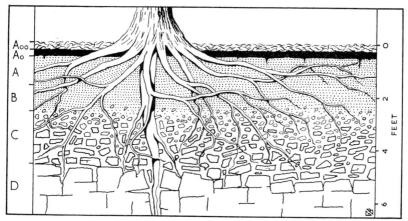

Fig. 16. A brown forest soil profile beneath oak forest.

layer whose colour gradually becomes paler with increasing depth. The brown colouration is due partly to the presence of humus which has been mixed downwards and, consequently, is less and less abundant with increasing distance from its source in the A_0 horizon. However, even if all the humus is removed chemically by treatment in the laboratory, the inorganic material itself is found to possess a brown or yellowish-brown coloration. This can be demonstrated to be due to the presence of the sesquioxides of aluminium and iron. Apparently the brown forest soil is sufficiently acid to stimulate the gradual decomposition of the clay minerals but leaching of the products of this decomposition is slight. If some eluviation of sesquioxides and unaltered clay does take place with subsequent redeposition lower in the profile the soil should be referred to as 'brown podzolic soil'. This type tends to replace the brown forest soil beneath deciduous forest where the parent material is poor in bases or where precipitation is heavy [8]. The mineral profile in this soil type comprises an upper A horizon and a lower B horizon even though this is not obvious in the field.

There can be little doubt that the thorough distribution of humus throughout these soils is due to the base-rich nature of the humus itself and, ultimately therefore, to the conservative nature of the deciduous trees. The base-rich medium attracts a fauna of soil organisms which is quite numerous as compared to that of the podzol. Organisms such as earthworms, beetles and rodents thrive and constantly carry humus downwards and mineral material upwards, thus offsetting horizon-forming processes. Soil bacteria also are much more numerous in the brown forest soil than in the podzol so that humus decay, with consequent release of nutrients, goes on more rapidly.

The people of Muscovy developed their agriculture in this environment. Here some of the soils were sufficiently rewarding and the original vegetation provided a wealth of different types of softwoods and hardwoods for houses, fences, implements and fuel. Here they built up sufficient strength to withstand the Tartar hordes and, from here, they were able ultimately to expand and occupy one-sixth of the land area of the earth.

Mixed forest ecotones of Europe

The mixed forest of European Russia is merely part of a forest belt which extends across Europe. This begins in the west with mere outliers in the Highlands of Scotland, the Massif Central and on the slopes of the Cantabrian mountains in northern Spain. It then becomes more continuous, extending eastwards from the slopes of the northern Alps, Alsace, central and north-east Germany and south-central Sweden. In lowland areas to the west and south of this belt the forest was almost entirely broadleaved. Since most of the mixed forest of central Europe has been swept away, it is very difficult today to determine its former nature and the former extent of its component associations. It seems certain however that the pine (*Pinus sylvestris*), spruce (*Picea excelsa*), silver fir (*Abies excelsa*), oak (*Quercus robur*), birch (*Betula spp.*) and beech (*Fagus sylvatica*) were all present in the mixed forests of Germany [45], but the extent to which the individual species were intermingled in the different associations is quite obscure. There is considerable evidence, however, to suggest that there was a close relationship between the distribution of some associations and that of particular types of soil-parent-material. The outwash sands of the North European Plain were almost certainly dominated by pine as in Russia and the richer soils were probably occupied in great part by broadleaved trees. It is possible that certain areas such as the loess belt of central Germany were so dominated by deciduous, broad-leaved trees, that they should be regarded as outliers of the deciduous summer forest formation rather than mere associations of the mixed forest ecotone. Here, however, almost complete deforestation took place so long ago in the Neolithic and Bronze Ages [10], that detailed reconstruction is impossible.

Eastern Asia

The mixed forest ecotone between the boreal forest and the summer deciduous forest does not reappear to any considerable extent in Asia until one reaches the middle Amur valley. It occupies much of northern and eastern Manchuria, northern Korea, southern Hokkaido and northern Kyushu. The species composing these forests are nearly all distinct from their counterparts in Europe although, in many cases, they are closely related. In northern Honshu, associations of broad-leaved, deciduous trees alternate with stands of both evergreen and deciduous conifers. Native species of maple, birch, beech, poplar and oak are found along with other species of particular economic importance such as the native ash (*Fraxinus mandschurica*) and the keyaki (*Zelkowa serrata*) while associated coniferous species include the Japanese cypresses (*Cupressus obtusa* and *C. pisifera*), Japanese cedar (*Cryptomeria japonica*) and the fir (*Abies firma*) [19]. A glance at the list of species in these forests indicates a flora which is far richer and more varied than its counterpart in Europe. The implications of this will be discussed later (Chapter VII).

North America

In North America also the lake forest and boreal forest give way to the summer deciduous forest formation through a zone of mixed forest. Indeed, throughout almost the whole of the areas of lake forest, occasional stands of deciduous trees were encountered by the first European settlers. It is probably because of this that the existence of the lake forest, as a distinct formation, has often been overlooked. However, the main zone of conflict between the lake forest conifers and the deciduous trees stretched from central Minnesota through Wisconsin and southern Michigan to the northern parts of Pennsylvania and New York. Throughout most of New Brunswick, Nova Scotia and northern New England also, mixed forest once covered the landscape but here it was mainly the dominant trees of the boreal forest formation which competed with the deciduous associations. As in Europe the two life-forms were rarely intermingled in the original associations [81]. Before clearance the sandy outwash areas in Minnesota, Wisconsin and Michigan were occupied predominantly by stands of white and red pine while the heavier, richer soils were mainly occupied by stands of oak (*Quercus spp.*) and hickory (*Carya spp.*). Relief also influenced the distribution of the two types of association; wherever the land in this western part of the ecotone was somewhat dissected, the north-facing slopes tended to be occupied by coniferous trees and the south-facing slopes by deciduous ones [81]. Further east, in Ontario, New York, New England and the Maritime Provinces, the oak-hickory associations gave place to associations dominated by maple (*Acer saccharum*) and beech (*Fagus grandiflora*) but the general space-relationship with the competing coniferous associations was maintained.

It is most interesting to note that, as in the case of Russia, the American nation also had its inception in the mixed forest environment. The Pilgrim Fathers and their descendants hacked their way into the hardwood stands to clear areas of brown forest soil for the purpose of cultivation. Near at hand were stands of tall, straight softwoods which were so convenient for building log cabins and for many other purposes. At a later stage the Canadian people also built a solid agricultural foundation for nationhood in a similar environment in the St. Lawrence lowlands and southern Ontario. It seems that, at the stage of economic development which preceded the machine age, with its mechanisation of agriculture, this type of environment offered the means for a relatively safe and stable rural economy. A large range of plant and animal raw material was available in the forest, and the brown forest soils, though not outstandingly productive, could be converted into grassland or arable soils which maintained their fertility or even increased it. In subsequent chapters a number of environments will be described which, at the outset, seemed to offer much greater rewards but which brought disaster to the communities which occupied them.

Quite distinct from these 'northern' mixed forests are the 'southern' mixed forests of south-eastern U.S.A. The southern 'piney' forests have already been described (Chapter IV) and these form a wide ecotone with the deciduous forest. This ecotone extends from New Jersey south-westwards along the whole of the Piedmont Plateau and then westwards across Alabama, Mississippi and northern Louisiana into Arkansas and Texas. In the east, associations of southern pine alternate with hardwoods such as maple (*Acer saccharum*) and chestnut (*Castanea dentata*) but in the west oaks and hickories become the commonest hardwood trees. As in the case of the 'piney' forests themselves, the status of the pine communities here is still doubtful. An increasing volume of evidence seems to indicate that the deciduous communities are the true climatic climax over the whole area, the pine being merely seral or sub-climax. If this is the case, the southern mixed forests have an entirely different ecological status from those further north.

2. EVERGREEN MIXED FORESTS

In several lowland areas of the world there are extensive mixed forests which appear to be true climatic climax formations and not mere ecotones. They are ecologically distinct in that their broad-leaved trees, as well as their conifers, are evergreens. Such forests are widespread in the southern hemisphere in Chile, New Zealand, Tasmania and South Africa. It appears that, even in historical times, they have also been extensive in the northern hemisphere but almost entirely within the confines of the Mediterranean basin. Some of the forests of southern Japan might conceivably be included in the same category but here the conifers seem to occupy only a relatively small proportion of the formation as a whole. Consequently it is probably most appropriate to regard them as broad-leaved forests (Chapter VI).

The Mediterranean mixed forest

To one who is familiar with the present-day landscapes around the Mediterranean Sea, a vision of almost unbroken forest from the water's edge up to the crests of all but the highest mountains may seem too fantastic for credence. Today, wherever cultivated fields and orchards do not occupy the ground, scrub, poor grassland or bare rock are almost universal. Ecological investigations, however, have indicated that the scantiness of this vegetation cover is mainly due to human activity; men have cultivated crops and herded animals around much of the Mediterranean Sea for several millennia.

Today, two main types of plant community recur with monotonous frequency (Chapter X) and, although many local names are applied to them in the countries where they occur, the French terms maquis (Italian: macchia) and garrigue are the two which have been most frequently borrowed in English literature. Nevertheless it seems clear that most of the area now covered with maquis was once forest. Forest trees of various kinds are found, often as isolated individuals, scattered over many areas of maquis and perhaps some of the species of shrub which are now dominant in the maquis communities were once no more than minor elements in the undergrowth of a vast evergreen forest. The exact nature of this forest is almost impossible to ascertain but, since the trees which are obviously natives of the Mediterranean lowlands are partly coniferous and partly broad-leaved, it seems likely that it was mixed. However, the extent to which the two life-forms were segregated into distinct associations under natural conditions, is obscure. It is also difficult to assess the extent to which the ranges of both types of species have changed because of man's activities.

Species of evergreen oak are characteristic and almost universal in the Mediterranean lowlands. The holm oak (*Quercus ilex*) is common west of the Dardanelles in Europe and westwards from Cyrenaica in Africa while the cork oak (*Q. suber*) is very common around the western basin of the Mediterranean particularly near sea-level. On the other hand, isolated individuals of the domestic, umbrella or stone pine (*Pinus pinea*) are found scattered throughout the maquis as well as in continuous stretches of woodland along the northern seaboard up to a height of about 3,000 feet. The maritime pine (*P. pinaster*) is similarly distributed along the northern side of the western basin but only up to about 1,500 feet, while the Aleppo pine (*P. halepensis*) is common in the north of the eastern basin.

Although it is almost certain that there are some areas of true climatic climax maquis, particularly of the wild olive-lentisk association, it is probable that these were confined to windswept areas where the rainfall was very low and drainage particularly free. Probably most of these communities of climatic climax maquis were in areas which are now occupied by garrigue since it has been shown that, given protection from grazing and firing, the dominants of the garrigue communities can be ousted gradually by taller plants. It is even possible that some of the land which is now under garrigue was originally under some kind of forest.

It is perhaps difficult to believe that such wholesale changes as those envisaged here have actually taken place. Nevertheless, forest communities do flourish in areas like south-western Australia which receive no more precipitation than many areas of maquis and garrigue in the Mediterranean basin. Furthermore, observations of plant succession have shown that trees can establish themselves in these apparently inhospitable places. One must remember that, when once forest trees have been almost annihilated over large areas, even if conditions again become favourable for re-invasion, this will inevitably be slow because of the scarcity of seeds. Not only has this impoverishment in seed parents been achieved over wide areas around the Mediterranean but conditions for regeneration remain most unfavourable. The maquis and garrigue have their place in the rural economy of the countries where they are found. Not only is the omnivorous goat still plentiful and almost universal—in many places the village communities have common rights to use the wood of any trees which manage to escape the maw of these voracious animals. Furthermore, fire is a common hazard wherever the vegetation becomes sufficiently thick to permit the flames to spread freely. In places it is even a common practice to use the maquis land in a kind of rotational agriculture, the land being cleared by burning, cultivated for a few years, and then permitted to revert. This kind of treatment favours shrubs like the lentisk and the cistus which can shoot freely from the base of the stems no matter how often they are burned. The saplings of trees, on the other hand, cannot survive.

Most important of all is the irreparable damage that was suffered by the soil when the forest was cleared. Even an Englishman knows how pleasant a haven is the shade of an isolated, spreading tree on a hot summer's day. Even more pleasant is the shelter provided by closed-canopy forest where negligible sun can penetrate to the ground. The type of day when such respite is valued is very much more frequent around the Mediterranean than in Britain. During the normal summer, one cloudless day follows another, the sun being almost overhead at noon. If human beings are susceptible to the contrast between sunshine and shade, soils are even more sensitive in this respect. For countless millennia[1] the soils around the Mediterranean had developed in constant shade. In any one place, only

[1] It must be remembered that forests in the Mediterranean lowlands were not swept away during the Pleistocene as they were further north.

once during a period of several centuries would an old tree die, fall and allow the sunshine to penetrate; even then the gap would be filled in a few years by natural regeneration. Under these conditions the air must have remained almost still even when a wind was sweeping over the tree tops. This still air would be of relatively high humidity because of water vapour received by evaporation and transpiration from the tree foliage, the undergrowth and the soil surface; the blanket of moister air would, in turn, prevent rapid evaporation from the ground. Beneath the tree canopy therefore, the soil would remain relatively cool and moist during the hot summer days even after weeks of drought. The water lost by the soil-vegetation system would be predominantly from the leaves of the tree canopy and this would have been obtained by the tree roots from the store of water, derived from the winter rains, deep down in the sub-soil and underlying rock.[1]

With the felling or grazing of the forest the whole hydrological system, along with the microclimate near the surface of the soil, was utterly changed. Soil which had never been thoroughly exposed during the whole of its evolution, could now be desiccated to a considerable depth by only a few days of summer sun and wind. Also the leaf-fall which had persistently supplied new humus to the surface layers was now cut off; no longer were mineral nutrients being drawn from deep within the weathering rock to be brought back to the soil by leaf decay. The nutrient cycle was thus altered as well as the hydrological one. The fertility of the soil fell rapidly as a consequence. The spongy, brown, deep forest soils lost their structure as the humus content decreased and, instead of being absorbed and passed downwards into storage, the winter rains ran off the surface in ever-increasing sheets and rivulets, carrying the soil particles with them. In a frighteningly short period of time, bare rock probably appeared at the surface on many slopes and, where pockets remained, these were very deficient in humus and prone to drought.

The 'terrae rossae' or red soils are often said to be characteristic of the Mediterranean region. They are certainly widespread there at the present day, not only on limestones but on a wide range of lithological types. The redness is due to the presence of sesquioxide of iron (Fe_2O_3), one of the products of mineral decomposition. Were there more humus in these soils this redness would be obscured and the whole aspect of many Mediterranean landscapes much altered.

The Mediterranean forest appears to have been a true climatic climax evergreen formation even though it was dominated by two distinct life-forms. Both appear to be very well adapted to the climatic conditions around the Mediterranean. Like the mixed forests of higher latitudes, but at a much earlier date, the Mediterranean mixed forests provided a habitat

[1] Both the broad-leaved oaks and the pines of the Mediterranean region have rooting systems which are amongst the longest and most efficient in the world.

in which great nations had their origin. In this case they nurtured the classical civilisation in which the whole of subsequent European civilisation had its roots. In the Mediterranean, however, the development and continuance of agriculture has resulted in the destruction of much of the former basis of subsistence. The climatic conditions here, associated with the prevalence of steeply-sloping land, entailed a most unstable ecological situation when once the original balance was disturbed.

3. EVERGREEN MIXED FORESTS OF THE SOUTHERN HEMISPHERE

Though occupying several types of climatic region, the mixed forests of the southern hemisphere are similar in many ways to those of the Mediterranean basin. The mixed forests of Chile, southern Brazil, northern New Zealand and Cape Province are all composed partly of coniferous trees and partly of broad-leaved, evergreen angiosperms. They occupy such diverse areas as the summer-drought region of south-central Chile, the very wet regions of southern Chile and the warm, moist area of northern New Zealand. Because of their separate evolution over a long period, the species of these forests, both coniferous and broad-leaved, are very different from their counterparts in the northern hemisphere but, in life-form and general appearance, they are similar.

South America

Passing southwards from the scrublands of north-central Chile, forests are first encountered at about 30° S. Here, however, they are confined to the middle slopes of the western Andes between 3,000 and 4,500 feet, the lowlands at this latitude still being very dry. One dominant species is the araucaria pine (*Araucaria imbricata*)[1] though this is normally associated with species of evergreen beech (*Notofagus spp.*) and other species like the Chile gumbox (*Escallonia spp.*), sumac (*Rhus spp.*) and the Chile soap tree (*Quillaja saponaria*) all of which remind one so vividly of the evergreen oaks of the Mediterranean area. From Valparaiso southwards these forests formerly occupied the lowlands as well as the mountain slopes. They were very dense and luxuriant. They appear to have attained their maximum luxuriance southwards from Valdivia to Penas Gulf; indeed large areas still remain, forming an almost impenetrable barrier from the coast up to about 1,500 feet. Species of evergreen beech are predominant at this latitude but the southern 'cypress' (*Fitzroya patagonica*) and the alerce (*Libocedrus tetragona*) are also important on wetter soils [19]. Formerly there were also immense stands of Chilean cedar (*Libocedrus chilensis*) but these have been very much reduced by felling. In some higher valleys, up to a height of about 3,000 feet and some distance away from the coast, where rainfall is not so heavy but where there is some shelter from the

[1] The 'monkey puzzle tree' of many suburban gardens.

wind, the forest is composed of different associations. Valuable species like the Chilean pine (*Podocarpus chilinus*) are still almost untouched and a considerable variety of broad-leaved evergreen trees such as the magnoliaceous 'Winter's bark' (*Drimys Winteri*) are also important. A similar type of forest extends southwards as far as the southernmost tip of Tierra del Fuego though with progressively diminishing luxuriance as the climate becomes cooler. The altitude of the upper limit of forest also decreases southwards. Species of evergreen beech, 'cypress' and magnolia are important throughout.

On the southern part of the Brazilian Highlands, where frosts are of frequent occurrence, tropical vegetation gives way to a discontinuous evergreen mixed forest which is similar in life-form to that of Chile, though by no means so luxuriant. This forest first appears at a height of between 3,000 and 4,000 feet at about 23° S. but gradually extends to lower elevations southwards. It is dominated by the Brazilian or Paraná pine (*Araucaria braziliensis*), a close relation of the Chilean araucaria pine. Broad-leaved evergreen low trees and shrubs such as the Paraguayan tea or yerba maté (*Ilex paraguayensis*) are associated with the Brazilian pine, often forming quite a dense undergrowth. Though this forest occupies an area which has a very different climatic régime from that in Chile, the two formations are obviously closely related.

Australasia

Forests of very similar structure are found in the North Island of New Zealand and in eastern Australia though they have been sadly reduced. In the northernmost part of New Zealand the kauri pine (*Agathis australis*) is the most famous tree. It was the main source of constructional timber for the first European settlers and, in consequence, can now almost be described as rare. At the outset it was of very limited range, being almost completely confined between latitudes 35° S. and 38° S. Apparently it never formed very extensive pure stands; on the other hand it was not found intermingled with other species. Typically it occurred in groves of between several score and several hundred specimens, these consociations forming part of a patchwork of forest communities [36]. Usually the kauri occupied heavy clay soils not far from the coast though in places it occurred in sheltered valleys up to an altitude of over 2,000 feet. Many of the areas formerly occupied by stands of kauri pine are now derelict. This is not merely because the trees have been swept away but also because much of the ground was subsequently turned over in the search for fossil kauri gum.

Mixed stands of many other species (Plate V*a*) which were originally interspersed with the stands of kauri, survive here and there in northern New Zealand as far south as the northern and north-western coastlands of the South Island. Conifers like the totara (*Podocarpus totara*), kahikatea

(*P. dacridioides*), miro (*P. ferrugineus*) and matai (*P. spicatus*) are found along with broad-leaved evergreens particularly from the *Proteaceae*, *Leguminoseae*, *Myrtaceae* and *Laurineae*. Species of tree fern and the Nikau palm (*Areca sapida*) are almost universal also. The forest is made to appear even more luxuriant by the presence of a great array of epiphytes growing on the trunks and boughs of the trees and numerous species of liane [22].

Equally varied floristically and with very similar structure, the rain forests of north-east New South Wales and south-east Queensland originally covered large areas of coastal plain and mountain slope. Numerous species of broad-leaved angiosperm form much of the canopy in the remaining forest, though several species of tall conifer are conspicuous as emergents above the general level; the crowns of bunyabunya (*Araucaria bidwillii*) and Moreton Bay pine (*A. cunninghamii*) are particularly obvious [83], giving the forest an appearance reminiscent of the Paraná pine forests of southern Brazil. In composition and general structure, both these Australasian forests are very different from those found in analagous situations in the northern hemisphere.

South Africa

The so-called 'temperate forests' of South Africa are now a mere remnant of their former selves. The narrow coastal belt between George and Humansdorp, a mere seventy miles in length, is now the only area where fairly continuous areas of mixed forest remain, though patches of similar forest are found between 3,000 and 4,000 feet on the seaward slopes of the mountains both to east and west. Very similar forests also fringe many of the rivers and streams far into neighbouring areas which are now covered generally with scrub. Though poorer in species than the equivalent formations in New Zealand and Chile, this forest still has the appearance of luxuriance; trees, shrubs, lianes and epiphytes are all represented by numerous species. Conifers (*Podocarpus spp.*) and broad-leaved angiosperms are both found in the most moist areas though the former tend to drop out completely on drier sites. The under-storey contains many ferns as well as broad-leaved shrubs. Many of these cannot survive competition when the tree canopy is removed and are normally displaced by more drought-resistant, sclerophyllous shrubs from the near-by scrub communities (Chapter X). There can be no doubt that felling, firing and grazing have caused this change to take place over large areas during the past two or three centuries.

CHAPTER VI

Broad-leaved Forests
of Middle Latitudes

In spite of the fact that the conifers have adapted themselves so successfully to a remarkable variety of climates, over large areas they have been unable to compete with broad-leaved angiosperms during post-glacial times. In the northern hemisphere much of this broad-leaved forest was of the deciduous summer forest formation-type with representative formations in Europe, eastern Asia and eastern North America. Certain specialised formations of this formation-type also appear to be climatic climax in between the boreal forest and the steppe in Saskatchewan, Alberta and parts of Siberia. North of the Tropic of Cancer it was only in Japan, south China and, locally, in south-eastern U.S.A. that evergreen broad-leaved trees had been able to retain almost complete dominance. In the extra-tropical regions of the southern hemisphere, however, the situation is reversed. Here the broad-leaved forests are almost entirely evergreen; it is only locally in Chilean Patagonia that there is a formation of deciduous summer forest.

I. DECIDUOUS SUMMER FOREST

The European formation

It has already been noted that an almost continuous belt of mixed forest extended westwards into Europe along the North European Plain as far as the Middle Elbe. West and south of this, in western and central Europe, the lowlands were occupied almost entirely by broad-leaved deciduous forests. These also extended around the coastlands of the Baltic as far as the Gulf of Danzig in the south and into Scania in the north. The formation extended as far north as the southern coastlands of Norway and eastern Scotland, and as far south as the Cantabrian Mountains, the slopes of the Ebro Lowland and the middle slopes of the Apennines and Dinaric Alps. It continued eastwards, south of the mixed forest, across the Danubian basins, the lower slopes of the Carpathians and the northern Ukraine. Here its southern boundary ran roughly from Chernovtsy eastwards to east of Kiev, north-eastwards to Tula, then irregularly eastwards to just south of Kuibishev on the Volga bend and, from thence, to the southern end of the Urals (Map 1). Along this southern boundary there was a great deal of inter-digitation with the steppe and the wooded steppe.

Much of this deciduous forest has now been swept away and what little remains has, nearly everywhere, been so profoundly affected by human activities that the original distribution of component associations and the details of their structure cannot be ascertained with precision. An attempt will be made later (Chapter XI) to envisage the original distributions over a relatively small part of the area formerly occupied by this formation. Generally speaking, forests dominated by the pedunculate oak (*Quercus robur*) covered the lowlands of the British Isles and northern France. The beech (*Fagus sylvatica*) and the ash (*Fraxinus excelsior*) were important or even dominant on calcareous soils or on particularly rich and well-drained areas, while the sessile oak (*Q. petraea*) and the birch (*Betula spp.*) were commonly dominant on shallower, more siliceous soils. Elm (*Ulmus spp.*) and lime (*Tilia cordata*) were also components in some of these associations, becoming generally more frequent southwards.

In southern Europe entirely different associations were predominant. In the lowlands of northern Spain and in the Po Basin other species of deciduous oak (*Q. lusitanica, Q. cerris* and *Q. pubescens*) displaced those of north-western Europe. The sycamore (*Acer platanoides*), chestnut (*Castanea sativa*) and ash-elm (*Fraxinus ornus*) were also plentiful here. At higher levels in the Mediterranean peninsulas and on the middle slopes of the Alps the beech is still predominant. Beech and chestnut forest occurs, interposed between the lowland evergreen forest and the uppermost coniferous forest, right across the northern Mediterranean lands at heights between 1,500 and 3,000 feet. It merges, often through extensive ecotones, with the remnants of the mixed evergreen forest beneath and with the coniferous forest above. It is noteworthy that the height of its lower limit rises southwards while the height of its upper limit remains fairly constant. As a consequence of this, the deciduous forest is ultimately squeezed out and, in the Atlas Mountains, the lowland evergreen forest, at a certain height, gives way directly to evergreen cedar forest (Chapter IV).

The great range of the European beech is emphasised by the fact that it is still the dominant tree at the northern edge of the formation in Scania and, in places, along the southern coastlands of the Baltic. Here it occupies the better soils developed in glacial clays and loams while the oak-birch association is found on the poorer, lighter soils developed in fluvio-glacial outwash sands and stony morainic material. The beech also forms extensive and almost pure stands between 600 and 2,000 feet on the flanks of the Hercynian block mountains in central Germany and in

Czecho-Slovakia, although west of the Rhine, at similar heights, it occurs predominantly in mixed stands with oak (*Q. robur*), hornbeam (*Carpinus betulus*) and chestnut (*Castanea sativa*). Further south, however, the deciduous forests of the Swiss Plateau and those of the lower slopes of the Alps are dominated by the beech [45]. The general level of beech forest falls eastwards so that originally, around the Pannonian and Transylvanian basins, it covered the foothills only. The centres of these basins were originally 'wooded steppes' and between these and the foothill beech forests there was an intermediate zone dominated by oak.

Eastwards and north-eastwards from these basins the beech rapidly becomes rare and it disappears entirely before one reaches the plains of Moldavia, Wallachia, central Poland and the lower Vistula. Thus, from the northern Ukraine across to the southern Urals, the forests are dominated by oak (*Q. robur*). This species was found on the rich loessic soils in this zone, the relatively infrequent sandy soils being partly occupied by pine (*Pinus sylvestris*) [44]. It seems probable that, had this area not had a rich covering of loess, the pine would have been much more common and, in consequence, true mixed forest would have extended right to the edge of the steppe.

The soils beneath the European deciduous forests vary a great deal, mainly because of the great range of parent-materials in which they are developed and the diverse demands of the different species composing the associations. Beneath the oak-birch forests on the sands of the North European Plain where an undergrowth of acidophyllous plants such as ling (*Calluna vulgaris*) is prevalent, podzols have developed with profiles which are almost as pronounced as those beneath the pines of the boreal forest. Beneath this oak-birch forest, the porosity and poverty of the parent-material, along with the moist climate and undemanding undergrowth, are sufficient to permit rapid podzolisation in spite of the products of leaf-fall from the trees. On the other hand many of the brown forest soils on marls and clays beneath deciduous forests in central Europe and the northern Mediterranean area are almost neutral in reaction; pH values of between 6 and 7 are of common occurrence [46]. Here the rate of leaching and dissociation of the clay minerals is very slow.

In most cases these soils appear to have passed their entire development beneath the one type of vegetation. It is only along the margins of the wooded steppe in the central European basins and in the Ukraine that there is a narrow zone of soils, known as the *grey forest soils*, which have obviously changed their mode of development in the not too distant past. Although the podzolisation process is now causing the development of a shallow, grey horizon near the top of the profile, the lower horizons in these soils are normally basic in reaction. Indeed they often contain nodules of calcium carbonate. It appears that these are soils which began their development beneath grassland (Chapter IX), but which were invaded by trees [44]. Beneath the tree canopy the microclimate of the surface of the soil was transformed. The rapid evaporation of the summer months was much reduced and so was the capillary rise of dissolved lime into the upper horizons. From thence, leaching exceeded capillary rise so that podzolisation began. This process was accentuated because the trees, though deeper rooting, were less demanding than the very *calcicolous* grasses; the lime was thus allowed to slip out of the nutrient cycle much more rapidly than formerly. Furthermore, whereas with a grassy vegetation, humus had been inserted directly into the soil profile by decay of the fine network of roots, it was now supplied primarily to the surface by leaf-fall; shallow-rooted plants were greatly reduced in number. Such a change in vegetation can obviously be responsible for a complete change in pedogenic processes without any change in rainfall or temperature.

The nature of the soil beneath the original wild vegetation is usually most difficult to determine, particularly in lowland areas in central and eastern Europe. It was here that sedentary agricultural communities established themselves thousands of years ago; in this forest environment they were able to protect themselves from the successive waves of nomadic invaders sweeping across the steppe from the east. The brown and grey forest soils have thus been transformed by many centuries of cultivation and it is only in small localities, by chance, that the evidence of past pedogenic changes has survived.

The American formation

Prior to the westward advance of European settlers, a great block of deciduous forest extended from the Appalachians to beyond the Mississippi. From the mixed forest ecotone of Minnesota, Michigan, Pennsylvania and New England it extended southwards to the southern mixed forest, interrupted only by a wedge of prairie which intruded from the west across Illinois and Indiana (Map 5). Projections from the main body of the forest extended even further west along the valleys of main rivers like the Missouri, the North Platte, the South Platte and the Arkansas as well as along outcrops of pervious Tertiary sandstones in Texas and Oklahoma.

Although much of this forest has disappeared, sufficient remnants can be seen and adequate records are available to provide a much better impression of its original nature than is the case in Europe. There can be little doubt, however, that in their structure the deciduous forests of the two continents were very similar. Each had its upper storey of trees which, generally speaking, allowed a good deal of direct sunlight to penetrate so that a rich shrub layer could flourish. Floristically, however, the American formation was very much richer than that of Europe. Whereas, in the latter, only a dozen or so species seem to have been common and wide-

A. MIXED EVERGREEN FOREST. Virgin forest in the Papakura District of the North Island, New Zealand. There is a rich undergrowth dominated by tree ferns while the large trees in the foreground are the totara (*Podocarpus totara*) on the left, and the rata (*Metrosideras robusta*) on the right. (From a sketch by C. Fischer presented as an engraving in von Hochstetter, 1867.)

B. DECIDUOUS SUMMER FOREST. Virgin stand dominated by white oak (*Quercus alba*) in Spring Mill State Park, Indiana. (U.S. Forest Service photograph, by permission of the U.S. Department of Agriculture.)

PLATE V

A

B

PLATE VI

A. BROAD-LEAVED EVERGREEN FOREST. Forest dominated by live oak
(*Quercus virens*) on Jeckyll Island off the coast of Georgia. The trees are festooned
with the inappropriately-named, epiphytic 'Spanish moss' (*Tillandsia usneoides*)
which really belongs to the *Bromeliaceae*.
(U.S. Forest Service photograph, by permission of the U.S. Department of
Agriculture.)

B. AUSTRALIAN SCLEROPHYLLOUS FOREST. Grassy eucalyptus forest in
Queensland.
(From a photograph by Semon in Schimper, 1903.)

spread, in the former several score species are in this category and many more are common locally. The American formation, in spite of its richness in species, shows great unity. Certain species of oak (*Quercus*) and hickory (*Carya*) are almost universal while the basswood (*Tilia americana*), maple (*Acer saccharum*) and beech (*Fagus grandiflora*) occur throughout the eastern part of the formation [81]. Other uniting factors are the trees which occur almost universally as subsere and subclimax species; birch (*Betula lutea*) and ash (*Fraxinus spp.*) are particularly common in the subsere on burned-over areas.

Within the general unity of the formation, however, three distinct sets of associations can be recognised. In the north-east, extending from southern Michigan, Ohio, New York and New England southwards into West Virginia and Kentucky, beech and maple are predominant. They also extend southwards along the lower slopes of the Blue Ridge and the other mountains of the southern Appalachians. These two main trees are in frequent association with several species of oak (*Quercus borealis*, *Q. alba* and *Q. bicolor*), sweet gum (*Liquidambar spp.*), chestnut (*Castanea dentata*) and the tulip tree (*Liriodendron tulipifera*). With increasing frequency from south to north the hemlock (*Tsuga canadensis*) is also found scattered amongst the deciduous trees, this outlier of the lake forest being particularly frequent on north-facing slopes and in deep, shady valleys [84]. These beech-maple associations form the most luxuriant forests in the whole formation. Not only is there a great wealth of species of tall tree but a large number of smaller trees closely related to the European hornbeam (*Carpinus*), wild cherry (*Prunus*), dogwood (*Cornus*), hedge maple (*Acer*), hawthorn (*Crataegus*) and alder (*Alnus*), as well as many peculiarly American ones, form a recognisable under-storey in most places [85]. Beneath these, equally rich shrub and herb layers must have thrived even under original, undisturbed conditions.

In Tennessee and the northern parts of the states of the Deep South the beech-maple associations gave way originally to those dominated by chestnut and species of oak. Unfortunately the chestnut has been almost eliminated over most of the area during the twentieth century by the virulent chestnut blight, so that now the chestnut-oak (*Q. montana*), scarlet oak (*Q. coccinea*), red oak (*Q. borealis*) and white oak (*Q. alba*) along with the tulip tree, sweet gum and shellbark hickory (*Carya ovata*) are the commonest dominants. As already stated (Chapter IV), many ecologists now hold strongly to the view that the chestnut and oak associations are the true climatic climax over much of the wide area in south-eastern U.S.A. which today is dominated by pines, though there seems to be little doubt that, fringing the Gulf Coast, the climatic climax is a broad-leaved evergreen formation (*vide infra*).

West of the Mississippi River the *mesophytic* chestnut and oak associations give way to more *xerophytic* oak-hickory associations. The latter are

Gvs

predominant in eastern Texas, south-eastern Oklahoma, Arkansas and southern Missouri and still form extensive forests on the slopes of the Ozarks and Ouachitas. They also extend, in a narrow fringe, around the edge of the prairie embayment in Illinois, Indiana and southern Michigan. Species which are only sub-dominant further east like the red oak, white oak and shellbark hickory, here become dominant along with even more drought-resistant species like the black oak (*Q. velutina*), bur oak (*Q. macrocarpa*), mockernut (*Carya alba*) and pecan (*C. pecan*). As the prairie is approached the latter, in turn, become absolutely dominant and compose those forest outliers which extend westwards up the river valleys and which form the famous 'Cross Timbers' forests which inter-digitate with the prairie on the sandstones of eastern Texas and south-eastern Oklahoma [75]. The flora becomes much impoverished in the westernmost outposts of the forest however; in the 'Cross Timbers' zones the post oak (*Q. stellata*) and the blackjack oak (*Q. marilandica*) are the only two common species and, whereas there are many dominants in south-eastern Nebraska, only a hundred miles further west in the valley of the South Platte, only the bur oak (*Q. macrocarpa*) persists [3]. Neverthe-less it is perhaps significant that this last most drought-resistant species is able to survive locally, in the river valleys, far westwards into Colorado and Wyoming.

The east Asian formation

Around the Yellow Sea in western Korea, Shantung and the Kwantung Peninsula, extending into the lowlands of Hopeh, Manchuria and the extreme south-east of the U.S.S.R., there are other areas which are dominated almost entirely by deciduous trees. Indeed this formation may have extended southwards as far as the Yangtse lowlands before human interference. Species which are co-dominant with conifers in the mixed forests of the mainland and Japan (Chapter V) become predominant in these areas of deciduous forest. The dominant trees here are nearly all of species belonging to the same genera as those which form the forests of the same formation-type in Europe and North America. Only the hickories (*Carya*) are notable absentees. The native oaks of Shantung were at one time economically important since their leaves provided food for the silk worms which produced the local 'wild silk'. The native beech (*Fagus crenata*) forms particularly magnificent forests and the ash (*Fraxinus mandshurica*) and birch (*Betula ermanii*) are almost universal.

The formations of the continental interiors

Between the boreal forests and the grasslands, in both Canada and Siberia, there is a narrow belt of deciduous forest. In Canada it extends in an arc eastwards from Edmonton along the North Saskatchewan River to just north of Saskatoon and then south-eastwards to the Winnipeg area and the Red River Valley. It varies from about forty to eighty miles in width, finally widening to over a hundred miles where it meets the Cordilleran forests of the Rockies. Two species are predominant through-out—the balsam poplar (*Populus balsamifera*) and the aspen (*Populus tremuloides*) [55]. The latter forms almost pure stands on the drier areas and the former is most common on moister land. It is also significant that the white spruce (*Picea albertiana*) is often associated with the balsam poplar throughout much of the northern part of the belt. It seems almost certain that, were it not for frequent fires and the nibbling of the snowshoe rabbit, this species of spruce would demonstrate that coniferous forest is the true climatic climax here. It is felt, however, that aspen forest is the true climatic climax in the southern part of the belt. Furthermore the range of the aspen extends, on the average, some sixty or seventy miles south of the southern edge of the continuous forest in the form of isolated groves intruding into the prairie (Plate X*b*). It seems possible therefore that the aspen would have formed a complete forest cover over much of what was prairie had it not been for periodic fires. More will be said on this subject in a later chapter (Chapter IX).

A similar narrow belt of forest extends from the Urals across the Siberian plain to the headwaters of the Yenesey at about 56° N. Although, generally speaking, this forest is much more ill-drained and discontinuous than its counterpart in Canada, it consists of very similar species. The Eurasian aspen (*Populus tremula*) is commonly dominant here along with the birch (*Betula verrucosa*) [31]. A further similarity between these two forest belts is to be found in their underlying soil. As in the Ukraine and the Pannonian Basin, the soils here were originally formed beneath grass-land. Almost universally beneath the continuous poplar forest in Alberta is an almost unaltered chernozem [55] and a similar soil is found locally beneath the stands of aspen in Siberia [31]. In both areas a distinct horizon of lime enrichment still survives—almost certain evidence that trees have invaded a former grassland area in the recent past.

Although these areas of deciduous forest are obviously much less attractive for settlement than the deciduous forest areas already described, it seems, nevertheless, that at certain stages in history, they were looked upon as being much more attractive than the boreal forests to the north and the grasslands to the south. The administrative capitals of two of the Prairie Provinces—Edmonton and Winnipeg—and the chief focus of communications in the third—Saskatoon—were all located within it. The Trans-Siberian Railway, following the line of old-established settlements through Omsk and Tomsk Provinces, also follows the belt with remarkable precision. It appears that, although the timber from the coniferous forests and the grain from the grasslands have been the main economic attractions in the twentieth century, the pioneer settlements were established in or near the moderately fertile soils and deciduous tree cover of these zones.

The Chilean formation

In the southern hemisphere it is only in the inland valleys and on lee-ward slopes in Chilean Patagonia that continuous areas of broad-leaved deciduous forest are to be found. The species composing these forests are much more closely related to those in the near-by evergreen forest than to the deciduous trees which dominate the deciduous summer forest formations in the northern hemisphere. It appears that, in the cooler, drier areas in Chile, trees of deciduous habit have evolved from local genera. The whole formation is dominated by deciduous beech of which the ñire (*Nothofagus antarctica*) and the lengue (*N. pumila*) are the most important [9]. The former forms graceful trees often reaching 100 feet in height and a base diameter of five feet, while the latter is much shorter and less valuable as a timber tree; it usually grows on higher and more exposed sites. It is the lengue which grows up to the timber line at about 1,000 feet in Tierra del Fuego where it frequently attains a stature of no more than four feet but has such densely interlocking growth that it almost bars the way to the alpine grasslands above. In the main, however, the deciduous forests are more easily penetrable than the luxuriant evergreen mixed forests nearer the coast (Chapter V).

2. BROAD-LEAVED EVERGREEN FOREST

Northern hemisphere formations

Isolated stands of evergreen oak and magnolia occur throughout much of Florida as well as locally along the coastal fringes of the Deep South of the U.S.A. and north-eastern Mexico. These are so limited in extent as compared to the subclimax pine forests and coastal grasslands, however, that in spite of the fact that they probably represent the true climatic climax of much of the coastlands of the Gulf of Mexico, they are too small to be shown realistically on a map of continental scale. The commones species in these forest communities are the live oaks (*Quercus virginiana* and *Q. virens*) and the evergreen magnolia (*Magnolia grandiflora*) (Plate VIa).

Outside the tropic in the northern hemisphere, it was only in southern Japan and southern China that extensive areas of broad-leaved evergreen forest flourished. Even here there was a small admixture of coniferous trees. In China this original vegetation has been either removed or completely transformed but in Japan sufficient of it remains to give a clear indication of its former luxuriance and variety. A large number of species of ever-green oak, peculiar to this part of the world, are predominant (*Quercus cuspidata, Q. glabra, Q. thalasica, Q. phylliraeoides, Q. acuta, Q. sessilifolia, Q. glauca* and *Q. gilva*) with large numbers of *Lauraceae* like the camphor tree (*Cinnamomum camphora*) and numerous *Magnoliaceae*. A dense underwood of shrubs and small trees along with many woody climbers are also characteristic.

Southern hemisphere formations

Passing southwards along the western side of the South Island of New Zealand the mixed forest gives way gradually to one which is predominantly broad-leaved. Along the southern part of the western coastlands and the windward slopes of the Southern Alps and across the lowlands of southern Otago, the forests are dominated by species of southern beech (*Notofagus spp.*). The size and shape of the leaves of these trees led the early European settlers to call this forest 'the black birch forest'. This must be regarded as quite different from the mixed forest to the north: *lianes* and flowering *epiphytes* are rare and, in all aspects, the community is less luxuriant. It has been suggested [12] that these *Nothofagus* forests should be grouped with similar stands found within the mixed forest of southern Chile and referred to as 'subantarctic rain forest', but it is doubtful if this term is very helpful or realistic.

Vast forests of broad-leaved evergreen trees also covered south-western and south-eastern Australia before the incursions of European settlers took place. Locally this forest bore a superficial resemblance to the rain forests of New Zealand; in well-watered coastal areas in Victoria, Tasmania and south-eastern New South Wales and on near-by seaward-facing mountain slopes, dense forests of tall trees, with a rich under-storey dominated by tree ferns and arborescent *Compositae*, covered the landscape. Most of the southern Australian evergreen forest was of a quite different kind, however, and must be regarded as a distinct formation. It has been referred to here as 'Australian sclerophyllous forest' (Map 10). It was much more open and had far fewer *epiphytes* and *lianes* than that already mentioned. Furthermore, the dominant trees in this more drought-resistant forest were nearly all of species belonging to the characteristically Australian genus of *Eucalyptus*. Species of *Eucalyptus* were also common in the local, wetter forests [30] but were here intermingled with a great variety of other broad-leaved species; indeed, the boundary between the rain forest and the sclerophyllous forest is everywhere very sharp [59].

Australia as a whole is a remarkably dry continent and, during certain periods in the past, it may very well have been even drier than it is at the present time. Most of the many species of *Eucalyptus* are strikingly drought-resistant so it is understandable that they should have risen to dominance in the forests of this continent. Great forests of these towering sclerophyllous trees extended far inland over large areas of Swanland, Victoria and New South Wales. In spite of their great height, however, the gum trees (*Eucalyptus spp.*) are remarkable in that they cast only a very slight shade (Plate VI*b*). Consequently they permit the development of a ground flora which may be quite light-demanding. The tall forests with an under-growth of grass and xerophytic scrub which were found by the first settlers from Europe, may very well have been the true climatic climax vegetation. On other continents, where trees with denser canopies are available, a grassy cover beneath forest would suggest that fire or some other deflecting factor had intervened (Chapter IX); in the sub-humid parts of Australia, where species of *Eucalyptus* are the only trees available, the existence of a 'grassy forest' may not require such an explanation.

The scores of species of tree-eucalypt in the forests of southern Australia have provided some of the most valuable hardwood timber on the earth. The Sydney blue gum (*Eucalyptus salignum*) of south-eastern Australia and the jarrah (*E. marginata*) and karri (*E. diversicolor*) of south-western Australia have become famous as timber trees. Though the record for the world's tallest known living tree is held by a Californian redwood (*Sequoia sempervirens*) with a height of over 360 feet, there is fairly good evidence for eucalypts of considerably greater height in relatively recent times. The tallest known eucalypt at the present time is a 'mountain ash' (*E. regnans*) in eastern Tasmania which was 322 feet high in 1956 [58].

CHAPTER VII

The Evolution of
Extra-Tropical Forests

A BRIEF examination of the distribution of all the main types of forest in middle and high latitudes reveals a pattern which is most confusing in its complexity. Although in some places there seems to be a close correlation between the distribution of a certain dominant life-form and a particular type of climate, in analogous climatic areas quite a different type of vegetation occurs. The relationship between climate and life-form is obviously far from absolute. Indeed, anomalies are so numerous that it is pointless, if not positively misleading, to make generalisations about such relationships. In the main an unsystematic patchwork of coniferous, broad-leaved evergreen, broad-leaved deciduous and mixed forests is revealed. Conifers are dominant on the edge of the tundra in northern Europe and on the fringes of the Sahara in Morocco. Mixed evergreen forests extend to the extreme tip of Tierra del Fuego in Chile, broad-leaved deciduous trees occupy analogous areas in Europe while an almost pure coniferous formation is found in western North America. Only the boreal forest, composed predominantly of evergreen conifers, seems to dominate throughout a whole climatic region with any degree of continuity and, even here, there are significant anomalous features. Some broad-leaved deciduous species are of frequent occurrence within the boreal forest; they even appear to be dominant in large areas such as the Kamchatka Peninsula (Map 2). More important still, where the boreal forest reaches its northernmost limit at about 72° 50′ N. in eastern Siberia, it is dominated not by *evergreen* conifers but by the *deciduous* Dahurian larch (*Larix dahurica*). It is quite clear from these selected examples that all the attempts that have been made to explain the distribution of forest types in simple climatic terms are premature and, ultimately, probably completely fruitless.

A little reflection on the nature and complexity of forest communities and on the length of time over which plant evolution has been taking place, is sufficient to convince one that any generalisations based merely on life-form must be misleading. Plants have been capable of changing their physiological resistances to drought, cold and other environmental characteristics without necessarily changing their more obvious structural features. Conifers have inhabited the earth for some 200 million years since Carboniferous times. During this period they have had sufficient time to diversify and to adapt themselves to a great range of climatic conditions. If they had not been capable of doing this they would probably not have survived. Although they are so very closely related and so similar in structure, the pines of the Mediterranean lowlands are obviously different in their physiological tolerances from the Scots pine (*Pinus sylvestris*) of the north European boreal forest. Where evolution has also brought about changes in structure, even greater contrasts in climatic tolerance have been made possible. The forests of northern New Zealand have been referred to as 'mixed forests' since they are dominated by conifers as well as angiosperms. It must be clearly understood, however, that the kauri pine (*Agathis australis*), the totara (*Podocarpus totara*), the kahikatea (*P. dacrydioides*) and other New Zealand conifers are different, structurally and physiologically, from the needle-leaved conifers of the boreal forests. Even within the northern hemisphere the coastal redwood (*Sequoia sempervirens*) of California is different in leaf structure and general appearance from the boreal forest conifers. In a similar way the silver birch (*Betula verrucosa*) and the Spanish chestnut (*Castanea sativa*), though of the same life-form, are obviously different in their tolerances of cold and shortness of growing season. Although not so ancient as the conifers, the angiosperms have had sufficient time to produce a range of woody deciduous plants adapted to very different types of environment. From these examples it is obvious that one should not expect to find a regular banding of the earth's forests according to life-form.

Some anomalies in formation distribution are so glaring however, that one is tempted to expect a fairly obvious explanation. Deciduous summer forest is obviously well-adapted to the lowlands of western Europe; with undisturbed conditions it flourished in many places there, almost to the exclusion of conifers. On the other hand, in western North America, where rainfall and temperature conditions are so similar, the forests are dominated by conifers with broad-leaved, deciduous angiosperms subordinate and local. In face of this anomaly one recollects that not only are the conifers and angiosperm trees ancient as groups but that individual species of both can be demonstrated to be millions of years of age (Chapter II).

It is reasonable to expect that, over such long periods, the different trees would have had sufficient time to meet each other in all areas where trees can grow, and there to compete with each other decisively. Over such periods, land connections have almost certainly existed between the continents and, failing this, agents of dispersal such as birds, wind and ocean currents have had sufficient time to effect a good deal of successful seed dispersal. It might therefore be expected that the same set of species would have demonstrated its superiority in all areas where the climate was similar. The distribution of the different types of forest at the beginning of historical times seems to indicate that this had not taken place however.

This problem has been partially resolved by the careful study of fossil floras. It has been discovered that, in early and middle Teritary times, the floras of mid-latitude areas in North America, in eastern and western Europe and in Asia were remarkably similar. In those times the climate of the whole earth was much warmer than now, so that the mid-latitude flora was similar to that in the humid sub-tropics at the present day. The forests in the London Basin in Eocene times must have been very similar, both in life-forms and superficial appearance, to those in northern New Zealand in the mid-nineteenth century. Indeed there is every indication that these New Zealand forests, along with other similar mixed forests in the southern hemisphere, are the true descendants of the ancient, and formerly almost universal, mid-latitude forests. The majority of present-day forests in middle and high latitudes in the northern hemisphere are very new and immature communities. In order to find the Tertiary ancestors of the dominant trees in our present-day deciduous summer forest, it is necessary to examine the fossil floras of Spitzbergen, Greenland and north-central Eurasia.

Though slight modifications in the vegetation of middle latitudes did occur as the Tertiary Era proceeded, it was not until the mid-Pliocene that the pattern began to change radically. Not only did the deciduous trees from higher latitudes begin to invade the mixed evergreen forests of middle latitudes, but conifers, many of which were almost identical with those composing the northern forests today, made their appearance in large numbers. The history of conifers like the Scots pine (*Pinus sylvestris*) and the European spruce (*Picea excelsa*) which now dominate in the European boreal forest is still obscure. Whether they were quite plentiful in the Lower Tertiary but flourished in areas whose fossil floras have not yet been described, or whether they had been only recessive or rare elements in the northern floras until the Pliocene, is not certain. That they did exist prior to the Pliocene is quite certain; their pollen has been found in numerous Tertiary deposits and, in any case, they are far too distinct to have evolved suddenly from other known coniferous stock.

During the late Pliocene Period, northern and sub-tropical species vied with each other for dominance in the forests of middle latitudes and many of the latter, including many species of palm and magnolia, disappeared completely. Nevertheless a complete transformation did not occur until the succeeding Pleistocene Period. It was then that the general cooling of the atmosphere reached its climax and great ice caps, probably unknown on the earth since Carboniferous times, formed over the arctic and over the northern parts of North America and Eurasia. Not only were vast areas, formerly occupied by forest, actually appropriated by ice, but a zone of varying width around the fringes of the ice caps had so rigorous a climate as to forbid the growth of trees; only dwarf shrubs and herbaceous plants survived in this tundra zone.

Vegetation and soil were obliterated by the advancing ice and large numbers of species were eliminated. There were only two ways in which survival could be achieved. Firstly if, at the beginning of the glacial epoch, a species had a wide north-south range, individuals at the extreme south of it might be able to survive the ice advance with no change in position. Secondly, if a species produced seed in sufficient quantities and of a suitable type to be carried southwards by agents of dispersal, it had a chance of migrating southwards ahead of the ice and of finding refuges in which it could survive until the ice retreated again. Neither of these characteristics was a complete guarantee of survival however. In Europe in particular the avenues of southward migration were almost completely blocked by the alpine ranges which had been thrust up during the Miocene orogeny. As the ice advanced from the north and temperatures fell, these mountain ranges developed their own ice caps, thus forming a most effective obstacle to southward migration. A few species escaped through the infrequent lowland gaps and were able to survive the glacial epoch in one or another of the Mediterranean peninsulas. Having survived this first advance however their Pleistocene misfortunes had barely begun! The ice, having achieved its maximum extent, began to retreat as world temperatures began to rise. The species that had escaped were now subjected to vigorous competition from evergreen sub-tropical species which were now able to migrate northwards again. Now that temperatures in the Mediterranean were again favourable for growth throughout the whole year, northern conifers and broad-leaved deciduous trees could not compete successfully with these sub-tropical species and, in order to survive, they had to retreat up the slopes of mountains like the Apennines and the Alps or to escape northwards again through the mountain gaps. Many species were sufficiently resilient to survive this double migration and to re-establish themselves in central and northern Europe. Indeed, in this first inter-glacial period, many trees flourished in Britain which are now absent from it [32]. This inter-glacial flora was not permitted to remain undisturbed for very long however; a second glaciation, more extensive than the first, soon followed and after this a third and yet a fourth. These four glacial epochs, known in central Europe as the Gunz, the Mindel, the Riss and the Würm, with their three intervening inter-glacials, occupied a period of about a million years. During this time the forest floras of Europe were forced into eight migrations and since, for many species, each migration entailed a crossing of the Tertiary mountain belt, it is small wonder that the present flora of Europe is remarkably impoverished as compared to that of the Pliocene.

Only a few hardy conifers and deciduous trees were able to survive these vicissitudes and to migrate back quickly in the rear of the last ice retreat. They re-formed the forests of central, western and northern Europe. It is probably less than 20,000 years since the ice retreated from the North European Plain however; consequently many species which, for one reason or another, are less 'mobile' than others, may not yet have had sufficient time to move to areas where they might very well be able to estabish themselves as elements in the true climatic climax communities.

There are thus two basic reasons for the floristic poverty of the forest communities of central and northern Europe; firstly there is the wholesale annihilation which took place during the Pleistocene migrations and secondly there is the fact that insufficient time has yet elapsed to permit many species to demonstrate whether or not they have the ability to invade northwards. In this context one may note the European sycamore (*Acer pseudo-platanus*) which, before its introduction by man, had not been present in Britain in post-glacial times. Since its introduction, however, it has proved itself capable of growing most aggressively and also of producing vast quantities of seed which germinate every year. One cannot be certain, however, that the sycamore would survive here if it had to compete with native trees in undisturbed forest.

It is clear from a case like that of the sycamore that the concept of 'climatic climax' may have some limitations. The pedunculate oak (*Quercus robur*) became the dominant tree in the clay lowlands of England in post-glacial times because, of all the species *which were present*, it competed most successfully. It cannot be asserted, however, that if all the species of tree *on the earth* (or even in Europe) had been present to compete, that the oak would have won in the struggle for dominance. It is possible that species exist which, in the prevailing conditions, could have beaten it into sub-dominance or even obscurity. The extent to which the sycamore would have been successful can never be known. It is only possible to carry out ecological experiments to discover whether it would be successful in our present climatic conditions.

In eastern North America during the Pleistocene, species mortality was by no means so great as it was in Europe. No east-west obstacle lay across the path of migrating species and a much larger percentage survived in consequence. Only one of the formations that had come to prominence in late Tertiary times suffered serious depletion. This was the lake forest. The towering conifers composing this formation were under severe pressure throughout the Pleistocene. During periods of climatic deterioration and ice-advance they suffered severe competition from the hardier conifers of the boreal forest advancing from the north. During periods of climatic improvement and ice-retreat they were at a disadvantage to the invading deciduous trees from the south. It is because of this that the lake forest, as a distinct formation, has often been omitted from vegetation maps of North America. When first seen by Europeans it was, throughout most of its range, intermixed with stands of trees belonging to either one or the other of its competitors.

In western North America, glaciation again effectively sifted the tree flora but here the ultimate result was very different from that in Europe. West of the Rockies the landscape is one of mountain ranges, high plateaux, intermontane basins and coastal plains. With the onset of each glacial epoch each mountain mass developed its own ice cap, and vegetation in the basins was often trapped: southward retreat could only be local and unimportant. Although large areas in the basins and on the plains were not over-ridden by ice they were, nevertheless, in close proximity to ice-covered surfaces and must have suffered severe weather conditions periodically. Late spring and early autumn frosts would be a great hazard, particularly for broad-leaved deciduous trees. It is probably because of this that, although numerous species of tall tree survived, they were nearly all evergreen conifers. Quite frequently, particularly in the montane forest on the middle slopes of the Cascades and Sierra Nevada, species of deciduous oak (*Quercus californica* and *Q. garryana*) and broad-leaved evergreen species like the madrone (*Arbutus menziesii*) are found intermingled with the conifers. These are either subordinate or in local stands however. After the Pleistocene Period the tall deciduous trees surviving in eastern North America were unable to migrate across the dry heart of the continent to compete with the conifers in the west.

Outside the tropic in the northern hemisphere, it was only in southern Japan and southern China that broad-leaved evergreen forests with some admixture of tall conifers, similar to the widespread forests of the mid-Tertiary, were able to survive over wide areas. Sufficient remnants of these eastern Asiatic forests have survived the onslaught of man for their original flora to be deduced. They appear to have contained representatives of nearly all the tree genera which characterised the Tertiary forests of both Europe and North America. Great ice sheets did not extend southwards in Asia as they did on the other two northern continents so that such drastic migrations were not necessitated. Consequently, even the more northerly deciduous and coniferous formations of eastern Asia are much richer in species than their counterparts in Europe.

The luxuriance of the mixed forests of New Zealand and their predominantly evergreen composition were permitted by the relatively slight interference suffered during the Pleistocene and, subsequently, by winters which are mild in comparison to those in comparable latitudes in the northern hemisphere. Although Patagonia and Tasmania suffered greater Pleistocene disturbance, the prevailing winter mildness, both then and now, has permitted the persistence of mixed evergreen forests. Although a great ice sheet has covered Antarctica since the beginning of the Pleistocene Period, the southern extremities of the other land masses of the southern hemisphere have been insulated from the effects of it by the vast extent of the southern ocean. The Tertiary floras here did not suffer total displacement though the mountains did carry their own local ice caps. It is difficult to appreciate that on the west coast of southern Chile, where fierce gales and bleak, wet conditions prevail throughout much of the year, frosts are, nevertheless, relatively rare; whole winters may pass without a single severe air frost being experienced. It is only in inland valleys, where night temperatures in winter can be much lower, that deciduous forest is found. It is primarily because of this maritime thermostatic effort that many genera which were very widespread in early Tertiary times are now confined to these southern lands. The araucaria pines (*Araucaria spp.*) and the kauri pine (*Agathis australis*) are in this category. It is possible also that the eucalypts were also of wide distribution in the northern hemisphere in the Tertiary though a great deal of doubt has now been cast on some of the European fossil plants which were formerly thought to belong to this genus. Even if these fossils are eucalypts, however, there is another possible reason why, ultimately, this genus became restricted to Australasia. These trees, though very tall, cast a very light shade; it is thus possible that they were unable to compete with the shadier angiosperms which evolved in similar sub-humid environments on other continents (Chapter VI).

In conclusion it must be re-emphasized that there is one outstanding contrast between the great forests which clothed middle and high latitudes in the mid-Tertiary and those which have occupied them in post-glacial times. The former were very varied whereas the latter are comparitively monotonous in their species-composition. Most of the Tertiary forests appear to have been climatic climax mixed forests very similar in their nature to those of the present-day in New Zealand and south-central Chile. Diversity of floristic composition is usually regarded as being an index of ecological maturity. If a plant community has occupied the same area for millions of years it might be expected to have become diverse because of the processes of invasion and competition. The tropical rain forest formations (Chapter XV) are the most complex and diversified on the earth; within one acre it is sometimes difficult to find two mature trees of the same species. On the other hand the separate communities of the boreal forest are often almost pure stands of single species. The probable reason for this evolution towards diversity is easy to appreciate. Each species of tree makes its own special demands upon the soil and the whole physical environment; these demands are different from those made by other species. In order to survive over a long period of time it may therefore find it necessary to form mixed stands with other trees. This combination of species will be able to keep the maximum bulk and the maximum variety of plant nutrients within the nutrient cycle.

In the light of this, the present segregation of coniferous and broad-leaved deciduous trees into distinct formations must not be regarded as an inevitable, long-term feature of the vegetation pattern of the northern hemisphere. It is possible that, given stable climatic conditions over a protracted period, mixed formations of both life-forms might very well evolve. Since any discussion of the details of such a change is entirely hypothetical as well as being highly involved, however, it will not be pursued any further here.

CHAPTER VIII

Tundra and Alpine Vegetation

DURING the Pleistocene glaciations large areas of land in lowlands fringing the major ice sheets and on the lower slopes of mountains fringing local ice caps had too rigorous a climate to permit the persistence of closed forest. Here herbaceous plants and dwarf shrubs were able to become dominant. The origin of these plants is still open to some speculation; it is possible, however, that many of them evolved from less hardy and less specialised species in order to inhabit the bleak mountain tops which came into existence during the alpine orogeny of mid-Tertiary times. They are adapted to an open environment with high light intensities and were therefore annihilated in mid-latitude lowland areas when the re-advance of forests took place in the wake of the retreating Würm ice caps. It was only on cliffs and very steep slopes, where absence of soil prevented the establishment of forest cover, that a few were able to survive. This probably accounts for their presence today in crannies in some steep limestone crags in the Pennines. In Teesdale the alpine saxifrage (*Saxifraga nivalis*) and spring gentian (*Gentiana verna*) are found, the dwarf willow (*Salix herbacea*) appears locally in Craven while mountain avens (*Dryas octopetala*) and a few other alpine and tundra species are found in both these areas [51]. The Pleistocene migrations also account for the fact that, in both Eurasia and North America, the arctic tundra vegetation is very similar to the vegetation of mountain slopes above the forest limits, not only in life-form but also in species-composition. It is clear that the Pleistocene tundra communities, originally occupying large, continuous areas, were able to survive only by migrating in diverse directions; there was a northward retreat along a continuous front and an up-slope retreat wherever tundra communities found themselves cut off on mountain flanks.[1] The present-day boreal and sub-alpine forests thus

[1] This point must not be over-simplified. There are numerous species which, before the end of the Pleistocene migrations, had become either definitely 'arctic' in type or definitely 'alpine'. It must be assumed therefore that each species was able to survive only where it could retreat in the appropriate direction.

give way ultimately to formations of low-growing plants which, though not identical, are very similar.

I. THE ARCTIC TUNDRA

The circum-arctic tundra can be regarded as comprising a single formation-type composed of Eurasian and North American formations. Even within the same formation, however, in spite of superficial similarities, considerable contrasts are evident in both life-form and general appearance. Although flowering herbaceous plants, particularly grasses and sedges, are almost universal throughout all communities, dwarf shrubs, mosses and lichens compete successfully in most places. Furthermore, there is a great contrast between the almost closed-cover tundra communities near the forest margins and the scattered, sparse communities on almost bare ground in the most inhospitable parts of north Greenland and Spitzbergen [58]. It is clear also that not all the extensive tundra communities are to be regarded as true climatic climax.

In the climatically more favoured areas of the tundra the true climatic climax communities are probably those in which grasses and sedges are dominant with a substantial under-storey of lichens and mosses. Communities of this nature are found on undulating areas with a considerable range of slope, aspect and lithology. The dominants in the American formation are various species of sedge (*Carex spp.*), cotton grass (*Eriophorum spp.*) and woodrush (*Luzula spp.*) along with species of many genera of grasses such as the bents (*Agrostis spp.*), foxtail grasses (*Alopecurus spp.*), fescues (*Festuca spp.*) and timothy (*Phleum spp.*) which are also of common occurrence in middle latitudes. The commonest associated lichens belong to the genera *Stereocaulon, Alectoria, Cetraria* and *Cladonia* while mosses of the genus *Polytrichum* are most frequently encountered. Along with these are a great variety of plants with showy flowers belonging to genera which are also distributed throughout the woodlands and meadows of middle latitudes such as the anemones (*Anemone spp.*), marsh marigolds (*Caltha spp.*), buttercups (*Ranunculus spp.*), avens (*Geum spp.*), cinquefoils (*Potentilla spp.*), campions (*Lychnis spp.*), catchflies (*Silene spp.*) and primroses (*Primula spp.*). These are accompanied by flowers which are more typically associated with high latitudes and high altitudes such as the mountain avens (*Dryas spp.*), saxifrages (*Saxifraga spp.*) and gentians (*Gentiana spp.*). A varied cover of predominantly herbaceous vegetation thus forms an almost complete turf over the ground.

Over very large areas, however, the grasses and sedges are much less important and dominance is assumed by foliose lichens and mosses, particularly the former. This cover of cryptogams, if ungrazed, may be anything from four to ten inches high and, in the more southern parts of the tundra, may be almost continuous. Further north similar

A. TUNDRA FROST SCARS. Completely bare patches of clay and stones
in a surface which, otherwise, is thickly vegetated.
(Photograph by J. Palmer.)

B. AN UPRIGHT STONE IN TUNDRA. In this relatively well-drained tundra
dominated by lichens and cotton-grass a fairly large stone has been raised into
a vertical position by frost action. Typical interfluve tundra
in north-west Canada.
(Photograph by J. Palmer.)

PLATE VII

A. STONE RINGS IN TUNDRA. These features have developed here on a
gentle slope covered by fairly luxuriant arctic scrub in north-west Canada.
(Photograph by J. Palmer.)

B. STONE POLYGONS. Sparsely-vegetated tundra in northernmost
Spitzbergen.
(From Polunin, 1960.)

PLATE VIII

communities to the two already described are much more discontinuous and a great deal of bare ground is found between individual plants or plant aggregates. This type of vegetation has been most vividly described in the Taimyr Peninsula by Middendorf [67]. Here, in many places, the ground is only about half covered by plants, permafrost is not more than three inches beneath the surface in many places and a varied patchwork of communities is found. In places mosses (*Polytrichum spp.*) dominate, elsewhere sedges or lichens (*Cladonia spp.*). Throughout all these communities, however, there is an admixture of flowering plants, creeping shrubs such as the dwarf willow (*Salix lapponum*) and xerophytic, heath-like shrubs such as the crowberry (*Empetrum nigrum*). Even at the height of summer, when many species are in flower, these communities give the landscape a drab or yellowish aspect. This is partly due to the colours of the mosses and lichens but partly to the fact that many of the grasses and other flowering plants, as an adaptation to reduce transpiration, retain the dead remains of last year's growth. Ultimately, around the very fringes of ice caps and on some exposed sea coasts, the vegetation dwindles and only an occasional tussock of sedge or cushion-shaped plant is to be found.

   Dwarf shrubs occur sporadically throughout the communities already described but locally shrubs become dominant. Usually, however, the shrubs which are subordinate elements in predominantly herbaceous communities, belong to different species from those which form arctic heath and arctic scrub communities. In arctic heath the species are low growing, the Canadian communities being dominated by small shrubs such as the arctic bell-heather (*Cassiope tetragona*), crowberry (*Empetrum nigrum*) and alpine bearberry (*Arctostaphylos alpina*). In arctic scrub, on the other hand, quite a bushy growth is attained which may reach a height of several feet (Plate VIIIa). It is dominated by broad-leaved shrubs such as the narrow-leaved Labrador tea (*Ledum palustre*), Lapland rose-bay (*Rhododendron lapponicum*), alder (*Alnus sinuata*), scrub birch (*Betula glandulosa*) and numerous species of willow (*Salix spp.*). Many communities are found containing both these types of shrub however.

   Usually the arctic scrub communities are found on sheltered slopes, particularly where the soil thaws unusually deeply in summer and where it is rich and moist [48]. It seems probable that they constitute a zone of ecotone between the more typical tundra communities on the interfluves and the outliers of the forest on the lowest slopes. There is some evidence that, where trees are near by, invasion takes place sporadically in a series of favourable years only to be rebuffed in a subsequent unfavourable period.

   Finally, there is another suite of communities, covering extensive areas in the tundra, which cannot be regarded as climatic climax; these are the tundra bogs and moors. They are very similar in appearance and species-content to the bogs or 'muskegs' which occupy ill-drained localities within

H s

the general area of the boreal forest. Very often in both Europe and America, they are dominated by cotton grass (*Eriophorum vaginatum*) along with many other species of sedge and some grasses. Usually also various species of moss, particularly bog moss (*Sphagnum spp.*), compose a sub-dominant under-storey. The surface of these bogs is usually 'tussocky' and the tussocks of cotton grass and moss are frequently invaded by dwarf willows. Indeed, at an advanced stage of development, when the peat has achieved considerable thickness, the vegetation may become quite luxuriant, similar in many ways to that of the better-drained arctic heaths. These 'heath moors', as they have been called, ensure a copious supply of available moisture throughout the growing season and, at the same time, thaw out to a sufficient depth to accommodate the rooting systems of shrubs. Apart from the heathy shrubs, some broad-leaved, deciduous shrubs also invade this kind of surface. In Canada, for instance, several species of birch (*Betula*) and many of willow (*Salix*) are of frequent occurrence. It seems likely that soil conditions here would be quite suitable for those species of tree which regularly invade drying-out muskegs within the boreal forest areas—species like the black spruce (*Picea mariana*) and the Canadian larch or tamarack (*Larix laricina*). Indeed, if these species of tree were able to advance uninterruptedly, along a solid front, across bogs in the southern parts of the Canadian tundra, they would probably demonstrate their ability to survive the climatic rigours there. Individual tree seeds which have been carried by the wind have been observed to germinate here quite satisfactorily but have subsequently been incapable of surviving the desiccating blizzards of winter [50]. Isolated saplings cannot survive but if they had the partial shelter of near-by parent trees, they would probably be able to do so; the forest would thus encroach gradually. This cannot take place, however, since the deep peat areas are separated from the forest edge by less favourable types of terrain.

The intense physiological drought of the winter months should be stressed as a limiting factor in the tundra far more than mere low temperatures. If the air always remained still, many of the northern trees could probably withstand even lower temperatures than they do. Strong winds, even at very low temperatures, are able to abstract water quite quickly from living plant tissues; if this proceeds beyond a critical point, death ensues, since no replacements are obtainable from the completely frozen soil. The plants of the tundra minimise the danger of this in several distinct ways. On the most exposed areas, all are very low-growing; many die down completely in winter and survive as underground rhyzomes, root stocks, corms and bulbs. Those herbaceous plants which do maintain organs above ground level in winter do so in the form of very tight-packed rosettes or cushions; many species of saxifrage do this very successfully. As a group, the lichens are noted for their ability to survive in an almost desiccated state for protracted periods. It is understandable therefore that they should form such an important element in the vegetation of many of the bleakest and baldest arctic interfluves. The arctic shrubs are also low-growing except in the most sheltered valleys. They have adopted different devices to prevent over-rapid transpiration. Many of the broad-leaved ones are deciduous (*Salix, Alnus, Betula*) while the heaths all have the characteristic small leaves with rolled-back margins.

The shortness of the growing season also imposes great restrictions on plant growth. It is mainly because of this factor that the flowering herbs of the tundra fall into two classes; there are those which flower at the beginning of the growing season and those which flower towards the end of it. The early-flowering species are typically plants with substantial underground food-storage organs; they use the stored energy from the preceding year to flower and produce seed and then utilise the remainder of the growing season to build up food for the following year. The late-flowering species use the first part of the season to build up a store of energy and then use most of it in producing seed; they then pass the winter as inconspicuous underground organs or, if annuals, survive merely as seeds.

The factors of winter frost, winter winds, short growing season and permafrost are responsible, collectively or separately, for precluding trees from the tundra. Because of contrasts in relief and aspect the relative importance of these different factors varies greatly and an evaluation of the exact reasons for the absence of trees in a particular place is usually obscure. However, the complex inter-digitation of the southern edge of the tundra with the northern edge of the forest indicates that, generally speaking, wind force along with shallowness of unfrozen soil are the operative factors in the southern parts of the tundra. Here, forests occupy the valleys while the interfluves are tundra. Wind force is obviously important on these interfluves but, whether forest would extend further up the slopes if the depth of the thawed layer in summer were deeper, is not clear. The existence of permafrost, as such, does not preclude forest since vast areas of the northernmost parts of the boreal forest are underlain by it both in Siberia and Canada; northern species such as the Dahurian larch (*Larix dahurica*) and Siberian dwarf-pine (*Pinus pumila*) are adapted to this soil condition in that they develop a root system which is entirely spreading with no tap root whatsoever [73]. In the northern archipelago of Canada and the arctic peninsulas and islands of the U.S.S.R., as well as in Spitzbergen and the north of Greenland, it appears probable that trees cannot possibly regenerate regardless of wind and soil depth. Though increase in the length of the season with continuous daylight does, to a certain extent, compensate for increased latitude, it is doubtful if the growing season here is sufficiently long for trees. Even if trees could survive the winter, it is very doubtful if they would be able to flower and produce seed. In these fringing areas, so near to perennial ice, there is no recognisable frost-free season. Everywhere in the tundra the mean length of the frost-free season is less than fifty days but here, in spite of continuous daylight, air frosts occur frequently in all the summer months. Returning to the southern fringes of the tundra, however, it is doubtful whether the length of the frost-free season has any bearing on the position of the limits of the forest. Probably the length of the frost-free season is sufficient here to permit forest regeneration but wind speed in winter and extreme shallowness of soil in summer forbid the spread of trees.

Fluctuations in the position of the northern edge of the boreal forest have occurred during the past few millennia. Locally, in European Russia, a distinctive fringing zone of 'tundra moor' has been described [44]. Here, hummocks of bog moss (*Sphagnum spp.*) peat, each several yards in diameter, with narrow, marshy channels between them, cover the landscape. The mosses on the hummocks are often still growing but the cores of the hummocks are permanently frozen. This permanently frozen peat contains the stumps of both birches and conifers, many of which are in position of growth. Clearly these areas were forested not so very long ago but the reason for forest recession is not altogether clear. Possibly a deterioration in climate was responsible; possibly the activities of reindeer herders along the former forest margins caused the saplings to be grazed off and thus prevented regeneration; possibly both climatic change and human activity played some part in the changes that occurred.

Soils of the arctic tundra

The peat soils of the tundra bog have already been described. Elsewhere soils vary greatly in depth and general appearance. Nevertheless, they have one characteristic in common which distinguishes them from the forest podzols. Whereas the latter are remarkable for their differentiation into distinct horizons, the tundra soils are markedly unstratified. The main reason for this is the thorough mixing which takes place, at regular intervals, because of frost-heaving. During the summer months the soil thaws to a certain depth but, because of the impervious frozen material beneath, it remains saturated apart from the top few inches. With the onset of the cold season the top part of the thawed layer freezes again so that, for a time, a rigid crust of newly-frozen soil traps a layer of unfrozen, saturated material beneath it. Since pure water increases its volume by between 8% and 9% when it changes from the liquid state into ice, some disruption is bound to take place as the entombed layer gradually freezes. Expansion can only take place upwards but it occurs very irregularly because of the heterogeneous nature of the materials composing the soil. Because of this the overlying crust is distorted and fractured and plastic or viscous clay (Plate VIIa), along with stones of various sizes (Plate VIIb), are extruded or wedged upwards. A great deal remains to be discovered about the exact nature of these frost-heaving processes but, whatever the details, a very uneven surface results. Many areas of tundra are characterised by a hummocky appearance and by the great number of stones which lie on the surface or project upwards from it.

The main reason why frost-heaving phenomena are much in debate is because there are two main periods of disturbance in the year. Expansion occurs in the autumn while shrinkage and subsidence of equal magnitude must take place during the spring thaw. It is not yet clear whether the surface unevenness is due primarily to differential uplift or to subsequent subsidence. The main argument has centred around the large stones which are brought up from below and which may ultimately project well above the general level of the ground (Plate VIIb). Some workers have inferred that these are actually thrust upwards an inordinately great distance during autumn freezes; however, the majority opinion now seems to favour the idea that the whole surface rises more evenly in autumn and that the stones are merely left behind when subsidence takes place. There can be no doubt that both periods of movement have some effect on the micro-relief of the surface but their relative importance has not been settled.

Quite often the large projecting stones are distributed sporadically over the surface but, over vast areas of tundra, they are arranged in a continuous pattern around the peripheries of irregular polygons (Plate VIIIb). In the light of the most recent evidence it appears likely that the stones come to the surface sporadically but, having arrived there, they are pushed outwards from 'centres of action' by some kind of lateral thrusting. Normally these 'centres' are so numerous that the circles of coarse material moving outwards soon interlock to produce the characteristic polygonal pattern. In some places, however, they are much more widely dispersed so that only the occasional 'stone circle' is encountered (Plate VIIIa). The reason why the 'centres of action' appear to remain fixed when once established is, as yet, unsettled. Most recent investigations seem to indicate that, when once a stone reaches the periphery of a polygon, it is effectively trapped and will remain there until finally destroyed by weathering processes. Stone polygons and stone circles are thus regarded as evidence of a particular stage in a soil-forming process; the materials of the whole layer above the permafrost are first sorted so that the coarse, unweathered fragments lie in an orderly pattern on the surface; here they are gradually broken down into sand, silt and clay particles. Since so much remains to be discovered about the nature of frost-heaving, however, it is possible that present views may be much modified, if not completely overturned, by future observations.

On slopes exceeding 5° to 10° in the tundra the nature of the surface phenomena is rather different; instead of occurring in circles or polygonal patterns, the finer and coarser materials tend to arrange themselves in parallel 'stripes' running up and down the slope. The general processes responsible for this pattern are similar to those which cause the patterns on the flatter ground but, because of relatively rapid down-slope movement, a banding of materials is produced. As well as the forces of frost-heaving, the rapid and protracted solifluction which takes place as soon as melting begins in the spring is also operative and produces a generally thin soil cover on quite gentle slopes. This is accentuated in the tundra because of the absence of substantial root systems such as are possessed by trees.

Although adverse climatic conditions are obviously responsible for the failure of the vegetation to form a completely closed cover over the surface of much of the tundra, the persistent mobility of the soil, just described, is also responsible to some extent. Even near the forest edge, in places where turf-forming grasses and sedges are able to thrive, bare stones and patches of blue or grey clay are frequent; constant re-invasion of fresh material thus goes on constantly. Blue and grey coloration is predominant in the mineral soil materials of the tundra mainly because of the prevailing water-logged conditions beneath the top inch or two. This maintains iron in its reduced or ferrous state so that the reds, browns and yellows of ferric oxide in its various forms, so common in the soils or lower latitudes, are here relatively rare.

Human interference

Under natural conditions the tundra was grazed by herds of wild animals of which the caribou or reindeer (*Rangifer tarandus*)[1] was the most prominent. This species grazes particularly on the lichens in the tundra vegetation. It is many centuries since the reindeer was domesticated by the Lapps in northern Europe but they have continued to utilise the same areas over which the animal had ranged in its natural state. In Asia and America, however, peoples like the Yukaghirs and Eskimos continued as hunters right up to the twentieth century. The history of the Eskimo economy in Alaska is interesting as regards the ecological status of much of the present arctic tundra in that area. Because of an increase in the use of firearms, the caribou herds became seriously depleted in the late nineteenth century and many Eskimo communities were brought to the verge of starvation. In an attempt to alleviate their plight, over 2,000 reindeer were introduced into Alaska. To ensure that these did not suffer the same fate as the caribou, rangers were employed to enforce some degree of conservation by controlling shooting and by killing wolves. So effectively did they carry out their duties that the herds increased enormously and it was felt that the basic problem had been solved. In effect, the result of

[1] These two animals are merely slightly different varieties of the same species, the reindeer being a native of Eurasia and the caribou of North America.

this 'conservation' proved to be catastrophic. Increasing numbers of animals were found to be dead and dying until, in one year alone, it was estimated that many thousands had perished. From an estimated peak population of something like 650,000 in the whole of Alaska just before 1940, there was a crash to below 50,000 only ten years later. On the island of Nunivak off the west coast, a peak population of 22,000 animals was reduced to little more than 4,000 in 1952 [20]. The reason for this whole-sale mortality was that the natural pastures had been seriously over-grazed by the increased animal population. The tundra lichens grow and regenerate very slowly; indeed it is estimated that half a century is required for the development of really luxuriant cover. It appears that, over large areas, these lichens had been almost eliminated and that the cotton grass (*Eriophorum vaginatum*), formerly co-dominant with the lichen, had come to form almost pure cotton grass moors. Unable to find sufficient forage, many reindeer perished. This is an excellent illustration of the suscepti-bility of some climax communities to indirect human interference.

2. SUB-ANTARCTIC AND ANTARCTIC TUNDRA

Owing to the fact that a great ice-capped continent occupies circum-polar regions in the southern hemisphere, the antarctic area is considerably colder than the arctic, particularly in summer. Whereas an almost closed cover of tundra plants, often dominated by flowering species, is of frequent occurrence around 70° N. and indeed considerably nearer the North Pole, not a single flowering species has been recorded south of 70° S. Throughout the whole of the continent of Antarctica, it is only on the Grahamland Peninsula that flowering plants manage to exist and, even here, only two species are found. The first of these is a grass, *Deschampsia antarctica*, closely related to the European wavy-hair grass (*D. flexuosa*), and the second is *Colobanthus crassifolius* which belongs to the order *Caryophyllaceae* [58]. Elsewhere in Antarctica, in sheltered places where the snow melts for a short period in the summer, brightly coloured mosses and lichens are found, but rarely, if ever, do they completely cover any con-siderable area. One has to move northwards to the sub-antarctic islands to find a flora in any way similar to that of the tundra of North America and Eurasia. Permafrost underlies the soils of all the sub-antarctic lands northwards to about the 60th parallel. On the more northerly islands such as Kerguelen (49° S. 70° E.) and Macquarie Island (54° S. 159° E.), where the vegetation is sufficiently rich to resemble that of the most luxuriant arctic tundra, it is unlikely that permafrost is to be found. Although these islands have a bleak, cool climate throughout the year, summer temperatures are almost certainly sufficiently high to support tree growth. One must therefore assume that the persistent strong winds are the factor which have permitted the continuing dominance of almost entirely herbaceous floras. Under this 'tundra' vegetation, pedogenic processes are very different from those beneath the arctic tundra. Instead of being much disturbed soils with universally impeded drainage they are almost certainly stratified soils with strong podzolisation. In the light of this contrast it is probably more desirable to refer to the vegetation on these islands as sub-antarctic *heath* and *moor* (Chapter XI) than as 'tundra'.

The main floristic differences between the vegetation of Kerguelen and that of the arctic tundra lies in the relative poverty of the former. It is particularly deficient in woody plants, the place of these being taken by a relatively large number of herbaceous species which adopt a cushion-like mode of growth. This last characteristic is a reflection of the extreme windiness of the habitat. The actual species composing the individual island floras are predominantly *endemics*—plants which are found nowhere else in the world. Thus, on Kerguelen, apart from the two antarctic species already mentioned, strange species such as the Kerguelen cabbage (*Pringlea antiscorbutica*) belonging to the *Cruciferae*, along with *Acaena adscendens* (*Rosaceae*), *Azorella selago* (*Umbelliferae*), *Lyallia kerguelensis* (*Caryophyllaceae*), *Agrostis antarctica* and *Festuca kerguelensis* (*Graminae*) and numerous species of cryptogam make up the vegetation. The striking fact about the Kerguelen cabbage is the fact that it is the sole species in a genus which is confined to this one island. Macquarie Island, South Georgia and the other oceanic islands in similar latitudes also have very unusual floras. All the islands south of 50° S. are covered with some kind of sub-antarctic heath and moor. It is only on Tierra del Fuego that forests extend to 55° S. and, even here, they are much stunted, most of the land being covered with a herbaceous or dwarf shrub vegetation.

3. THE ALPINE FORMATIONS

The reasons for the basic similarity between the alpine vegetation on mid-latitude mountains and the arctic tundra have already been given. Although a large number of quite distinct species are found only in either one or the other, large numbers of species are common to both. Indeed, on the same continent, there is often almost as much difference between the alpine floras of two separate mountain ranges as there is between the flora on one of those ranges and that in the related arctic areas. Thus, in North America, the Rocky Mountain alpine formation contains numerous endemic species but, of the remaining plants, just as many are found in the Canadian arctic tundra as are found on the near-by Cascades–Sierra Nevada Range. The most significant point, however, is that most of the species in all three areas belong to common genera and that nearly all those species which are endemic in any one area have sprung, in relatively recent times, from species of wider distribution [81].

The same kind of relationships have been discerned in Eurasia. Because these relationships are so close, some ecologists [81] regard all the arctic and alpine communities of the northern hemisphere as being divisible into only two formations—a Eurasian one and a North American one. They then divide each formation into associations—an arctic one and an appropriate number of alpine ones. Whether this classification is realistic or helpful is doubtful. Its main advantage is that it draws attention to the great similarities, both in life-form and species-content, between arctic and alpine vegetation.

Environmentally, alpine vegetation stands in just the same relationship to arctic tundra vegetation as sub-alpine forest does to boreal forest (Chapter IV). Alpine vegetation experiences relatively high light intensities with consequent day-time warming throughout the whole year. Except in very well sheltered places, it also has to withstand very high wind speeds. Also, whereas permafrost is almost universal beneath arctic tundra, it is normally absent beneath the generally steeply-sloping areas of alpine vegetation. Because of these differences, physical drought is a much more obvious and frequent hazard to alpine communities than to their arctic equivalents; fortunately for many of the widely-dispersed species, the same adaptations which help to resist the great cold and physiological drought of the arctic winter are also efficacious in lowering transpiration rates in the drying winds and hot sun on high mountain slopes.

On the lee side of mountain ranges and, more generally, in continental interiors, an entirely different kind of alpine vegetation can be found, particularly on south-facing slopes. Because of frequent and intense drought, these communities are similar in many ways to the semi-desert vegetation of the continental lowlands with which they are contiguous (Chapter X). The 'pamirs' of the mountains of central Asia are a good example of this vegetation type. Quite often, however, even in these generally dry mountain areas, the north-facing slopes carry a more typical alpine vegetation with cushion-like and rosette plants along with alpine sedges and grasses. Even within the more usual type of alpine communities, however, regional contrasts can be found. The western slopes of the Cascades, for instance, carry an alpine vegetation which is rather different from the equivalent on the western slopes of the Rockies. In the case of the former, a heavier precipitation and more humid atmosphere permits a much more luxuriant growth and peat bogs occur more frequently.

Alpine vegetation is usually even more gay with showy flowers in the growing season than is the arctic tundra and, as in the latter, two distinct aspects can again be distinguished. It is often possible to see the flowers of the spring aspect pushing through the edges of the melting snow-drifts and to return, later in the year, to find masses of flowers of the summer aspect only just making seed in time to escape the first withering blasts of oncoming winter.

The height at which sub-alpine forest gives way to alpine vegetation rises gradually with decreasing latitude and general statements can be made for each latitudinal zone. It must always be remembered, however, that many factors are capable of giving rise to local anomalies. Generally speaking, at a given latitude, the larger the mountain mass, the higher the tree line. Wetness is also of some significance; on the western side of the Cascades Range, for instance, the forest is higher and the snow line is lower than in the equivalent position on the Rocky Mountains. Greater wetness thus not only causes the lower boundary of the alpine communities to be displaced upwards, it also brings down the upper limit.

Generally speaking, the lower limit of the alpine zone is about 4,000 feet above sea-level in southern Alaska, 7,000 feet in southern British Columbia and 10,000 feet in the Sierra Nevada, although inland, in Colorado and New Mexico, it is often as high as 12,000 feet. It is found at about 15,000 feet on the mountains of south-central Mexico. In Eurasia it occurs at about 3,000 feet on the Cairngorms in Scotland, 7,000 feet in the Swiss Alps and the Caucasus and 12,000 feet in the western Himalaya.

On the mountains of middle and high latitudes there is often a characteristic type of vegetation in the transition zone between the forest and the alpine zone. This is a narrow belt of sparse woodland known as 'elfin woodland' (Plate XXIIb). It is composed of different species on different mountain masses; in the Alps one finds the mountain pine (*Pinus montana* var. *pumilio*). All these trees are able to adapt themselves to extreme windiness by spreading their branches at, or just above, ground level. Growing in this way, they are able to attain considerable bulk, thus demonstrating that climatic conditions here would be quite favourable for more normal forest were it not for wind speed. Here again, however, just as in the case of the northern parts of the boreal forest, one is faced with the very difficult problem as to which factors are ultimately responsible for preventing trees from occupying the upper regions. The zones of elfin woodland are critical zones of competition in which this problem can be studied. Unfortunately many of them, particularly in more accessible regions, have now been modified or completely removed to make way for 'alpine pastures'.

CHAPTER IX

Grasslands

I. THE NORTH AMERICAN PRAIRIES

WHEN the first European settlers saw the North American continent, it appeared to them that a great sea of forest stretched endlessly westwards. In Canada, New England and Virginia the further inland they penetrated, generation after generation, the more boundless forests they discovered. It was not until the nineteenth century that, ultimately, they emerged at its western edge and discovered that the forest was not endless; an equally vast sea of grassland lay beyond it. The American and Canadian peoples who had achieved nationhood in a forest environment had now to master a very different kind of landscape if their territories were to expand still further. The grassland lay in a continuous north-south belt from the forest edge in Alberta and Saskatchewan to the Gulf Coast in south-eastern Texas; it also had extensive outliers in the basins and on the plateaux to the west and in the Central Valley of California as well as smaller outliers to the east, particularly in the Kentucky Blue Grass Region and the Nashville Basin of Tennessee. Apart from these outliers it also bulged far eastwards on to the Middle West plains in Illinois and western Indiana. Westwards from the forest edge the grassland stretched, without interruption, to the foothills of the Rockies.

Over such vast areas, as one might expect, considerable differences in component species were to be found. Consequently the prairie 'formation' has been sub-divided into several categories; one authority recognises seven distinct 'associations' [81].[1] It seems that three quite distinctive general types have been recognised however: (a) *true prairie*, (b) *mixed prairie* and (c) the *Pacific* and *Palouse prairies*.

The true prairie extended in a great arc from south-central Alberta and Saskatchewan through southern Manitoba and the Middle West to east-central Texas, ultimately giving way to the closely-related coastal prairie in south-eastern Texas. The western edge of this type of grassland ran from

---

[1] The terms 'formation' and 'association' have been indicated as quotations here to imply the doubt that has been expressed about the climatic climax status of at least some of these communities by some authorities.

central Texas through central Oklahoma, central Kansas and Nebraska and across the Dakotas into southern Saskatchewan and Alberta, though recent observations have demonstrated that considerable fluctuations in its position must have occurred at frequent intervals, shifts eastwards taking place during sequences of dry years, and westwards during periods of more plentiful moisture [82]. The eastward bulge of the prairies across Illinois into north-west Indiana (Map 5) was also of this type of grassland. Agriculture has almost completely removed these communities but old photographs (Plate X*a*) and descriptions along with modern observations of vegetation relicts, leave no doubt about its former nature. The dominant grasses mostly grew to a height of three or four feet during the summer season. They were a mixture of sward-forming grasses like *Agropyron smithii*, *Andropogon scoparius* and *Bouteloua curtipendula* and bunch grasses like *Sporobolos asper* and the common stipa grass (*Stipa spartea*). In spite of a general 'tussocky' appearance therefore, particularly in the winter season, the true prairie did form a continuous sward over the ground.

Generally speaking, the true prairie gave way westwards to mixed prairie. This occupied most of the Great Plains west of the line already indicated. The name 'mixed prairie' arises from the fact that it is composed of grasses of two distinct life-forms; there are those growing to a height of two or three feet and there are dwarf grasses whose fruiting stems achieve a height of only a few inches. Common throughout the entire area formerly there were mid-height grasses like *Stipa comata*, *Sporobolos cryptandrus*, *Agropyron smithii* and *Koeleria cristata* while the buffalo grass (*Buchloe dactyloides*) and grama grass (*Bouteloua gracilis*) were the most widespread, sward-forming dwarf grasses [82]. In some of the driest areas of south-western Texas, southern New Mexico and Arizona [35], extending locally on to the plateau of northern Mexico, the true mixed prairie gave way to an even more xerophytic community dominated almost entirely by the dwarf species alone. Here several species of grama grass (*Bouteloua*) along with species of *Aristida* are dominant and the areas they cover are often spoken of as 'The Desert Plains'. Extensive areas further to the north, in Kansas, Colorado, Wyoming and Montana, have also been referred to as 'short grass prairie' in twentieth-century literature. This is because the over-grazing to which much of the Great Plains was subjected periodically from 1870 onwards was effective in killing out the mid-grasses and leaving just a turf of grama. Indeed even the latter was seriously reduced over wide areas leaving bare soil into which the woody or succulent species of near-by scrublands (Chapter X) were able to invade. Species of cactus (*Opuntia spp.*) and sagebrush (*Artemisia tridentata*) were thus able to spread far beyond their former ranges. The main reason for this disturbance is that many of the grasses of the mixed prairie are annuals so that over-grazing which is so intense as to prevent seeding, can eliminate a species almost within a single season.

The Pacific prairie was confined to the floor of the Central Valley of California and some coastal lowlands of Lower California. In general appearance, species and life-form it was very similar to the Palouse prairie which was found over extensive areas of basin and plateau country in eastern Washington and Oregon, southern Idaho and northern Utah as well as on the Palouse itself. These western grasslands differed from those already described in that they were dominated by bunch grasses whereas the true prairie and mixed prairie were characterised by a definite turf or sward. Thus the blue bunch wheat grass (*Agropyron spicatum*) was a common dominant in the Palouse prairie and a species of stipa grass (*Stipa pulchra*), of pronounced bunch form, was almost universal in the Pacific prairie. Again, over-grazing and cultivation have almost removed these grasslands in their original form; the floor of the Central Valley is in great part cultivated and vast wheatlands extend over the Palouse. Grazing has taken place in the drier and least fertile areas, interfering with the original vegetation to such an extent that wholesale invasion by sagebrush has occurred.

Soils

A soil developing beneath grassland is subjected to conditions which are very different from those experienced by a developing forest soil. It has been shown in previous chapters that, beneath much natural forest, most of the organic material which comes to the soil is supplied, by the process of leaf-fall, to a surface with only a discontinuous covering of herbaceous plants. A loose, friable and relatively well-aerated layer of material thus accumulates which can only be incorporated into the predominantly mineral horizons beneath by purely mechanical mixing processes. These are caused almost entirely by the activities of the larger members of the soil fauna. The aerial parts of grasses, on the other hand, form a much denser 'sod' or 'turf' as they die down and accumulate. Furthermore, grasses form a dense rooting network throughout the entire depth of the soil beneath. These die away and decay in an almost continuous process so that humus, in a finely-disseminated form, is actually implanted in the soil by the vegetation. This intimate penetration by roots and humus not only ensures deep inherent fertility but also gives the soil a fine crumb structure and perviousness, regardless of differences in original parent-material. Generally speaking, grasses also take up greater quantities of mineral nutrients, particularly calcium, than do forest trees. The humus returned to the soil by grasses is therefore commensurately richer in nutrients. In turn, as the humus decays, the nutrients are released only to be caught up again by the efficient rooting systems. In this way a very rich nutrient cycle is maintained.

The microclimates experienced by the soils beneath the prairie grasses are also very different from those experienced by forest soils. Wind and sun cause more rapid evaporation from the actual surface of the soil beneath grassland and, even more important, the rapid transpiration from the grasses themselves causes them to draw on the soil-water around their roots. Even though summer is the wettest season over most of the prairies therefore, the soil is subjected to most efficient drying processes and, over vast areas, rain rarely manages to percolate beneath root depth. Indeed, beneath some of the driest areas, the sub-soil can be regarded as almost permanently dry.

One of the most widespread soils beneath the prairies of North America is usually referred to as the *black earth* or *chernozem* (Fig. 17). It extends in a wide belt from central Alberta and Saskatchewan southwards through the eastern part of the Dakotas and Nebraska and across central Kansas and western Oklahoma to central and southern Texas (Fig. 12). It is thus found beneath the western part of the true prairie and beneath the eastern edge of the mixed prairie. Typically it has a brown coloration which is so dark as to appear blackish when seen fresh in the field (as in the top foot of the mollisol shown in Plate IX*a*). This deep coloration is due almost entirely to the humus content; if this is removed chemically in the laboratory, the inorganic fraction of the chernozem is usually found to be a pale grey. The universal dark coloration of the chernozem humus has not been satisfactorily explained but it seems likely that it is in some way due to a combination of the physical nature of the humus supplied by grasses, the high base-status and the high soil temperatures during the summer [63]. This dark-coloured humus is so evenly distributed throughout the chernozem profile that quite small amounts give rather a misleading impression. The average chernozem has an organic content of only 8%–10%; more than 16% is quite exceptional. Indeed chernozems with as little as 6% still have the typical 'black' coloration.

Unlike forest soils, chernozems belong to the great soil group of *pedocals*. They have developed in sub-humid conditions so that evaporation with consequent upward movement of the soil solution preponderates over downward movement and leaching. Only rarely is precipitation so prolonged or snow-melt so copious as to cause percolation to beyond root range. As a consequence, although the very soluble salts of potassium and sodium are leached from the profile by the occasional downpour, the less soluble calcium salts, particularly calcium carbonate ($CaCO_3$), accumulate. Indeed one of the diagnostic properties of chernozems is that they contain an excess of lime and therefore have a pH value of more than 7; in other words they have a basic reaction. Because of this, the clay compounds in the profile have no tendency to dissociate; the clay minerals remain completely stable and the podzolisation process cannot begin. The average chernozem consists of an almost homogeneous, dark-coloured layer which may be up to six feet or more in depth but which is normally two or three feet thick. This soil cannot be differentiated into eluviated and illuviated horizons; the only distinction that can be made is that the upper part of this soil is often darker than the lower part.

There is usually a distinct A_{00} horizon at the top of the chernozem profile in which the yet-undecomposed dead leaves and leaf-bases of the grasses accumulate. This may be several inches in thickness and grades

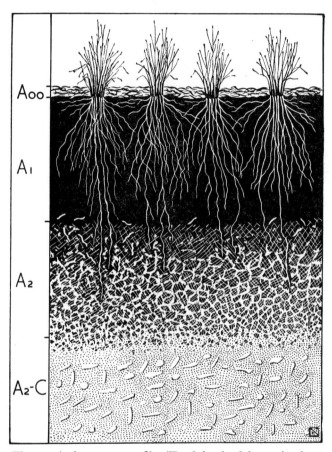

Fig. 17. A chernozem profile. (Total depth of the section is 4½ feet.)

downwards into a predominantly organic, almost structureless, A_0 horizon which is usually of negligible thickness however. As the organic material at the top of the soil humifies, the plentiful soil fauna mix it downwards very rapidly. The A horizons sometimes give way quite sharply to the underlying C horizons of partially-weathered parent-material but usually this junction is obscured because of the development of a thick horizon of calcium carbonate accumulation (A$_2$—C). This substance occurs in the form of irregular concretions or nodules; it is a feature in nearly all chernozems and obviously forms at the maximum depth to which water penetrates except on the infrequent occasion. Less numerous lime concretions also occur irregularly throughout the lower part of the dark-coloured horizons (A$_2$). Most usually they are in the form of fine tubes or filaments[1] which are obviously the former burrows of the soil fauna. Air can penetrate downwards most freely in these burrows so that evaporation and re-deposition of lime is rapid.

Westwards, beneath the even drier areas of the mixed prairie, chernozem grades first into *dark-brown* or *chestnut-brown soils* and ultimately, in Alberta, Montana, Wyoming, Colorado, New Mexico and Arizona, into *brown soils*. The chestnut-brown soils are closely related to the chernozems. They owe their paler coloration merely to the fact that they contain, on the average, only 3%–5% organic material. This is due to the fact that, in the drier climate, the annual production of vegetable matter per unit area is reduced. The brown soils are even poorer in organic matter and quite often contain no more than 1%. The actual depth or true soil also decreases with increasing aridity and the horizon of lime accumulation thickens and occurs at progressively shallower depths. There is also a general tendency for the pH value to increase with increasing aridity (though pH8 is not exceeded if calcium carbonate is the only free base present). In all other respects the chestnut-brown and the brown soils are similar to the chernozems.

All the moister, eastern parts of the true prairie were underlain by an entirely different kind of soil. In southern Minnesota, Iowa, Illinois, north-western Indiana, northern Missouri, eastern Nebraska and Kansas, east-central Oklahoma and Texas and the coastal prairies of Texas and Louisiana, the grassland soils were pedalfers not pedocals and are referred to as *prairie soils* (Fig. 12). Here, in spite of the wind and sun of the prairie environment, precipitation is sufficiently heavy and frequent to cause frequent percolation right down into the sub-soil beneath root range. Because of this, all free calcium carbonate, along with more soluble salts, were entirely leached from the soil; there was no horizon of accumulation and there were no disseminated nodules. Nevertheless, because of the lime-demanding grasses and the rich nutrient cycle, these soils were little less inherently fertile than the chernozems. The pH was normally between 6 and 7—only just on the acid side of neutrality. Because of this slight acidity there was some slight tendency for the liberation of sesquioxides even with undisturbed conditions, but this was completely obscured by the dark brown coloration imparted to the soil by the abundant humus. As far as can be ascertained no clear differentiation into eluviated and illuviated (A and B) horizons had taken place in any natural prairie

[1] Referred to in Russian literature as 'krotowinas'.

B

PLATE IX

A. A TOP CATEGORY MOLLISOL. This soil from South Dakota, formerly classed as a chernozem, is similar to the chernozems except that there has been considerable accumulation of clay, as well as organic matter, between the depths of one and $2\frac{1}{2}$ feet. The very light spots below $2\frac{1}{2}$ feet are roughly spherical accumulations of calcium carbonate; these are the *beloglazka* or 'white eyes' of the Russian literature on chernozems. (Photograph by Roy W. Simmonds, by permission of the United States Department of Agriculture.)

B. 'MEADOW STEPPE'. This steppe, not far from the forest edge in the Voronezh District, is dominated by a sod-forming species of feather-grass (*Stipa lessingiana*).
(Photograph by Keller, 1927.)

A

A. TRUE PRAIRIE. An old photograph of the Woodburg area of western Iowa in
its 'original' state. Isolated trees occur in folds in the ground along the valley
sides and an almost continuous fringe of trees follows the valley bottom.
(Photograph by J. M. Coulter in Schimper, 1903.)

B. THE PARK BELT. Aspen groves and prairie near Vermilion in southern
Alberta.
(Photograph by J. Arthur Herrick in Henry A. Gleason and Arthur Cronquist,
The Natural Geography of Plants, 1964, Columbia University Press, New York.)

PLATE X

soils though it is now extremely difficult to find an area which is known, with certainty, never to have been disturbed for agricultural purposes. In spite of these vital characteristics which confirm the prairie soil as a pedalfer, its fine crumb structure and high base-status made it similar to the chernozem. Indeed, because of its generally greater depth and more humid climate, this soil type came to be regarded as agriculturally more valuable than the chernozem.

The grassland-forest boundary

From Alberta to Texas there extended one of the most interesting and complicated zones of ecotone ever observed on the earth. Forest trees here met prairie grasses in a struggle for dominance. Since the forest tree is the taller of the two life-forms, however, it is more realistic to view this boundary as the place where trees were forever attempting to invade the prairie; if at any time or in any place they were successful, they soon shaded out the grasses and precluded them. The main difficulty is that of assessing which factors had been responsible for preventing this invasion long before the advent of European settlers. There can be no simple answer since, in different parts of the ecotone, different species of tree, probably with quite different tolerances, composed the forest edge. In Alberta it was predominantly the aspen (*Populus tremuloides*) [55], in Minnesota it was aspen along with species of willow, poplar, cherry and oak (*Q. macrocarpa*) [25], while in central Texas it was the two most xerophytic oaks—the post oak (*Quercus stellata*) and the blackjack oak (*Q. marilandica*) [81].

Although human interference has now destroyed most of the evidence, the historical record indicates quite clearly that the boundary between forest and prairie was remarkably abrupt in many places. One could pass from closed-cover forest to grassland almost, if not entirely, devoid of trees, in the space of a few yards [7]. The boundary was, nevertheless, a most sinuous one. Not only was there complex inter-digitation, but numerous islands of prairie were found within the generally forested area and many outliers of forest lay in the prairies. Although the prairie embayment in the Middle West extended, as a fairly continuous area, only as far east as north-west Indiana, quite large islands of prairie were scattered throughout the forests of eastern Indiana; the continuous poplar forest of Alberta gave way southwards, in a zone up to eighty miles in width, to the Park Belt in which isolated groves of aspen were surrounded by prairie (Plate X*b*); in central and southern Texas, isolated outliers of dwarfed, xerophytic oaks were completely surrounded by short grass prairie.

Though the general picture in North America is of a humid, forested east and north giving way to a dry, grassy interior, it has, so far, been found impossible to state, even in general terms, which attributes of the

Ivs

physical environment were responsible for holding back the advance of
trees. Careful examination has revealed that there are areas of former
prairie which, on almost any conceivable climatic criterion, are no less
favourable for tree growth than near-by areas of former forest. It is not
remarkable therefore that numerous workers in this field of ecology have
begun to seek for explanations other than simple climatic ones. Indeed,
some have reached the conclusion that vast areas of prairie were not true
climatic climax vegetation at all; they think that trees would have
occupied this land had it not been for certain deflecting factors. In North
America, destructive fires used to sweep over the prairies at very frequent
intervals. Many were started by the Indians as a deliberate policy but
there is some evidence that others were quite natural fires, mainly
started by lightning. The seedlings and saplings of woody plants are
destroyed by frequent fires whereas grasses are quite unharmed when the
dead remains of last year's growth are removed by a fire in the dormant
season. There are also eye-witness accounts to show that some of the vast
herds of bison spent at least part of the year grazing in the forest margins
[7]. Since these animals had very regular habits and seem to have grazed
over their territories systematically, there can be little doubt that they
would be quite capable, not only of preventing the advance of forest, but
also of preventing its regeneration even in areas which, at one time, it may
have occupied. If man-made fires were responsible for keeping back the
forest, then the prairies on the areas concerned must be regarded, not as
climatic climax vegetation, but as plagioclimax (disclimax).

Carl Sauer [66] has drawn attention to a significant correlation between
relief and the former distribution of forest and prairie in the Middle
West; it points most suggestively to the efficacy of fire as the main ecological
factor. The details of his argument are best expressed in his own words:

'I grew up in the timbered upland peninsula formed by the junction of
the Missouri and Mississippi rivers. The prairie began a few miles to the
north and extended far into Iowa. The broad rolling uplands were prairie,
whatever their age and origin, the stream-cut slopes below them were
timbered; river and creek valleys and flanking ridges were tree covered,
be they formed in bed rock or on deep loess mantle. From grandparents I
heard of the early days when people dared not build their houses beyond
the shelter of the wooded slopes, until the plow stopped the autumnal
prairie fires. In later field work in Illinois, in the Ozarks, in Kentucky, I
met parallel conditions of vegetation limits coincident with break in relief.
I gave up the search for climatic explanation of the humid prairies.'[1]
In the same work Sauer points to other facts which he considers to be
significant. Firstly he notes that, when once the frequent spread of fire
was prevented by the advance of cultivation on to some former prairies,
natural invasion of trees took place on the odd corners of land not actually

[1] *Op. cit., p.* 16.

cultivated. Secondly he draws attention to the fact that woody plants are dominant in the scrub communities which cover even drier areas than those occupied by grassland. His main conclusion is that fire was not so important as an ecological factor in the wet forested lands of the continent where the vegetation was too frequently moist or green; neither was it important in the driest areas because insufficient organic waste is produced to permit fire to spread; it was in the intermediate areas, wherever the relief permitted fire to race before the wind, that woody plants were precluded by it and heliophilous, herbaceous plants were allowed to dominate.

This theory is so grandiose in its perspective that it cannot fail to stimulate the imagination. In the absence of clear climatic correlations it is very persuasive. Indeed, for those few areas where invasion by trees has been shown to have taken place after fire protection was given, there is every reason to suppose that the truth lies in this direction. Nevertheless, for most of the prairie areas, moist as well as dry, many competent ecologists would contest these sweeping assertions. It has been pointed out that the true prairie communities [81], as well as those of the mixed prairie [82], are so complex and well-integrated for it to be inconceivable that they are merely *ad hoc* collections of plants which have recently come together from a variety of sources. The true prairie shows a consistency in structure and species-composition over very large areas that argues very forcibly in favour of its antiquity. One of the logical inferences which may be drawn from Sauer's theory is that the extensive grasslands which appear to have existed in Tertiary times, probably could not have done so had not natural fires been a frequent phenomenon to hold back the encroachment of forest trees. This may very well have been the case since very little is known about the frequency and history of natural fires. Nevertheless this puts fire in a position of importance as an ecological agent which it has not formerly occupied.

However strong the arguments against the pyrogenic theory, it cannot be denied that the Middle West prairies occurred in an area which was remarkably humid. Although the landscape has now been transformed, there seems to be no good reason why ecological experiments and observations should not ultimately demonstrate whether or not the native trees of the Middle West will germinate, grow and reproduce themselves in the areas of former prairie. Further north, in Alberta, research has certainly indicated that fire is a most potent ecological factor [55]. The existence of separate groves in the Park Belt demonstrates that the seeds of the aspen (*Populus tremuloides*) could germinate, survive, and mature right up to the thirteen-inch isohyet even under the conditions of frequent firing which existed before farming spread into the area. Even today, in those places where undisturbed groves remain, fire scars on the trees indicate that fires have swept through the vegetation at frequent intervals. The groves are primarily in hollows or on north-facing slopes (Plate X*b*), the aspen having been able to survive here in spite of firing. It is possible that this species would be able to spread into even drier regions if protected from the debilitating effects of fire. Its ability to encroach on grassland is demonstrated by the fact that, even beneath the continuous forest to the north of the Park Belt, the soil is one which formed beneath prairie. Not so many centuries ago, trees encroached upon grassland in this area and the chernozem is, even yet, little altered. The remains of a lime-accumulating horizon are still recognisable and podzolisation of the uppermost horizons is only at an early stage. If undisturbed, it seems likely that these will develop into *grey forest soils* of a type very similar to those in the Danubian Basin and in the Ukraine. Beneath the individual groves of the Park Belt there are *chestnut-brown soils* which are almost indistinguishable from those beneath the contiguous prairies; again, podzolisation of the uppermost horizons is only just beginning and it is clear that invasion by trees took place only in the relatively recent past.

It is quite certain therefore that trees can invade areas of prairie underlain by typical grassland soils. Since aspens are now growing in Alberta and Saskatchewan where, formerly, there was unbroken prairie, it cannot be argued that the soils occupied by grassland have innate qualities which forbid the advance of trees. It is important to recognise this since it will be demonstrated later (Chapter X) that relatively fine-grained soils, such as those containing a high percentage of loess, do certainly favour shallow-rooted plants like grasses by preventing rapid percolation. Since many of the soils in the Middle West and the Great Plains, as well as those of the steppes of Russia, are rich in loess, it has frequently been argued that this is the main factor preventing the spread of forests. Nevertheless, if aspens can spread into dry prairie in Alberta, there seems to be no over-riding climatic or pedological reason why other species of tree should not have been able to invade westwards further south in the U.S.A. The aspen has one salient morphological feature which favours it as an 'invasion' species however; it propagates very rapidly by means of *suckers*. These can survive the fiercest fires in which all those parts of trees which are above ground are destroyed, and can subsequently send up numerous secondary growths to re-establish the forest.

Observation and logical inference may thus lead one back to the per-suasive arguments which postulate fire as a primary factor influencing the distribution of the North American prairies. It must be emphasized, however, that the available evidence for north American grasslands is complex and sometimes confusing; what may be true of one area need not, of necessity, apply in the case of another. There is certainly no agreement on the matter, indeed some ecologists who have worked in Arizona and northern Mexico [35] clearly favour the view that the 'desert grasslands' are climatic climax communities and that their recent contraction and deterioration is more likely to be due to climatic change and the introduction of large herds of cattle than to a decrease in the frequency of fires.

2. THE EURASIAN STEPPES

Though the name 'steppes' had been applied customarily to the wild grasslands of Eurasia, this must not be permitted to obscure the fact that these were composed of grasses of identical life-form to those which formed the 'prairies' of North America. A very similar change in grassland type occurred as one passed from the forest edge to the semi-deserts of the Caspian area and central Asia. The 'meadow steppes' [44] just beyond the forest edge were very similar to the true prairies of North America, being composed predominantly of sod-forming grasses such as *Stipa pennata*, *S. lessingiana*, *S. joannis*, *Festuca sulcata* and *Koeleria gracilis* along with subordinate tussock grasses (Plate IX*b*). Not only did these grasses grow luxuriantly to a height of four feet or more, but numerous broad-leaved herbs were associated with them providing attractive, flowery, natural meadows throughout the summer season. This type of grassland occurred in a narrow belt extending from the lower Dniester Valley east-north-eastwards across the Dneiper Bend, the Saratov Region and the southern end of the Urals, and from thence eastwards across the Akmolinsk area to the Kulunda Steppe and the foothills of the Altai. It also inter-digitated with the forest in the forest-steppe belt just to the north.

With slight increase in drought southwards, this meadow steppe gave way to another type of grassland dominated by grasses of almost equal height but much poorer in species. This 'large-tufted *Stipa* steppe', like the western grasslands of North America, was dominated by tussock grasses such as *Stipa stenophylla* and *S. capillata* with turf-formers like *Festuca sulcata* and *Koeleria gracilis* as a subsidiary element. The dense rooting systems of the tussock grasses at a shallow depth are able to absorb available water so efficiently that the competition appears to have been too rigorous for most of the flowering herbs which flourished in the better-watered meadow steppes to the north.

Southwards again another narrow belt of a distinct type of grassland was found. This was very similar in life-form to the grasslands of the 'desert plains' of south-central U.S.A. Here the tussock grasses became less important again and fairly dwarf, turf-forming grasses like *Festuca sulcata* became the dominants. Associated with these were semi-desert shrubs like *Artemisia maritima* (Plate XII*b*) which are the dominants still further south in the semi-desert scrub around the northern end of the Caspian Sea and, indeed, in many semi-deserts in central Asia.

Although these wild grasslands have now been either removed or much altered by cultivation and grazing, sufficient evidence remains to reconstruct their former distribution. A very similar series of communities was also found in the lowlands of Manchuria occurring in north-south belts with decreasing luxuriance from east to west. Short grass steppes are still very extensive on the eastern edge of the vast semi-desert scrublands in Inner Mongolia and the eastern part of Outer Mongolia.

Soils

There appear to be no extensive counterparts of the North American *prairie soils* beneath the steppes of Russia. Generally speaking the meadow steppe and the tussock steppe are underlain by chernozem and the short grass steppe partly by chernozem and partly by chestnut-brown soil. Descriptions of the chemistry of some of the Russian chernozems, however, make it quite clear that they have some affinities with the prairie pedalfers of the Middle West. The deep chernozems of the forest-steppe and meadow-steppe zones frequently attain a depth of five or six feet. Although they possess lime-accumulating horizons at the base, they are normally quite devoid of free calcium carbonate throughout the uppermost two or three feet of their profiles. It appears as though they might once have been typical chernozems which have been partially converted into prairie soils by an increase in the rate of leaching. Though there is no exact correlation between vegetation and soil types, the more leached chernozem normally gives way to the normal, completely pedocalic chernozem beneath the tussock steppe. The profile becomes shallower southwards until ultimately the chernozems pass into chestnut-brown soils.

The forest-steppe transition

As in North America the boundary between the forest and steppe was a complex one. The Park Belt of Canada has its counterpart in the 'forest-steppe' of European Russia. It is probably of some significance that in the latter, though outliers of oak (*Q. robur*) are the dominant features, groves and thickets of aspen (*Populus tremula*) are of very frequent occurrence. Furthermore, this European species of aspen is very similar indeed, morphologically and physiologically, to its American relative. Further east, in Siberia, the aspen becomes an even more obvious element in the vegetation of the forest-steppe transition. Since there is very strong evidence that forests have invaded former steppes in European Russia even more extensively than in Canada, the aspen may again have performed the important function of pioneer invader.

Beneath many of the southern parts of the oak forests around Kaluga, Tula and Tambov, and south of Kazan, relics of chernozem have been found. Unleached free calcium carbonate, and even complete horizons of calcium carbonate nodules, have been identified in what are otherwise typically podzolised forest soil. The remains of steppe rodents, probably the mole rat (*Spalax microphthalmus*), have been found associated with these, the hollows in the bones often being filled with dark chernozem soil [44]. Not only are the isolated groves and southward-extending tongues of forest on areas formerly occupied by steppe, but the southern part of the continuous oak and pine forest is also of recent origin.

It is in the area of the oak forests and the forest-steppes of European Russia, probably more clearly than anywhere else, that the close relationships between soil and vegetation have been clarified. Soil maps on a world or continental scale have seemed to emphasise the essential relationship between soil and the elements of macroclimate. It seems to have been assumed that certain mean values of precipitation and evaporation (as measured by standard meteorological instruments) would automatically

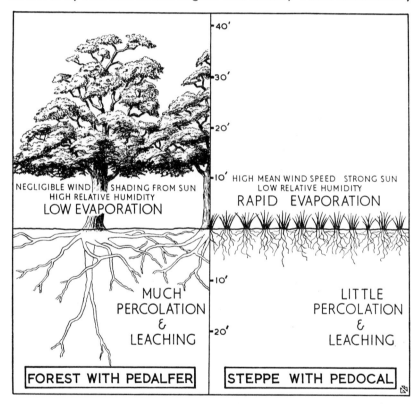

Fig. 18. The contrast between hydrological conditions beneath trees and grasses in the forest-steppe region.

call into existence a pedocal like a chernozem whereas other values would result in the formation of a pedalfer. The general assumption has been that if mean annual precipitation exceeds mean potential evapotranspiration a pedalfer will result but if the reverse obtains, a pedocal will form. As long as this principle is applied with some refinements it is probably acceptable; the main point to be remembered is that the very process of invasion of a grassland area by trees will much reduce the rate of evaporation from the surface of the soil; this is particularly so in summer when the trees cast a deep shade and the sun is at its hottest. This reduction is so great that it usually changes the water balance completely *so far as the upper soil horizons are concerned*. Water will still be lost to the atmosphere in the same or even greater quantities than before, but this water will be drawn from the sub-soil by the tree roots (Fig. 18). Although the amount of precipitation reaching the ground must be reduced to a certain extent, being intercepted by the foliage of the trees and re-evaporated, this is more than compensated for by the enormous reduction in evaporation from the ground and in reduced transpiration from shallow-rooted herbs.

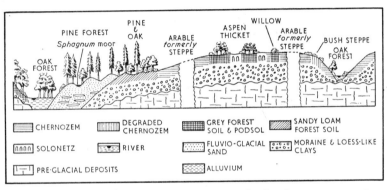

Fig. 19. Diagrammatic section across an area in the forest-steppe region showing the relationship between the distribution of relief, soil and vegetation types. (Adapted from Boris Keller, 1927.)

It is primarily because of this, that unaltered chernozem was found beneath the steppe grasses in the forest-steppe region, whereas almost completely podzolised soils could be found, often not many yards away, beneath contiguous areas of old-established oak or pine forest (Fig. 19). Within such small areas, quite obviously, macroclimatic differences could not possibly be held to account for the soil contrasts. Furthermore, particularly in the northern part of the forest-steppe, patches of forest soil extended on to exposed flat or gently-undulating areas, exactly similar in relief to those occupied by chernozem. Both forest soils and chernozem could be found on thick loess deposits; both were also found on more pervious, poorer glacial outwash materials. Neither relief nor lithology can therefore be said to have exercised absolute control. It is quite clear that the only close correlation is between vegetation and soil; the vegetation was so effective in its control of microclimate and type of humus as to over-ride other factors.

Ultimately therefore, the factors which limited the spread of forest indirectly controlled the extent of podzolised forest soils. Relief also had an indirect effect. As in the Middle West in the U.S.A. there was the same general relationship between relief and forest-grassland distributions; the forests were more and more restricted to valley sides and dissected terrain the further one advanced into drier areas. Oak forests were almost continuous along the banks of the lower Volga and lower Dneiper, while the flat interfluves at the same latitude were almost continuous steppe.

3. EXTRA-TROPICAL GRASSLANDS OF THE SOUTHERN HEMISPHERE

Vast areas of grassland were found south of the tropic in the southern hemisphere though, here again, as on the northern continents, the original wild vegetation has now been removed or much changed by cultivation and grazing. The Veld of the eastern interior of South Africa, the Pampas of Argentina and the Canterbury Plains of New Zealand were predominantly grass-covered. Extensive areas in central New South Wales, northern Victoria and south-eastern South Australia, though liberally scattered with trees and shrubs, were also dominated by grasses. Indeed, even the driest areas in these parts of Australia, where drought-resistant shrubs are generally dominant, contain quite frequent patches of grassland. The distribution of vegetation types on the southern continents is thus even more complex and difficult to explain than in North America and Eurasia and vegetation maps on a continental scale can be very misleading.

The Pampas

When European immigrants first saw Uruguay and the north-eastern part of the Pampas south of the River Plate, the land was covered with tall prairie dominated by species of feather grass (*Stipa*) and melic (*Melica*). Only to the north of the site of Buenos Aires, along the lower reaches of the River Uruguay, was there any considerable extent of forest. Nevertheless, small patches of forest or thickets of small trees were of frequent occurrence in moister hollows throughout. The dominant species of this moist Pampa were bunch grasses, so that a large percentage of the soil surface lay bare between the individual tussocks. Westwards and south-westwards, the long grass prairie gave way to a mixed vegetation of short grass and xerophytic shrubs, the short grasses predominating in a belt extending northwards from Bahia Blanca, but the shrubs becoming more and more prevalent as one approached the Andean foothills. The shrubby vegetation also dominated locally even in the east, wherever hilly or dissected land rose above the general level of the plains. The two largest areas of this were the Sierra de la Ventana north of Bahia Blanca and the Sierra del Tandil running westwards from Mar del Plata. Because of its general association with more hilly land, the Spanish settlers of South America applied the name 'monte' to this shrub vegetation [39]. Since the eastern parts of the tall prairies of the Pampas receive more than thirty inches mean annual precipitation, well distributed throughout the year, it is small wonder that, as in the case of the Middle West, numerous ecologists have debated the ecological status of these grasslands. Again it has been maintained [68] that monte, or even forest, formerly extended to the east but was pushed back by Indian fires.

The Veld

Though grasses are of widespread occurrence on the Karroos and in Botswana, it is only in the eastern parts of the interior of South Africa that extensive grassland, with little or no admixture of trees or shrubs, is to be found. Northwards from the Orange River in Griqualand and in the southern Kalahari there is a region known collectively as the 'Sand Veld'; treeless grassland is found here but it gives place, at very frequent intervals, to tree savanna and bush savanna. The 'Eastern Grassveld Region' proper extends from eastern interior Cape Province northwards to the Limpopo River (Map 9). Northwards in the Transvaal and eastwards, below 3,500 feet, in Natal, however, it is increasingly sprinkled with trees and shrubs and becomes more and more like tropical savanna in appearance [2]. The latter is usually referred to here as 'bush-veld'. It is noteworthy that the majority of species composing this eastern 'bushveld' belong to tropical genera so that it can be regarded as a south-ward extension of the tropical savannas of East Africa.

The true 'grassveld' is thus confined to the eastern mountain slopes and high plateaux. In its original state it appears to have been dominated over large areas by the red grass (*Themeda triandra*) which formed a fairly level sward and is a really nutritious grass. Unfortunately, bad management of these natural grasslands has much reduced the range of the red grass. It has been discovered that heavy burning at the wrong time of the year so weakens it that more xerophytic and much less nutritious grasses such as the wire grass (*Aristida junciformis*) and the 'blousaadgras' (*Eragrostis spp.*) are able to oust it almost completely. The grazier's problem is a difficult one, however, since the dead growth of the previous year must be removed in order to ensure palatable pastures of young growth in the spring. It is obvious that very careful management indeed is necessary, and carefully controlled grazing and cutting are now being practised in order to maintain the red grass community. A further problem facing graziers, particularly on the 'High Veld' above 5,000 feet, is that much grassland is obviously plagioclimax. If the grass is left unburned or uncut for a number of years, the early stages of a forest subsere make their appearance. Taller grasses, particularly of the genus *Andropogon*, invade first, to give a type of grassland which is known locally as 'Tambookie'. This may be followed quite soon by a dense cover of bracken (*Pteridium aquilinum*) which in turn gives way to scrub and forest. On a farm at Nottingham Road in west-central Natal, over sixty acres of initially almost pure red grass veld was systematically protected from fire over a period of thirty-five years. At the end of this period no veld remained, the whole area being occupied by dense scrub into which yellow-wood saplings (*Podocarpus sp.*) were rapidly intruding [5]. Forest was obviously the climatic climax vegetation here as indeed it probably is over all the higher parts of the veld where mean annual precipitation exceeds twenty-five inches. The extent to which forest or scrub are the true climatic climax in even drier parts of the Veld is much debated.

Australian savanna

There appear to have been no extensive areas outside the tropic on the Australian continent where grasses were completely predominant but, just west of the Great Dividing Range, from the tropic southwards to the coast in south-eastern South Australia, there was a wide belt of tree savanna. This savanna varied a great deal from place to place; locally there were wide areas of tall grass, dotted here and there with isolated trees; elsewhere giant gum trees (*Eucalyptus spp.*) formed an almost continuous canopy (Plate VI*b*) but, because of the very slight shade cast by these species, they were underlain by an almost continuous grassy or shrubby carpet (Chapter VI). The areas of scrub in western Victoria and in South Australia also contained frequent patches dominated by grasses. Many genera, characteristic of both middle latitudes and tropical regions, were represented in these Australian 'savannas', species of *Poa*, *Panicum*, *Festuca*, *Agrostis*, *Aira*, *Andropogon*, *Stipa* and *Bromus* being particularly important. It is also significant that many of these species are very nourishing pasture grasses.

The Canterbury Plains

The only other extensive grasslands in the southern hemisphere were the 'tussock' grasslands of the South Island, New Zealand. They occupied an area of at least 25,000 square miles on and around the Canterbury Plains and were, apparently, remarkably uniform in structure and general appearance throughout. The formation was dominated by tussocks mainly of species of *Poa* and *Festuca* and, although it is now difficult to assess details of the original communities, it seems certain that the sea of waving bunch grasses was dotted sparsely with shrubs and cabbage trees (*Cordyline australis*) [13]. In the sheltered microclimates between the tussocks, there was quite a rich flora of low-growing grasses, sedges and other herbs. Today almost the whole of this grassland has been replaced by cultivated vegetation and, even where cultivation has not taken place, heavy grazing by rabbits and domesticated animals has almost completely transformed it. The former distribution and nature of this grassland has aroused great interest. In places it extended into areas with more than sixty inches mean annual rainfall, in spite of the fact that its dominant grasses have marked xeromorphic characteristics. As in the case of other long grass prairies, many explanations have been suggested for its exist-ence, and recent observations, archaeological and ecological, seem to indicate quite clearly that prehistoric man may again have played a leading role.

The earliest human inhabitants of the South Island, from the ninth century onwards, seem to have based their economy on the hunting of moas (*Dinornis spp.*)—the giant, flightless birds which formed the most conspicuous element in the fauna of New Zealand. From the very frequent occurrence of the remains of charred tree trunks which have been found in the soils of the Canterbury Plains, it is apparent that forests, dominated by conifers as well as hardwoods (Chapter V), were formerly very much more extensive than they were at the beginning of the nineteenth century. Furthermore, carbon 14 dating has indicated that this destruction took place in the thirteenth and fourteenth centuries. It may well be that these primitive hunters used fire, at certain times of the year, to drive their quarry, and that the area covered by forest was reduced as a consequence. Indeed the evidence seems to suggest that the grasslands may have trebled in size during this period [18]. It should be noted however that the very presence of these giant grazing birds implies the existence of some kind of grassland from Tertiary times onwards: these animals were structurally adapted to an open environment if not to an almost treeless one so that one cannot postulate a universally forested landscape under natural conditions. In fact it is possible that, during the early phases of their activity, these hunters may, inadvertently, have stimulated an increase in the numbers of moas by increasing the area of the type of habitat to which they were adapted.

Whatever may be the exact history of the vegetation in the South Island, there can be no doubt that a number of the native forest trees of New Zealand, conifers and angiosperms, can be seen to be thriving on numerous scattered sites which have a mean annual precipitation of no more than 25 inches.

Soils of the southern hemisphere grasslands

In any general consideration of the grassland soils south of the Tropic of Capricorn, one should emphasise the dangers of seeking direct analogies with counterparts in North America and Eurasia; it should not be assumed that any black soil found beneath grassland on the Veld or in Australia must, of necessity, be a chernozem or that one will, inevitably, find 'prairie soils' or 'chestnut-brown soils' anywhere beneath these southern grasslands. Amongst other things, most of the prairies of North America and Eurasia suffer far colder and more protracted winters than are experienced anywhere in the grasslands of the southern continents. In point of fact a great variety of soils is found in these southern areas and the modes of formation of most of them are, as yet, obscure. There are points of similarity between the northern and the southern hemisphere counterparts, but these may be quite overshadowed by points of contrast. It is clear, for instance, that some kind of pedocal soil has developed beneath the veld grasses on the plateau west of the Drakensberg in South Africa and that, in places, this soil has a very dark coloration [53]. It would be quite premature to equate this soil with the northern chernozems, however, just as it would the black soils of interior Queensland and New South Wales. It would also be premature to equate the 'tosca' soils of the southern and western Pampas [39] with the chernozems; the former are certainly characterised by a lime-accumulating horizon within their profiles but whether or not this came into existence in the same way as did the analogous horizon in the chernozem is doubtful.

Nor are the soils beneath the more humid grasslands in the southern hemisphere exactly similar to the prairie soils and degraded chernozems of North America and Eurasia. In the humid Pampas of Argentina, though much of the area is underlain by deep, dark-brown soils, considerable tracts of yellowish-brown soil have been described which seem to bear witness to a degree of podzolisation. The soils beneath the more humid parts of the Veld in South Africa are predominantly pedalfers, being laterised in the north and podzolised in the south. Nevertheless, there are patches of a much richer, black soil within this general area of 'sour veld' and these have been found to correspond almost exactly with outcrops of basic igneous rocks such as norite (Chapter XIX). It is quite clear therefore that soil-parent-material as well as climate has been operative in giving rise to soil differences beneath these grasslands. As yet, however, far too little is known about pedogenesis in these less intensively surveyed regions even to speculate on the relative importance of the soil-forming factors involved.

CHAPTER X

Woodlands and Scrublands

THE unforested areas of middle latitudes were not all dominated by grasses and herbaceous vegetation. Areas equal to, if not greater than, those of the grasslands were occupied by a vegetation which was composed mainly of woody plants. Indeed, over great expanses of territory, low-growing woody plants are still completely dominant. In the previous chapter it was noted that, in many cases, these scrublands occupy even drier areas than those dominated by grasses.

This woody vegetation varies from place to place in its structure and general appearance. In places it is composed of small trees forming a more or less open canopy over the ground so that it has been referred to quite widely as 'woodland'; elsewhere shrubs form a more dense but lower-growing cover; in other places the main shrub layer is discontinuous with the intervening ground occupied by lower-growing shrubs, grasses or other herbaceous plants while, finally, there are places where large areas of completely bare ground intervene between sparse clumps fo low-growing vegetation.

These woodland and scrub communities are clearly of two distinct kinds. Firstly, there are those which have developed in the semi-arid areas of the continental interiors where drought is persistent throughout most of the year; here precipitation is unreliable as well as light and winter frosts, even protracted cold, have to be contended with. Secondly, there are those which are adapted to sub-tropical latitudes, mainly on the western sides of continents, where there is pronounced summer drought but where winters are relatively mild with only slight frost or no frost at all. These latter communities are dominated by *sclerophyllous* shrubs of predominantly evergreen habit which are able to make full use of the relatively short periods favourable to growth in spring and autumn but which have morphological adaptations enabling them to withstand the summer drought. The scrub of the continental interiors is normally even more xerophytic; a larger proportion of the plants here are deciduous while, in areas of very low rainfall, many of the species have all their foliage leaves reduced to mere spines or scales.

I. SCLEROPHYLLOUS FORMATIONS

The Mediterranean area

Two main types of vegetation, already referred to (Chapter V), are widespread around the Mediterranean basin. Of these the maquis is undoubtedly a type of sclerophyllous vegetation. It is dominated by shrubs and, although these vary very much in height and species from place to place, they are predominantly of the same life-form: although some are near-leafless, spiny plants, the great majority are evergreens with leathery, drought-resistant foliage. Species like the wild olive (*Olea europaea*), the carob (*Ceratonia siliqua*), the lentisk (*Pistacia lentiscus*), the Kermes oak (*Quercus coccifera*), the cistus (*Cistus spp.*) and the arbutus (*Arbutus unedo*) are typically sclerophyllous, but they are often mixed with heathers (*Erica spp.*), gorse (*Ulex spp.*) and broom (*Genista spp.*) which are of different life-forms. Many types of maquis have been recognised, often within the same general area, each being named after its dominant species [1].

Garrigue also covers large areas in the Mediterranean peninsulas and islands, particularly on more pervious outcrops such as limestone. Although some species of typical maquis reappear in garrigue along with a number of really low-growing, prickly shrubs, typical garrigue communities are characterised by aromatic, herbaceous plants and should not therefore be classed as true sclerophyllous vegetation. Species of the deadnettle family (*Labiatae*) and thyme (*Thymus spp.*) are characteristic in many places.

Though there is no doubt that much of this present-day maquis and garrigue occupies the territory of former forests, it is also certain that, over limited areas, these scrub communities are true climatic climax. This is probably true of considerable parts of southern Italy, south-eastern Spain, the Mediterranean islands and the coastal lowlands of north-western Africa, particularly where very pervious rocks and wind exposure give exceptionally dry habitats. Similar communities are found around coastal Asia Minor and eastwards across Syria although in the latter, with increasing dryness and winter cold, these ultimately give way to the continental type of semi-desert scrub community.

California

Another formation of the same formation-type as the maquis is found on the lower flanks of the Coastal and Santa Lucia Ranges in southern California and, with increasing altitude southwards, on mountain slopes in Lower California. It is usually referred to as 'coastal chaparral' and, although its component species are *phylogenetically* related to those of scrub communities in the continental interior to the north-east, its evergreen, sclerophyllous habit makes it a close ecological relation of the maquis.

Only one common dominant in this formation, a species of dwarf oak (*Quercus dumosa*), is deciduous. The coastal chaparral is normally up to ten feet in height though, in slightly more favoured places, individual trees or clumps of trees of greater height may be found. These are usually evergreen oaks or a species of drought-resistant conifer called the piñon (*Pinus cembroides*) [17]. On the other hand, at the drier end of its range, the chaparral may be much less than the height of a man. It is composed of species which, in spite of their ecological similarity, are only distant relatives phylogenetically. The commonest are bush oaks (*Quercus dumosa*, *Q. agrifolia* and *Q. chrysolepis*), *Adenostoma fasciculatum* belonging to the rose family (*Rosaceae*), species of *Ceanothus* (*C. cuneatus*, *C. oliganthus*, *C. spinosus* and *C. divaricatus*) belonging to the same family as the British buckthorn (*Rhamnaceae*), *Arctostaphylos pungens* and *A. tomentosa* belonging to the heather family (*Ericaceae*) and *Prunus ilicifolia* belonging to the same genus as the wild cherry (*P. avium*). The flora of the formation is particularly rich and these dominant species are associated with a very large number of other less frequent ones [67].

Some of the species of the coastal chaparral are particularly resistant to fire; when their stems above ground are completely destroyed they are able to send up new shoots and thus to form even denser bushes. It is probably because of this that the formation has been able, not only to maintain its range over uncultivated land, but also to expand at the expense of grassland and forest. Consequently, though it is probably climatic climax only where the mean annual precipitation is between ten and twenty inches,[1] it now occupies mountain slopes in northern California and Oregon where as much as fifty or sixty inches is received [17].

Australia

In the southern hemisphere three more formations of sclerophyllous scrub are found. The Australian one is particularly extensive and the South African and Chilean formations also cover considerable areas. The Australian formation extends from around Shark Bay in Western Australia south-eastwards in a wide belt and then swings eastwards in a narrower belt along the immediate coastlands of the Bight. It then broadens out again to occupy the whole of Eyre's Peninsula and Yorke Peninsula, most of the south-eastern corner of South Australia and the north-western quarter of Victoria (Map 10). This belt of mallee scrub thus extends along the dry side of the forests of south-western and south-eastern Australia in areas where a sparse rainfall is concentrated in the winter season. Once again, although it is dominated by plants of very

---

[1] It is only on sites very close to the sea, where atmospheric humidity is prevailingly high, that the coastal chaparral can survive with as little as ten inches mean annual precipitation.

A. SCLEROPHYLLOUS SCRUB. Typical 'Cape vegetation' in the nature reserve in the Cape of Good Hope Peninsula south of Cape Town. The dominant shrub with the very showy flowers is a species belonging to the *Proteaceae* but species of many other families are present.
(Photograph by A. F. Braithwaite.)

B. XEROPHYTIC WOODLAND. Woodland dominated by the evergreen blue oak (*Quercus douglasii*) west of Williams, Colusa County, California.
(Photograph from Lyman Benson, *Plant Classification*, © 1957 D. C. Heath & Co., Boston.)

PLATE XI

A. SAGEBUSH (*Artemisia tridentata*) west of Kamloops, British Columbia.
(Photograph from Lyman Benson, *Plant Classification*, © 1957 D. C. Heath & Co.,
Boston.)

B. RUSSIAN SEMI-DESERT. The dominant plants here are a wormwood
(*Artemisia maritima*) and a species of fescue (*Festuca sulcata*).
(Photograph by Keller, 1927.)

PLATE XII

similar life-form to those in the other sclerophyllous formations, its plants are close phylogenetic relatives of Australian forest species. Many of the mallee species belong to the genus *Eucalyptus; E. dumosa, E. oleosa, E. uncinata* and *E. bicolor* are particularly frequent, originally forming an almost pure dwarf *Eucalyptus* scrub, six to eight feet in height, extending over hundreds of miles. Elsewhere, in spite of its similarly monotonous appearance, the mallee is very rich floristically, containing shrubs from dozens of families most of which are not represented in the British flora [67]. In this context it must be remembered that the south-west Australian area is particularly rich in genera which appear nowhere else in the world. The predominant colour of the leaves of the mallee scrub is grey-green, this giving a most monotonous aspect to the landscape. The shrubs, in their original state, formed a dense, almost impenetrable scrub which was avoided as much as possible by early travellers. In the sclerophyllous nature of their foliage and the aromatic, indeed pungent odours that many of them emit, these shrubs are remarkably like the typical shrubs of the Mediterranean maquis.

South Africa

Though much of the uncultivated land in the south and south-west of Cape Province in South Africa which is now occupied by scrub must formerly have been forested, there can be little doubt that large areas of climatic climax sclerophyllous scrub are to be found there. Apart from the better drained and more exposed areas in the coastal strip from Cape Town eastwards, large parts of the Karroo and the eastern inland counties of Cape Province as far north as Calvinia, almost 200 miles north-north-west of Cape Town, were covered with a scrub formation very similar in structure to those already described [2]. This South African formation varies in height between two and eight feet, its bushes having markedly sclerophyllous foliage of a predominantly bluish-green hue. If anything, the average size of these leaves is even smaller than that of their counterparts in the Mediterranean and Australia but, like these, they are usually hard and in-rolled at the margins and possess numerous other adaptations to lower the rate of transpiration. Though this formation, even in its original state, appears to have been more open than the mallee, few herbaceous plants seem to have taken advantage of this. Underneath the predominantly shrubby vegetation, apart from occasional tussock grasses, the ground was almost bare (Plate XIa). The formation is also noteworthy for the absolute predominance of perennials; it has been said that there is probably not a single indigenous annual in the flora of south-western Cape Province. Most of the species composing this scrub flora are endemics and although widespread families such as the *Cruciferae*, *Rosaceae* and *Leguminosae* are well-represented, a large number of the species belong to families which are not represented in Europe.

Kvs

Chile

The sclerophyllous scrub formation of Chile occupied much of the coastal fringe and the Central Valley northwards from Concepcion for about 450 miles almost to La Serena, and about 150 miles even further north along the westward-facing slopes of the Andes between about 3,000 and 6,000 feet. Again, it is remarkably similar in appearance to the analogous formations except that, locally, it is quite frequently dotted with taller-growing trees. Where typically developed, it is even more difficult to penetrate than the Australian mallee. Again, it is dominated by species quite peculiar to itself of which three are particularly common: *Quillaja saponaria* (*Rosaceae*), *Kageneckia oblonga* (*Rosaceae*) and *Rhus caustica* (*Anacardiaceae*) are almost universal [67].

The evolution of the sclerophyllous formations

One cannot but be impressed by the close structural and physiological similarities between the five formations of the sclerophyllous scrub formation-type. It is clear why the so-called 'Mediterranean type' of environment has attracted so much attention. In this environment of winter moisture and summer drought, hundreds of species from scores of only distantly related families have evolved, probably through millions of years. They would not have survived here had they not, through a series of *mutations*, conformed to a particular type of general norm. The countless species which, during this period, have not conformed, have either become extinct or have had to survive elsewhere. The vegetation as a whole has thus undergone a process usually referred to as 'convergent evolution'. The solidity of some of these formations and the richness of their floras is even more remarkable when one remembers that frequent climatic changes have occurred, particularly during the past million years. These formations must have been forced to migrate backwards and forwards with the summer-drought—winter-moisture climatic zone. They had survived these fluctuations remarkably well; mankind has done much more to reduce and impoverish them in the course of only a few centuries.

It would be very easy, on the basis of the correlation between the distribution of the sclerophyllous scrub formations and that of the 'Mediterranean' climatic type, to arrive at a philosophy of vegetation distribution based wholeheartedly on climatic determinism. This would be rash and unwarranted however. Although plant geographers and climatologists in the past have accepted the basic premise that there is an absolute and inescapable relationship between climatic distributions and the distribution of dominant life-forms, this is by no means justified. In the case of sclerophyllous scrub and the 'Mediterranean climatic type' the correlation is close but, as has already been seen, in other cases it is almost non-existent.

2. SEMI-DESERT SCRUB AND WOODLAND

North American formations

The mid-latitude semi-desert scrub communities of North America have been studied more systematically and intensively than those of either Eurasia or Patagonia. This is fortunate because, within a relatively small area, North America possesses an interesting range of xerophyllous woodland and scrub communities. These can be regarded as a series occurring between either forest or prairie on the one hand and arid desert on the other. Generally speaking, these communities were originally concentrated in the south-western interior of the U.S.A.—in the basins and on the plateaux and lower mountain slopes between the Sierra Nevada in the west and the Rockies in the east; there was, however, some extension southwards, particularly of the most drought-resistant communities, into the Sonora Desert and on to the northern part of the Mexican Plateau. Even under undisturbed conditions there were also outliers to the north and east beyond these areas.

At an average height of about 7,000 feet in the southern Rockies, the Sierra Nevada and the ranges of the Great Basin, as well as on plateau surfaces at about the same height in Arizona, Nevada and Utah, the sub-alpine and montane forests give way downwards to a distinct formation dominated by low trees. This is usually referred to as the 'woodland' formation. It is dominated by its own distinctive species which typically grow to a height of between twenty and forty feet (Plate XI*b*). As a climax formation it reaches the eastern limits of its range on the Edwards Plateau and on the flanks of the Davis and Guadalupe Mountains in western Texas and its northernmost limits in Wyoming. It is also found between the sub-alpine forests and the sclerophyllous chaparral between 8,000 and 9,000 feet in southern California and Lower California. Though some are coniferous and some broad-leaved, the woodland dominants are all evergreen. On the Rockies, in Texas and in southern New Mexico and Arizona the formation is composed of species of evergreen oak (*Quercus reticulata* and *Q. hypoleuca*) [35] and juniper (*Juniperus pachyphloea, J. occidentalis* and *J. sabinoides*) with sub-dominant pine (*Pinus cembroides*). The more typical piñon-juniper woodland occurs throughout the Great Basin dominated by the piñon (*P. cembroides*) and species of juniper (*J. occidentalis* and *J. californica*). In southern California and Lower California pine (*P. sabiniana* and *P. cembroides*) and evergreen oak (*Q. douglasii* and *Q. wislizeni*) predominate with juniper only subordinate.

Though it is probable that the woodland formerly formed an almost complete canopy throughout much of its range, it is now mainly reduced to more open woodland, being much invaded by bunch grasses where it is contiguous with prairie and by shrubs where it meets chaparral. Indeed, over wide areas, the woodland has been removed altogether and replaced by either prairie or chaparral. The term 'woodland' is therefore perhaps a misnomer for the true climax formation; though quite xerophilous this formation should really be regarded as a type of low forest [81].

In many places in the Great Basin and on the slopes of the Rockies from Colorado southwards, the 'woodland' was succeeded down slope, somewhere between 7,000 and 5,000 feet, by the deciduous chaparral formation. Frequently this occurred in quite a narrow belt, usually where the mean annual precipitation was between fifteen and twenty inches. Like the sclerophyllous chaparral of California, however, because of its tolerance of fire, the deciduous chaparral has extended its range greatly, primarily at the expense of woodland and forest. This plagioclimax chaparral is now found far beyond the range of the climatic climax formation, having extended as far north as the foothills of the Rockies in Wyoming. It is normally much lower-growing and poorer in species than the true climatic climax formation however. Two of the commonest dominants in this formation are deciduous bush oaks (*Q. gambelii* and *Q. undulata*) whose acorns formed such an important part of the diet of the original primitive food gatherers of the Great Basin. These oaks are associated with a variety of other shrubs belonging to a number of different families. The most frequent are *Cercocarpus parvifolius, C. ledifolius, Rhus trilobata, Prunus demissa, Amelchier alnifolia* and *Robinia neomexicana*.

East of the Rockies the chaparral gave way downhill to prairie, but between the Rockies and the Sierra Nevada, in most places, its place was taken at lower levels by sagebrush. This was also the case locally on dry slopes in California and Lower California. This formation covered most of the surface in many of the basins and on the lower slopes of mountains, its dominant plant, in most places, being the sagebrush (*Artemisia tridentata*). The greyish-green, hairy leaves of this low shrub gave the landscape its characteristic dusty and dreary appearance over very wide areas (Plate XII*a*). It is extremely xerophytic and, as a true climatic climax dominant, it appears originally to have been confined to areas with only five to ten inches mean annual precipitation. In these areas, however, it is co-dominant with considerable numbers of other shrubs of similar height mainly belonging to the genera *Artemisia, Atriplex, Salvia, Chrysothamnus, Pentstemon* and *Grayia*. With increasing soil salinity, species such as shad scale (*Atriplex confertifolia*) and greasewood (*Sarcobatus vermiculatus*) predominate over the sagebrush. Today the sagebrush is dominant over large areas far beyond its original range. Over-grazing of the annual grasses in the western Great Plains and on the Palouse Prairie has permitted this shrub to invade areas with as much as twenty inches mean annual precipitation. Technically this is therefore a kind of plagioclimax but the extent to which this change in vegetation is an irreversible one, does not seem to have been ascertained. Whether or not the prairie

grasses would be able, universally, to reassert themselves if human interference were terminated, is not known.

In the driest areas of Lower California, south-east California, south-
west Arizona and southern New Mexico, as well as in the northern part of
the Sonora Desert and the Mexican Plateau, the sagebrush gives way to an
even more xerophytic formation. This 'desert-scrub' climax [35] is similar
in many ways to the American tropical semi-desert scrub (Chapter XVII).
Though the individual plants of the 'desert scrub' are, on the average,
three to six feet high—taller than those in sagebrush—the former are
usually much more widely spaced. Though these bushes normally stand
several feet apart, this does not mean that the apparently untenanted
ground between them is unused. These shrubs differ from the ones
in communities already described in having predominantly spreading
root systems. The bare ground between the bushes is normally underlain,
at no more than a few inches depth, by a universal, aggressive network of
roots. In this way the plants are able to make the maximum use of whatever rain does fall. This is important since the mean annual rainfall over
much of the range of this formation is only three to six inches. Although
species of cacti such as the chollas (*Opuntia spp.*) and the saguaro (*Carne-
gaeia gigantea*) [35] are often the most conspicuous elements in these
communities, they are not normally the most numerous species. The main
dominant in Arizona and California is the creosote bush (*Larrea mexicana*)
which is frequently associated with other evergreen, spiny bushes such as
white bursage (*Franseria dumosa*). Within the general range of this climax,
deeper valleys often have a richer, denser vegetation in which taller bushes
such as acacias (*Acacia spp.*) and mesquite (*Prosopis spp.*), as well as various
cacti, are found. In these valleys and hollows the water-table is nearer the
surface, consequently plants with inherently penetrative rooting systems
are able to become dominant. Both the acacias and the mesquite are noted
for the enormous depths to which their roots have been observed to
penetrate, the latter sometimes having achieved as much as fifty feet. This
deep-rooted vegetation must be regarded as merely an edaphic climax,
however, in an area where, in most places, the water-table is far out of
reach and the sub-soil almost permanently dry.

Eurasian formations

There appear to be no systematic descriptions of the semi-desert
formations of Eurasia as a whole in the languages of western Europe.
Apart from the case of Soviet Asia [73], all that are available are scattered
descriptions of plant communities in separate regions. These deserts
and semi-deserts cover the vast area from the plains east of the Caspian
to the Great Khingan Mountains on the eastern edge of the Gobi Desert
(Map 2). From the various descriptions available several general points
can be made. Firstly, herbaceous plants are of much greater frequency

than in the semi-deserts of North America; grasses in particular are much more important. Secondly, the Eurasiatic semi-desert is generally much more open, isolated bushes or tussocks of grasses existing, often in a half-submerged condition, in a sea of sand or loess. So great is the contrast between this sparse vegetation and that in areas of similar precipitation in North America that one cannot help suspecting that successive millennia of exploitation by nomadic pastoralists have left a profound mark on the vegetation cover. It is probably significant, however, that some of the species which are dominant and widespread here, belong to the same genera as some of the common dominants of the North American semi-desert; the wormwoods (*Artemisia maritima* and *A. pauciflora*) (Plate XII*b*) are very close relatives of the American sage-brush. In wetter and more sheltered places, larger shrubs and even small trees are found; particularly widespread in central Asia is *Haloxylon ammodendron* (*Chenopodiaceae*), a leafless tree which may grow to a height of almost twenty feet [73]. Poplars (*Populus euphratica*, *P. alba* and *P. nigra*), elm (*Ulmus campestris*), willow (*Salix alba*) and ash (*Fraxinus spp.*) as well as many shrubs such as the wild rose (*Rosa canina*), raspberry (*Rubus idaeus*), hawthorn (*Crataegus pinnatifida*) and honeysuckle (*Lonicera sp.*). can all be found growing in favoured oases in spite of the extreme winter cold and summer heat. Human disturbance and seed transference must have been going on for so long in this area, however, that it is hazardous to try to reconstruct the original flora without much more evidence than is available at present.

The Patagonian formation

The semi-desert of Patagonia stretches southwards on the eastern side of the Andes from the Rio Negro (40° S.) to Tierra del Fuego. Even the driest parts of this area carry a shrub vegetation and, although this becomes somewhat discontinuous locally, it normally reaches a height of between three and nine feet. The shrubs are predominantly deciduous but in most species the leaves are so small that they are barely noticeable. Mainly they are reduced to long thorns or small scales. In spite of its fairly low growth and monotony, this scrub has quite a complex structure when fully developed; so much so that it can be most difficult to penetrate. Some species push upwards, others form an under-storey, others creep along the ground while another group, particularly in more open areas, form hummocks or cushions. In their shape and in the nature of their bark and leaves, most of the species are remarkably well-adapted to the cold winters and dry climate of this region.

True deserts of middle latitudes

On relatively few areas in middle latitudes is precipitation so infrequent as to permit complete desert. In parts of Death Valley in California and locally in Mongolia and Sinkiang, no higher plant life exists; the surface is either a rock pavement or an unstable, sandy plain. Even in these absolute deserts, however, it is difficult to assess whether it is *drought*, as such, which precludes plant growth or whether it is the nature of the surface. In these areas where, at best, vegetation can only be sparse and intermittent, the wind is able to move any friable material which comes into existence. Some areas are thus maintained as completely bare rock while others are characterised by persistently shifting sand-dunes which smother or destroy any plants which try to invade. Bearing in mind the remarkable powers of drought-resistance of some xerophytes and the almost unbelievable ability of others to send down roots to tap deep-seated ground water, however, one is tempted to the conclusion that many of these deserts would carry some vegetation if only a stable rooting medium could be established over their surfaces.

Rooting habits in semi-arid regions

During the past century a great deal of information has been obtained about plant-soil relationships in semi-arid lands. In America it has been noted that the shallow rooting systems of the plants in the most xerophytic 'desert scrub' formation are entirely different from those of the other scrub formations. The plants of chaparral and sagebrush have roots which penetrate deeply into the sub-soil [81]. It is significant that the climatic climax sagebrush of the Great Basin occupies land which, in great part, is underlain by very pervious, sandy or gravelly material. The chaparral also is primarily a formation of mountain and hill slopes and appears to have achieved its most luxuriant development and its greatest penetration into really dry areas on deep, stabilised screes. Both these formations are thus rooted in material into which the water from slight or short showers can penetrate with maximum efficiency. Such a medium is obviously most suitable for plants which produce inherently long root systems. It is also most favourable for water conservation generally; if water can escape downwards quickly, it has the best chance of escaping the fate of rapid evaporation. A greater proportion of rainfall and snow-melt is therefore available to plants on permeable soils, provided these plants have deeply-penetrating roots.

On the other hand, the heavier, less permeable soils in the semi-arid regions of North America seem to have favoured the dominance of grasses. In freely-drained areas where loess or fine-grained alluvium predominated as the soil-parent-material, grasses often became dominant even on the dry side of the ten-inch isohyet. This was often found to be the case on the plateaux and in the basins to the west of the Rockies. The earliest white settlers in these dry areas began by using the heavy, grassland soils on which to grow their wheat and other crops. Here the soil was retentive and rich in minerals and, in relatively moist years, produced heavy crops. In the frequent years when drought prevailed, however, total crop failure was incurred and destitution often followed. It was almost as a gesture of desperation, in one period of protracted drought, that some farmers cleared a patch of sagebrush and sowed it with wheat. It was discovered that this seed survived and yielded a crop when that on the near-by heavier grassland soil failed completely. Wheat, though a member of the grass family (*Graminae*), is quite capable of sending down roots to a depth of six feet or more; it is thus able to utilise the water which percolates deeply into porous soils. On the heavier grassland soils only very rarely does precipitation percolate deeper than a few inches so that most of the incoming water is lost very quickly by evaporation. Thus only shallow-rooted plants which require very little water may be able to survive as the climatic climax vegetation; the short and mid-grasses of the mixed prairie and the 'desert plains' are predominantly annuals who complete their life cycle in the spring and lie dormant as seeds during the heat of the summer and the cold of the winter.

It is clear, therefore, that the distributions of sagebrush and prairie are related to clearly recognisable hydrological conditions. As Carl Sauer has shown, however [66], this is not a full explanation for the relative distributions of grassland and scrub because, in the driest areas of all in North America—the 'desert scrub' areas—the vegetation is not dominated by grasses but by shallow-rooted shrubs.

3. SOILS

Freely-drained soils

Though a great range of contrasts is to be found in the soils which develop beneath scrub, certain characteristics recur frequently; some are almost universal. All the free-draining soils beneath semi-desert scrub, and most of those beneath sclerophyllous scrub, are pedocals of one kind or another. Because of the vast range of geological materials in which development has taken place, however, this statement requires qualification. In South Africa for instance, even where the climate is very dry, true lime-accumulating soils are rare. This is because of the extreme poverty of the soil-parent-materials, however, and not because of leaching. Many of the 'Brown Soils of the Cape' which underlie true climatic climax sclerophyllous scrub, though they contain no free lime, cannot be regarded as pedalfers; in spite of the winter rains here, upward movement of soil water must predominate over leaching. Beneath the Australian mallee, wherever salt does not predominate (*vide infra*), the soil is found to be predominantly calcareous, particularly where extensive limestone outcrops are the parent-material. Indeed, over large areas, a thick, rock-like 'pan' of calcium carbonate concretions has formed at quite shallow depth. On the other hand it is not certain how much of the true climatic climax sclerophyllous scrub developed on leached soils. Today in South Africa, the Mediterranean regions and the United States, this type of vegetation can be found on soils which are podzolised but, in most cases, these are areas where scrub is known to have replaced forest due to human interference.

Beneath all freely-drained areas of semi-desert scrub the soils are found to be some type of very light-brown or grey soil. These are like the soils of the drier prairies in that they are pedocals but, generally speaking, they differ from the grassland soils in structure. The chestnut-brown and light-brown soils which develop beneath even the driest grasslands have a characteristic crumb-structure imparted to them by the finely-divided and well-disseminated humus which they derive from the grass roots. This is normally undeveloped in the scrubland soils where the humus is, in great part, the product of a very sparse leaf-fall. The roots of the shrubs do not form such a fine network and, except in the case of the 'desert scrub', they penetrate far beneath the true soil. The hard leaves form a thin, loose mulch on top of the soil. This does not form a dense mat as do the dead blades of grass beneath prairie; the layer of hard leaves remains well aerated so that most of the organic material decomposes completely and is never incorporated into the soil. The very pale, greyish-brown or grey colours of these scrub soils are evidence of their poverty in organic material. The lime-accumulating horizon which is normally present is usually quite close to the surface. Locally, where the rainfall is very low, this horizon actually emerges at the surface and attains considerable thickness. This rock-like crust of lime ($CaCO_3$) or gypsum ($CaSO_4$) is referred to as 'caliche' in North America and as 'tosca' in Argentina. Quite apart from this, however, the whole body of these soils is normally markedly alkaline and pH values of about 8 are usual.

Soils with impeded drainage

Wherever the water-table is deep beneath the surface in these semi-arid regions, the occasional rainfall is quite sufficient to leach very soluble materials out of the soil and carry them down into the sub-soil out of reach of vegetation. Materials such as common salt (NaCl) and the salts of potassium (KCl, K_2SO_4) are thus removed from freely-drained soils. It must be realised, however, that the basic difference between humid and dry areas is that, in the former, precipitation gain exceeds potential evapotranspiration and in the latter the reverse is true. It follows from this that, in humid regions, there must be an overall excess of water which will flow off and ultimately reach the sea. If, at the beginning of an erosion cycle, there are any enclosed hollows or basins in the landscape, these will be filled to overflowing, water will escape over the brim at the lowest point, and the escaping water will ultimately cut a gorge so that the former basin will ultimately become part of a graded valley system. No such thing occurs in a dry region. Here the water, which may form shallow lakes or ponds during a wetter period, is all evaporated during the succeeding dry period long before the basins can begin to overflow. Any soluble materials which are dissolved in the ephemeral streams which supply the shallow lakes during wet weather, will be re-deposited as solids during the dry weather. In this way, year after year, the ground-water in these enclosed areas will become a more and more concentrated solution of salts. On the lower slopes of inland drainage basins and on flat plains where the water-table is not far beneath the surface, this concentrated solution may be drawn upwards by capillarity during the protracted periods of drought and be evaporated at the soil surface. Inevitably the salts which were dissolved in it will be deposited on the surface in a fine powder or efflorescence.

If the rocks forming one of these regions of inland drainage are predominantly limestones, the accumulating salts in the ground-water may be dominated by calcium carbonate ($CaCO_3$) and gypsum ($CaSO_4$) and re-deposition will be in the form of an extra-thick pan of caliche. It frequently happens, however, that salt (NaCl) predominates; as a consequence saline soils develop wherever the water-table is sufficiently near the surface.

There are obvious reasons for the frequent occurrence of saline soils in dry continental interiors. Firstly, since sodium chloride is so much more soluble than the commonly-occurring salts of calcium, the former will always preponderate in the solution if it is available in only small quantities, even though the surrounding rocks are predominantly lime-stones. Secondly, although sodium is of much less frequent occurrence in rocks than is calcium, there are other ways in which sodium chloride can find its way into the ground-water in a continental interior. In the case of the Caspian area in Eurasia or the Lake Eyre depression in Australia, former connections with the oceans are an obvious explanation for their saltiness; the Caspian Sea and Lake Eyre were cut off from the ocean, they have subsequently been reduced in size by evaporation and the salt from the former sea water has been left impregnating the deposits and rocks thus laid bare. There are extensive areas in central Asia, North America and central Australia which probably did not obtain their salt in this way however. Recent research has indicated that, in these cases, the atmosphere is the probable source. Salt derived from evaporated spray over the oceans is universally present in the atmosphere. Though the quantity in a cubic foot of air is minute, nevertheless it has a significance quite out of proportion to its bulk. Studies of the mechanism of atmospheric condensation have indicated that each cloud droplet forms around a *hygroscopic nucleus* and that one of the commonest kinds of nucleus is the particle of sodium chloride. Thus every shower of rain that falls in a dry continental interior brings with it a tiny quota of salt. Even in a lifetime the quantity falling per square yard will be small but over a period of a million years the result is bound to be great. In a humid region the effect on soil salinity is negligible but in a dry area, where any soluble solid is irrevocably trapped, the gradual build-up in the soil and the ground water is inevitable. The rate of salt accumulation may be even greater in dry areas near the coast where salt spray direct from the ocean, may be blown many miles inland. Thus, in spite of underlying limestone, most of the soils beneath the mallee scrub around the northern side of the Great Australian Bight are very saline.

Soils impregnated with sodium chloride have frequently been referred to in British literature as 'white alkali soil' and in Russian literature as 'solontshak,' though the latter has gained world-wide currency during recent decades. It is, in fact, rather misleading to refer to a saline soil as an 'alkali' soil because usually it has a neutral reaction (pH 7) and not an alkaline one. This is because the clay-humus complex is base-saturated (Chapter III), almost entirely with sodium, and the substance which is present in great excess in the soil solution is a neutral salt, sodium chloride, not an alkali. Like all soils whose clay-humus complex is base-saturated, solontshak has a good crumb-structure; even if very rich in clay it is not sticky or intractable. In spite of this, however, because the majority of crops are poisoned by a high concentration of salt, these soils, in their original state, are not agriculturally useful. It is necessary to remove the salt in some way before they can become so. Unfortunately it is not easy to de-salinise these soils without bringing about changes in their chemistry and structure which render them even less desirable. Such induced changes occurred in many places before modern techniques were devised. If the excess salt is washed out by the introduction of irrigation water, the soil may become both structureless and poisonously alkaline. This is because, when once the excess salt is removed from the chemical complex, ordinary rain-water, which is really a dilute solution of carbonic acid (Chapter III), is able to react with sodium from the clay-humus complex. A considerable proportion of the remaining sodium goes to form the strong base, sodium carbonate (Na_2CO_3), because the carbonate radicle (CO_3) is chemically stronger than the clay-humus radicles:

$$\text{Clay-humus} \Big\langle \begin{matrix} Na \\ Na \end{matrix} + H_2CO_3 \rightleftharpoons \text{Clay-humus} \Big\langle \begin{matrix} H \\ H \end{matrix} + Na_2CO_3$$

This soil is referred to in much British literature as a 'black alkali soil' and in the Russian as 'solonetz'. Quite often this is only a transition stage however; ultimately the sodium carbonate may be washed out of the soil leaving it structureless and acid. This is the soil referred to as 'soloti' or 'soloth' by the Russians.

Solonetz and soloti soils are not all produced by human activity. Any slight fall in the water-table within an area of solontshak can produce the changes described. Thus a change to a slightly drier climate or the encroachment of a nick point into an inland basin may produce a lowering of the water-table. This results in the highest areas of saline soil being just too far above the maximum height of ground-water to receive a periodic supply of excess sodium chloride. Subsequent showers will wash out the excess salt which is present and solonetz will be produced. Subsequently there may be even further change to soloti. Because of this kind of change, these soils are often arranged concentrically around the centres of inland drainage areas, with a salt lake or pan in the centre, solontshak around it and solonetz and soloti, in turn, even further out (Fig. 20).

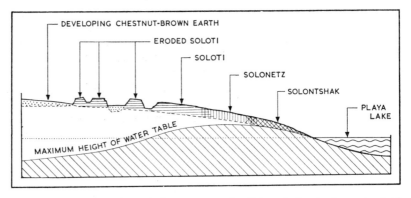

Fig. 20. The disposition of different types of soil in relation to the water-table around a playa in a semi-arid basin.

The characteristics and origins of these soils with impeded drainage have been described here, at some length, because of their increasing importance in world affairs. In dry areas, food production can only be increased materially by irrigation and, quite obviously, large-scale irrigation schemes can only come into existence on plains or in large basins; it is here that the large areas of saline soils are to be found. Techniques for removing the salt without losing structure or incurring solonetz conditions are now being practised. They involve, amongst other things, the addition of a heavy dressing of lime or gypsum to the soil before the irrigation water is introduced. This ensures that, as the sodium is removed from the clay-humus complex, calcium takes its place so that a fertile soil with a good structure results.

PART THREE

THE BRITISH ISLES

CHAPTER XI

The Climatic Climax Vegetation and its Associated Soils

SOME world vegetation maps show the British Isles to be covered mainly with deciduous summer forest, parts of northern Scotland only being covered with mixed forest and alpine vegetation. Similarly, world soil maps show a continuous sweep of brown forest soils on most of the area with a zone of podzols across the north-west. It is clear that these maps have serious limitations. Hikers and amateur naturalists, in planning their weekend excursions, are guided by the knowledge that, within only a few square miles, they can find enormous contrasts in scenery; the soils, vegetation and associated fauna are an integral part of this varied scenery. The world maps merely show the type of vegetation and soil that would be found over *much of* the area if it were *naturally well-drained, of only gentle gradient* and *completely undisturbed by man*. Since hardly any of the present surface of Britain fulfils all these requirements, it is small wonder that the world maps are of little practical value. Apart from the fact that the compilers of such maps intended to show nothing but the theoretical climatic climax vegetation and the zonal soil types, the limitations of scale are also obvious. It is because of the small scale that relatively small areas, even of climatic climax, have to be omitted from such maps. It is because of the basic premises upon which the maps are constructed, that no indication at all is given of the nature of the numerous subclimax, plagioclimax and seral communities which cover even the uncultivated parts of the landscape.

To a certain extent, a book which dealt only with vegetation and soil belts on a continental scale would also, of necessity, obscure a great deal of detail. Indeed, it might leave the reader with an entirely erroneous impression of the true landscape. For this reason three chapters here are devoted to one relatively small part of the earth in an attempt to show what great variations can occur over short distances. It must not be thought that the British Isles have been chosen for any more significant reason than that they are well-known to the author. Many other parts of the earth show just as great a variety within small areas; this example is being used, not specifically to provide a preponderantly large volume of information about the British Isles, but in an effort to put into proper perspective all the rest of the land areas of the earth which are covered in general terms in the rest of the book.

The age of the climatic pattern

Like many areas, particularly in middle and high latitudes, the British Isles have experienced many profound changes in climate during the past million years. Consequently, although the concept of climatic climax forms the basis of the ecological classification of vegetation, it must be remembered that the present climate of the British Isles, with which climatic climax vegetation is in equilibrium, has only held sway for about 2,500 years. Some 20,000 years ago, the Pleistocene ice sheets finally retreated from the lowlands of Britain but, for another 10,000 years or so, an ice cap persisted over Scandinavia. This maintained a cold climate over much of western Europe so that a tundra vegetation remained, even in southern England. After one or two minor oscillations, a rapid increase in temperature took place between about 7500 and 6800 B.C. [32]; during this time the birch (*Betula spp.*) and the pine (*Pinus sylvestris*) rapidly invaded the lowlands. The climatic amelioration continued, with increasing temperatures and dryness, until about 5000 B.C. and there is good evidence that, round about this date, July temperatures were 3° to 4° F. higher than they are at the present day. During this warm, dry Boreal Period (6800–5000 B.C.) pine forest preceded by birch and hazel (*Corylus avellana*) invaded almost to the crests of even the highest mountains in the British Isles though, at the highest levels and in the Shetlands and the far west of Ireland, pine probably never became dominant over the birch. Many of the marshes and glacial lakes dried out or became silted up during this period. At first these extensive, wetter areas would be dominated by a scrub of alder (*Alnus glutinosa*) and willow (*Salix spp.*) but as the Boreal Period progressed these were displaced over large areas by hazel, birch and pine.

A significant change in climate occurred about 5000 B.C. [32] when, quite suddenly, a large increase in rainfall took place. Many areas that had been drying out through the Boreal Period now became marshy again and the pine receded rapidly. Its position of importance was taken, at least for a time, by the alder which appears to have invaded and even become dominant on quite well-drained slopes. It is apparent, however, that the period of extreme wetness did not last for very long and, although the climate remained considerably wetter than it had been in the Boreal

A. LOWLAND OAKWOOD. Valuable stand of mature, planted oaks, with thick field layer of bracken, at Hovingham, North Riding of Yorkshire.
(Photograph by S. R. Eyre.)

B. UPLAND OAKWOOD. Closed-cover woodland at just below 1,500 feet above sea-level at Keskadale, near Keswick, Cumberland. Though not exceeding twenty-five feet in height many of these trees are centuries old, each clump of small trunks probably originating from a single individual, the original trunk of which long since decayed and disappeared. The sparse ground flora and thick mulch of dead leaves are notable.
(Photograph by S. R. Eyre.)

PLATE XIII

A. ASHWOODS ON LIMESTONE. These steep, rocky slopes in Dovedale, Derbyshire, are mainly clothed with ashwood which ceases abruptly at the edge of the plateau. (Photograph by W. J. B. Blake in Tansley, 1953.)

B. BEECHWOOD ON CHALK. Winter aspect of a pure beechwood at Oakham Bottom. A thick mulch of dead leaves covers the ground and the young trees in the foreground show that a certain amount of natural regeneration is taking place. (Photograph by C. J. P. Cave in Tansley, 1953.)

PLATE XIV

Period, rainfall gradually decreased again as the centuries passed. Temperatures remained relatively high in spite of the wetness and, after the first outburst of the alder, other deciduous trees gradually gained dominance. It was at this time that the pedunculate oak (*Quercus robur*), the wych elm (*Ulmus glabra*) and the lime (*Tilia cordata*), for the first time in post-glacial times, became dominants in British forests. This period between 5000 and 3000 B.C., in which a deciduous summer forest formation established itself, is usually known as the Atlantic Period. It was followed by a period, usually known as the Sub-Boreal, in which, if anything, the climate gradually continued to become drier but which remained quite warm. The pine extended locally, particularly on to drying out bogs, during this period and the beech (*Fagus sylvatica*), ash (*Fraxinus excelsior*) and sessile oak (*Quercus petraea*) attained some importance. It was at this stage in pre-history that man himself began to complicate the ecological picture however. Neolithic herdsmen and cultivators entered Britain in the latter part of the third millennium B.C., and were followed, about 1900 B.C., by Bronze Age peoples of similar culture. From this time onwards it is not always clear whether vegetation changes are due entirely to climatic change or whether pastoral and cultivating activities are partly or wholly responsible.

In spite of this, however, there can be no doubt that a marked climatic deterioration took place over a period from about 1200 B.C. until about 500 B.C. [16]. During this time the climate not only became wetter again but summer temperatures became much lower until, by the end of the transition period, they appear to have become very similar to those at the present time. This marked the beginning of the Sub-Atlantic Period and, although there have been some minor climatic fluctuations, it appears that no radical changes have occurred between then and the present day. The onset of these cooler, wetter conditions caused rapid change and impoverishment in the forests. The pine and beech became much more restricted in their ranges while the elm and the lime became rarer. Birch again increased in importance, particularly in the upland forests. Most obvious of all, the forests as a whole contracted in area; they retreated downhill on mountain slopes; they retreated from flat or rolling plateau surfaces in Scotland, Wales, the Pennines and on Dartmoor; they retreated from areas of bog and fenland in the lowlands of Ireland, the glens of Scotland and on some flat areas of eastern England.

Even if man had not interfered, the vegetation in the British Isles today would have had only 2,500 years in which to come to equilibrium with the cool, moist Sub-Atlantic climate. When many of the forests were cleared in the thirteenth century they had had a mere 1,800 years of evolution in this type of climate. Since individual trees may live for many centuries and since it may take many generations of competition to demonstrate which species are truly dominant, one cannot be certain

Lvs

that the forests that were cleared were any more than an approximation to true climatic climax. Furthermore, the evolution of soil profiles may be a very slow process; one must anticipate therefore that some of the uncultivated soils we see today might carry some characteristics which they obtained before Sub-Atlantic times. Indeed it is quite obvious that, where climate has changed considerably at quite frequent intervals, the concepts of climatic climax vegetation and zonal soil are subject to certain limitations. Added to this there is the fact that there can be very few square yards of soil or vegetation in the British Isles today which have not, consciously or inadvertently, been modified in some way by human action.

In the light of these problems it is small wonder that no detailed reconstruction of the 'original' vegetation has been achieved. Even if one had a complete historical record of the wild vegetation which was originally removed to make way for cultivation and grazing (and nothing of the kind exists) this would still not tell one for certain the nature of the present theoretical climatic climax vegetation. For the reasons already given one cannot be certain what the ultimate pattern of plant communities would be if man and his works could be completely eradicated for several centuries. Nevertheless certain broad indications can be given with a high degree of confidence (Fig. 21).

The climax formations

The former existence of five distinct climax formations can be postulated in the British Isles. Nearly all the deciduous trees which arrived back in post-glacial times survived the onset of the Sub-Atlantic climate with a greater or lesser degree of success. These species integrated themselves into a number of associations to form the deciduous summer forest formation; this covered most of the naturally drained lowlands and slopes up to 1,500 feet or more, throughout England, Wales, Ireland and southern Scotland. The summers were sufficiently long and warm to permit these trees to regenerate, though some seem to have survived only in the south.

The second formation, an outlier of the Eurasian boreal forest, appears to have survived the wet Sub-Atlantic climate with any degree of solidity only on intermediate slopes in the Highlands of Scotland. There is some debate as to whether its chief dominant, the Scots pine, survived in England and Wales; if it did so, the evidence suggests that it must have been very local and rare. It appears that the pine has a competitive advantage on upper slopes in Scotland, particularly in the eastern Highlands, where late spring and early autumn frosts can impair flowering and seed-ripening in trees like the oak. The reasons why this formation was not more substantially represented in the rest of the British Isles in historical times, particularly on mountain slopes, is not altogether clear.

On gently-sloping lands on the upland plateaux of the Pennines,

Dartmoor, Wales and western Scotland, and even in lowland areas in western Ireland, forests appear to have failed to regenerate in Sub-Atlantic times. The persistent wetness of the ground and the vegetation led to a combination of circumstances which debilitated the trees and prevented their regeneration. It also caused rapid leaching and increased soil acidity so that bog and moor species came to dominate and a layer of peat built up on the surface of the mineral soil. In this way climatic climax 'blanket bog' came into existence.

Though probably not so extensive as the last, there can be no doubt that a fourth formation existed on many exposed slopes between 1,000 and 3,000 feet above sea-level, particularly where the gradient was steep and the aspect was westerly. Indeed, it must have extended right down to sea-level on exposed ridges and on cliff tops along the western seaboard. In areas like this, forest was precluded by persistent strong winds. A high light intensity was thus permitted to penetrate to low levels near the soil surface and light-loving plants, both herbs and dwarf shrubs, were able to become the dominants. This formation can be most conveniently referred to as climatic climax 'heath'. Where it was the wetness of the climate which caused the existence of bog instead of forest, it was wind speed which caused the displacement of forest by heath.

Finally, at relatively high levels, there was a fifth group of communities known as the 'arctic-alpine' formation (Chapter VIII). This was dominated by plants of alpine and tundra origin which were only able to survive in the British Isles in places where taller and more aggressive plants were unable to invade. This type of vegetation is dominant above 3,000 feet. It can be found at lower levels but, beneath this height, it tends to give way gradually to heath. On some mountains in western Britain and western Ireland, however, arctic-alpine vegetation may be dominant everywhere above 2,000 feet, indeed, with extreme steepness and exceptional windiness, isolated representatives of it can be found even in lowland areas. Generally speaking, this formation is found where low temperatures and short growing season are sufficient to prevent forest regeneration and the growth of trees. High wind speeds are also a prohibiting factor in most places at these high levels but trees, tall shrubs and the more bulky herbaceous plants almost certainly could not grow here even if wind speed were less.

Although most of the dominant species of all these formations have survived human supremacy, in most cases the structure of the communities in which they originally lived has not done so. Consequently, when the primeval oak forest of this country is referred to, one should not envisage a typical oakwood as seen today in the English lowlands (Plate XIIIa). Even in those places where the land has been maintained in a wooded state, its products have been used almost as intensively as those on land which has been tilled or grazed. Some species have been cut out and

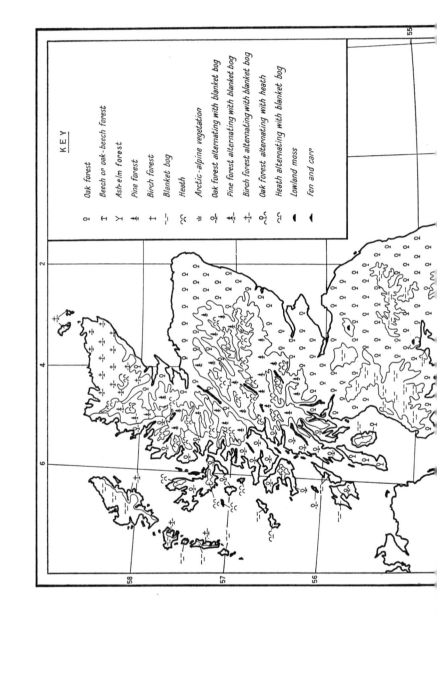

KEY

ọ	Oak forest
Ⲧ	Beech on oak-beech forest
Ƴ	Ash-elm forest
҂	Pine forest
†	Birch forest
⊹	Blanket bog
⟨⟨	Heath
⩊	Arctic-alpine vegetation
ọ̣	Oak forest alternating with blanket bog
⧾	Pine forest alternating with blanket bog
⟊	Birch forest alternating with blanket bog
ọ̣	Oak forest alternating with heath
⟨⟩	Heath alternating with blanket bog
◖	Lowland moss
◖	Fen and carr

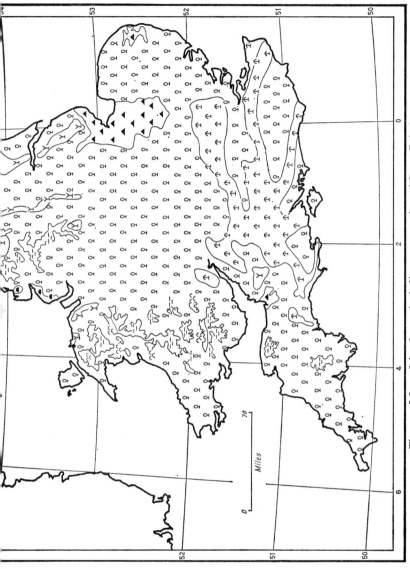

Fig. 21. Map of the theoretical climax vegetation of Great Britain.

reduced in number and others, because of their greater usefulness, have been encouraged if not actually planted. The species-composition has thus changed and, more noticeably, the nature of the community has often changed from a closed-canopy forest of tall, straight trees to a much more open woodland with a much richer ground flora. The surviving non-forest climax formations have suffered similarly, though the resulting changes have not been so spectacular.

I.   THE DECIDUOUS SUMMER FOREST FORMATION

Wherever marshy conditions or frequent floods did not prevent the establishment of true climatic climax, the lowlands of England, Wales, Scotland and eastern Ireland were originally covered with forest. Much of this was dominated by the pedunculate oak (Q. *robur*); with few exceptions it was only where the soil was underlain at shallow depth by unweathered rock or where the soil was calcareous that this species was displaced by others. The pedunculate oak thrives on deep clays [40]. Normally it sends a tap root down into the unconsolidated sub-soil; this reaches the zone of permanent saturation (water-table) so that an unlimited supply of water is available throughout the growing season. Nevertheless this tree cannot tolerate permanently waterlogged conditions in the upper parts of the soil profile. With increasing height of the water-table, in the undisturbed state, it would almost certainly give way to the alder (*Alnus glutinosa*) which is tolerant of marshy conditions.

Though the primeval oak forest would cast a much deeper and more complete shade than most of our present-day oakwoods (Plate XIIIa) it must have possessed clearly-marked shrub and herbaceous layers. The oak tree does not cast a really heavy shade and it seems clear that shrubs like hazel (*Corylus avellana*) and hawthorn (*Crataegus oxyacantha*), though they thrive in open country, would be scattered in a discontinuous stratum in the original forest. Also, numerous herbs found in present oakwoods have obviously evolved in a shady forest environment. The majority of them, like the bluebell (*Endymion non-scripta*), dog's mercury (*Mercurialis perennis*) and wood sorrel (*Oxalis acetosella*), flower in the spring before the tree foliage gives heavy shade. Even this fairly rich ground flora would be discontinuous in primeval forest, however, where there would be large patches bare of herbs but covered by dead leaves.

On shallow soils overlying siliceous rocks, oak forest still predominated but here the place of the pedunculate oak was, in great part, taken by the sessile oak (Q.*petraea*). This species is forced to develop a predominantly spreading root system on the shallow truncated soils on hill slopes on the Millstone Grit of the Pennines and on slates, shales, grits, schists, gneisses, and acid igneous rocks in the Lake District and the Southern Uplands of Scotland. Though it can develop a deeper rooting system and occur along with the pedunculate oak on deep, unconsolidated materials, generally speaking, though by no means universally, it replaced the latter in these upland areas.[1] These upland oakwoods extended above the 1,500 foot contour in many places and fragments of them are still to be found. The Birkrigg and Keskadale oaks (Plate XIII*b*) south-west of Keswick in the Lake District have been described in some detail [47] and the latter certainly extend above 1,500 feet. By an accident of history these sessile oakwoods have survived, possibly in a form very similar to their original one, though it is suspected that, from time to time, they may have been pollarded to supply firewood, charcoal or tannin. Slopes of similar height, gradient and aspect elsewhere in England, Wales and southern Scotland are mainly or entirely treeless today but, generally speaking, there can be no doubt that their true climatic climax vegetation is forest.

In more sheltered localities and on shaly or slaty rocks, the undergrowth of the upland oakwoods is normally very similar to that in the pedunculate oakwoods. Bracken (*Pteridium aquilinum*) and soft grass (*Holcus mollis*), along with numerous flowering plants, cover much of the ground. Originally, though the herbaceous cover would be sparser, there is little reason to suspect that its floristic composition would be very different. In the more exposed upland oakwoods the place of the soft grass is usually taken by wavy-hair grass (*Deschampsia flexuosa*) and the herbaceous flora is more impoverished. In the highest woods of all, particularly on gritstones, the ground flora is normally composed almost entirely of plants which are normally associated with upland heaths. Ling (*Calluna vulgaris*), purple bell-heather (*Erica cinerea*), bilberry (*Vaccinium myrtillus*) and bent grass (*Agrostis spp.*) are the commonest dominants. Usually, however, wherever woods are found on very poor soil-parent-material with a heathy ground flora, the oaks are mixed to a greater or lesser extent with silver birches (*Betula spp.*) This applies equally on both Millstone Grit uplands and sterile glacial sands in lowland areas.

The soils beneath oak forest

Since oakwoods are found on such a variety of rock materials, it is not surprising that they are underlain by a considerable variety of soils. Though most of them belong to the brown forest soil zonal group, they vary a great deal in depth, fertility and general appearance. At the one extreme there are very acidic soils with a pH of about 4 which have developed in materials derived from rocks like Coal Measures sandstones on the flanks of the Pennines and Cretaceous sandstones in the Weald— materials which are highly siliceous and poor in mineral nutrients. At the other extreme there are extremely rich brown forest soils formed in silty

[1] The fundamental differences in the environmental requirements of these two species have not been ascertained [74].

alluvium on river terraces or loessic materials like the 'brick earth' of the London Basin. These are rich in minerals and have a pH value between 6 and 7; very few are now to be seen in their original state since, because of their great fertility, they have been converted into agricultural oils.

On the most sterile types of parent-material, oakwoods can be found associated with well-developed podzols. This is particularly so on very pervious gritstone screes in areas of high rainfall in the north-west, where leaching is very rapid and effective, but it also occurs on sterile sands and gravels in the south-east. The oakwoods on the Bagshot Sands in the London Basin are underlain by some of the best developed podzols in Britain [63]. It is perhaps difficult to appreciate how a podzol can come into existence beneath oak trees whose leaf-fall supplies relatively base-rich, mild humus or mull to the soil surface. It must be remembered, however, that several factors can intervene to offset this. Firstly the forest undergrowth on these podzols is nearly always heathy and undemanding; it requires little nutrient material from the soil and the humus it produces is consequently a base-deficient, acid mor. Secondly, there is normally, at the present day, a strong admixture of birch in the woods on these poor soils and this tree is less demanding than the oak itself. It appears therefore that on these podzols, acid humus production predominates, any nutrients which are released from humus being leached away rapidly through the porous sub-soil. The mature woodland on these soils is pre-dominantly of a very open type today and it is possible that this was the case even under natural conditions; the oak may never have been able to thrive here. On the other hand it is possible that some of the extreme podzolisation may only have occurred since mankind thinned out the forest and permitted the development of a thick, heathy undergrowth. Until this time the soils may very well have been a fairly poor type of brown podzolic soil—maintained as such in a much richer nutrient cycle than exists today. A great deal remains to be discovered about the origin of the inland heaths and heathy woodlands and their associated soils.

The ashwoods

Wherever the soil contains any free lime, oaks tend to become unim-portant or to disappear altogether from woodland communities. It is only in regions of heavy precipitation such as western Ireland and north Lancashire that oak trees appear to grow freely on limestone outcrops; here leaching is sufficiently rapid to remove all free calcium carbonate from the soil in spite of the nature of the underlying rock. Under natural conditions there can be little doubt that the ash (*Fraxinus excelsior*) would be an infrequent tree on siliceous soils; it casts a much lighter shade than the oaks and would thus be crowded out from true climatic climax forests. On many limestone soils, however, oak growth would be much weakened and other species of tree must have been dominant in the original forests. This applies to the large areas of old, hard limestone in the north and west such as the Derbyshire Peak District, the Craven Highlands of Yorkshire and Wenlock Edge in Shropshire, as well as to the Magnesian Limestone, Oolitic Limestone and Chalk outcrops in northern England (Fig. 21). Although these areas are now largely treeless (Chapter XIII), there seems to be little reason to doubt that they were once forested; indeed fragments of woodland are still to be found in many places where steep valley sides are incised into the limestone plateaux (Plate XIVa). These woodlands are nearly all dominated by the ash and it is tempting to infer from this that the original limestone forests which were swept away by man were of a similar nature. All recent research, however, indicates quite clearly that the present nature of these woodlands is very misleading; pollen analysis [29] in all peat bogs which lie near the limestone areas in Britain reveals hardly any ash pollen [16] in peat which pre-dates the arrival of man as a herdsman and cultivator (Chapter XIII). The only inference that one can reasonably draw from this is that the ash, in quantity, returned to Britain at a relatively late stage in the post-glacial period; indeed its achievement of dominance in limestone areas may have become possible only because human communities swept away pre-existing forests. The nature of these pre-existing forests thus remains obscure. It may well be that the wych elm (*Ulmus glabra*), which can thrive on limestone soils and which is known to have become common and widespread along with the pedunculate oak, may have been the main dominant here. Until much more palaeo-botanical research has been carried out however, no statements on this subject can be more than speculative.

Today the ash is also a widespread co-dominant with the oak on many areas of intermediate lithology [77]. On many outcrops of basic igneous rock, marl and calcareous sandstone, sufficient free calcium carbonate is present in the soil to weaken the growth of the oak, thus enabling the ash to compete on more or less equal terms. Again however, one can only speculate regarding the nature of the forests here before the return of the ash.

The ash woodlands of today have an extremely rich flora and luxuriant undergrowth. Not only do they occupy relatively rich soil but their canopy allows much more light to penetrate to lower levels than does that of the oak forest. The shrub layer is usually almost continuous with species like hazel (*Corylus avellana*), dogwood (*Cornus sanguinea*), spindle tree (*Euonymus europaeus*), buckthorn (*Rhamnus catharticus*), privet (*Ligustrum vulgare*) and elder (*Sambucus nigra*) all frequent. The field (herbaceous) layer is also rich and luxuriant; indeed the present-day ashwoods are the happy hunting grounds of plant collectors probably more than any other type of plant community. The vast majority of oakwood herbs are present, along with many more.

Limestone soils

Much of the original forest was removed from flat or gently-sloping limestone surfaces some thousands of years ago; one cannot there-fore be certain of the exact nature of the original soils. Most of the present-day ashwoods, since they grow on predominantly steep slopes, are underlain by permanently immature, *truncated* soils. Because of the persis-tent mixing due to soil creep and the plentiful soil fauna, calcium carbonate is being mixed upwards almost continuously in these soils, thus off-setting the leaching process. In consequence, the pH is usually between 7 and 8. These basic soils occurring on limestone in a humid climate are referred to as 'rendzinas' or 'grey limestone soils'. On moderate slopes they are normally of a homogeneous blackish or dark-grey coloration while on very steep slopes, apart from being very thin, they are paler grey in colour. This difference is due entirely to the slower rate of down-slope movement on the moderate slopes; these soils are thus not only thicker—they have had time to accumulate humus from the overlying layer of leaf litter.

On some of the flatter limestone surfaces, from which forest has long since disappeared, it is likely that natural forest rendzinas were originally to be found. This is almost certainly true of the younger limestones—the Magnesian Limestone, the Oolitic Limestone and the Chalk. Being fairly soft and friable, these rocks weather into finely-divided material which can be mixed, evenly and quickly, into the body of the overlying soil by earthworms and the rest of the soil fauna. An excess of calcium carbonate is thus maintained throughout all the mineral horizons of the soil, keeping the pH above 7 and preventing any podzolisation or even the slightest tendency towards dissociation of the clay minerals. On the other hand, soils on flatter surfaces on the Palaeozoic limestones had probably developed differently even before man's intervention. Here the rock weathers into much larger fragments which cannot be disseminated and intimately incorporated into the soil. Consequently, when a depth of several inches of soil has been achieved, only a moderate precipitation is sufficient to leach away any calcium carbonate which dissolves from the hard limestone fragments and thus to maintain an acid soil reaction. In this acid medium the dissociation of the clay minerals can begin and is normally evidenced by the red coloration of the ferric oxide (Fe_2O_3) so released. Because of this, these soils are referred to as 'terrae rossae' or 'red limestone soils'. Where the residual soil has achieved considerable depth, in extreme cases, it is possible to find true podzols with clearly marked, bleached A_2 horizons; this has occurred locally on the Carboniferous Limestone plateaux in the Peak District and in

Craven. It is impossible to assess the extent to which these terrae rossae and podzols had developed beneath the original forest; possibly many of them have only achieved their present pronounced characteristics since the forest was replaced by a grassy turf (Chapter XIII). Before this, the trees and their associated shrubs may have been able to maintain sufficient calcium and other bases in the nutrient cycle to keep the base status considerably higher than it is at the present time. It is even possible that soils much more like true rendzinas were widespread where terra rossa is now found.

The beech associations and their associated soils

The beech (*Fagus sylvatica*) is the only other deciduous tree which has achieved the status of a common climatic climax dominant in the British Isles. Though it had a wider range in the Sub-Boreal Period, it appears to have survived the rigours of the Sub-Atlantic climate only in southern and south-eastern Great Britain. Because of widespread planting during the past five centuries it is difficult to assess its natural range with exactitude, but the evidence indicates that this extended only as far west as the eastern side of Salisbury Plain and no further north than the Wye Valley, the Cotswolds, the Chilterns and Epping Forest (Fig. 21). The reason for this very restricted range is not obvious since beech plantations as far north as Aberdeenshire produce viable seed in good years which subsequently germinate and produce healthy saplings. There are two possible explanations. It is conceivable that the first few centuries of the Sub-Atlantic Period had a climate so much worse than that at the present day that beech regeneration was quite impossible except in the south-east; between that time and the time of the great mediaeval forest clearances, the beech had not had sufficient time to re-encroach northwards and westwards into areas where it was now possible for it to grow again. In this context, it must be remembered that the beech, like the oak, produces heavy seeds which are not carried long distances by the wind; only chance, sporadic transport by animal agents can carry seeds more that a yard or two away from the parent tree. An alternative explanation is that the beech only regenerates in northern and western Britain today because man has eliminated the competition of other species like the oak. The beech mast from plantations often falls into near-by clearings or on to open heath where the beech saplings do not have to compete with those of other trees. In completely natural forest it is possible that they would not survive the vigorous competition of those species which can tolerate a cooler and less sunny growing season. It is noteworthy that the beech, even in southern England, only produces large quantities of mast about once in every twelve years while the oak produces acorns plentifully almost every year.

The soil requirements of the beech have frequently been misinterpreted.

Because most of the 'natural' beech woods occur on the Chalk, it has been said that this species requires a highly calcareous soil in order to achieve its optimum development. All recent ecological work has indicated that this is a complete misinterpretation of the facts [74]. It has been pointed out that the beech grows fastest and achieves greatest stature on deep clay-with-flints or on consolidated scree material overlying the lower chalk slopes. These deposits are often leached and quite acid in reaction. Much of the rooting system of the beech occurs at remarkably shallow depth so that, in these localities, the root tips must be almost entirely in this acid medium and not in the underlying chalk. The beech tree's adaptation to really acid conditions is even further emphasised by the fact that natural beech forests are to be found on very poor sands and gravels in the New Forest, Epping Forest and at Burnham Beeches. The inescapable conclusion is that the beech grows widely on the chalk out-crops, not because this rock is calcareous, but because it is naturally very porous and well-drained. Careful analysis of the distribution of beech-woods indicates that, more than anything else, these avoid poorly-drained areas.

The actual soil requirements of the beech are finally emphasised by those almost pure beechwoods that are found growing on shallow, chalky soils. In spite of the fact that abundant calcium is available in the rooting medium of these trees, it is quite apparent that they actually absorb very little calcium. Indeed, chemical analysis of their foliage reveals that they absorb relatively little mineral nutrient. As a result of this the humus supplied to the soil surface is a really acid mor [78].

The impoverished nature of the beechwood humus coupled with the fact that beech trees cast a denser shade than any other British deciduous tree, results in a very sparse shrub and herb population. In pure, dense beechwoods, shrubs are almost lacking and the ground is strewn with beech litter almost bereft of herbs. In mixed beech-oakwoods like those in the New Forest or in many places on the clay-with-flints, the shrub and herbaceous flora is much richer and similar to that in oakwoods.

The reason why oak and beech seem to compete with almost equal success on many acid soils in the south-east is, as yet, obscure. Though the possibility must not be ruled out, it must not be assumed that one of them would inevitably prove to be the absolute dominant if human interference were to be excluded. It is possible that the beech needs the oak for its own survival. The beech would perhaps be the cause of its own ultimate elimination if it formed a pure stand on a fairly level area over any con-siderable period. The acid mor that it manufactures certainly seems to create soil conditions which prevent the germination of its own seeds; a certain 'leavening' of the humus by the more base-rich oak leaves may be a necessity in those areas which are too flat to allow soil-creep constantly to mix calcium carbonate upwards. It is fairly obvious, however, that the ash cannot survive in competition with beech though the former would possibly be a dominant here if the latter did not exist. It is note-worthy that, when a piece of chalk grassland is permitted to revert, the ash is normally the first tall tree to re-invade.

A variety of soils are found beneath the beech associations. On steep slopes on the chalk, soil-creep ensures that a shallow rendzina comes into existence. On gentler slopes beneath pure beechwood, however, the nature of the beech humus initiates a vicious circle of interactions. When once the soil has weathered to a certain minimum depth, the upper horizons begin to be podzolised under the influence of the acid mor. Because of its acidity, this mor also fails to attract a soil fauna of earth-worms, beetles and other insects and, because these are absent, burrowing mammals such as wood mice and voles are also lacking. The organisms which are normally present to disturb and mix the horizons of the brown forest soils are thus not present here and persistent leaching prevents free calcium carbonate from reaching the uppermost mineral horizons. Podzolisation thus proceeds and distinct, bleached A_2 horizons can be detected in places. Even more pronounced podzols have developed beneath beechwoods on the sands and gravels in the London Basin. Beneath the oak-beech association on the other hand, almost typical brown forest soils are usually found. The admixture of oak, along with the rich shrub and herb flora, is sufficient to maintain a richer nutrient cycle.

The variety and complexity of soils beneath deciduous summer forest

Not only do all grades of brown forest soil occur beneath the deciduous forest associations of the British Isles; entirely different classes of soil, some of which are commonly found in association with quite different plant formations, also occur. Though brown forest soils are most commonly associated with deciduous summer forest, the existence of the one in no way implies the inevitable existence of the other. Problems of soil classification are also illustrated very well by the soils beneath deciduous forest. It is customary (Chapter III) to classify soils into three distinct groups—zonal, azonal and intrazonal. Over quite wide areas beneath deciduous forest, however, there are soils which, quite clearly, must be regarded as intermediate types. Limestone soils such as rendzinas and terrae rossae are classed as intrazonal since they occur anomalously within areas which are generally podzolised; they have as much in common with chernozems as they have with podzolised soils. It is quite clear, however, that every gradation exists between a typical terra rossa and a typical podzol and between a typical rendzina and a typical brown forest soil. Forest soils on calcareous sandstones or marls frequently cannot be classed categorically either as zonal or intrazonal. In just the same way many soils beneath forest on moist clays cannot be said to be zonal brown forest soils or gleys (Chapter XII); they have attributes of both. Though the basic classification of soils is most useful when considering general aspects of pedogenesis, it cannot be adhered to rigidly when classifying soils in a small area.

2. THE NORTHERN PINE FOREST

The zone of ecotone between the European deciduous summer forest and the boreal forest has its westernmost outlier in the Highlands of Scotland. Here, generally speaking, deciduous trees, particularly the pedunculate oak, occupied the floors of the straths and the glens up to 500 or 600 feet above sea-level, while the mountain-sides up to the tree limit were covered by forests of pine and birch. These reached their most massive development in the eastern and central Highlands, often extending upwards to a height of 2,000 feet, but they also covered extensive areas of moderate slope in the western Highlands as well (Plate XV*a*). Wherever climatic conditions permitted, the pine dominated the birch but the latter, being a little more tolerant of windy and wet conditions, appears to have composed the highest and most exposed forests. The spruce (*Picea excelsa*), which is one of the commonest dominants in the mixed forests of Germany and southern Scandinavia, never re-entered Britain in post-glacial times and so was completely absent from the Scottish forests.

The Scottish pine forests survived the period of mediaeval forest clearance to a much greater extent than the forests of England and Wales; indeed vast areas of pine forest remained almost intact up to the time of the '45 Rebellion. Shortly after this, however, they were in great part removed, partly by the English soldiery who were intent on destroying all possible cover for recalcitrant clansmen, partly to make way for expanding sheep grazing activities and partly to provide fuel for small iron industries which were brought into places like Glen More, merely because the forests were there. Pine forests survived only as isolated fragments and, though these are numerous and widely scattered, there remain only two or three fairly substantial areas. Ballochbuie Forest, on the royal estates between Balmoral and Braemar, extends upwards from 1,000 to about 1,750 feet and Rothiemurchus Forest occupies a large basin of glacial sands and gravels in the Cairngorms, extending from 1,000 feet to over 2,000 feet in places.

Because of human interference, most of these wild pine woods must have a very different structure from the one they possessed under natural conditions. Due to heavy grazing by deer, sheep and rabbits, most of them have not been regenerating for some time and are still failing to do so. Consequently the canopy of the forests has opened out in many places and the birch has been able to invade. This tree is therefore much more important today than it was in the true climatic climax forest. The undergrowth also is much more luxuriant than it would be before heavy grazing began. The more intense light, streaming on to the forest floor, has permitted heath plants like the ling (*Calluna vulgaris*) and the bilberry (*Vaccinium myrtillus*) to form an almost continuous cover over the ground in many parts of the forest. This dense, shrubby growth, in its turn, possibly contributes towards the pine's failure to regenerate. It is only in certain small pinewoods, for reasons which quite often are not at all obvious, that wild pine can be seen freely regenerating. One example of this is to be found in the Applecross peninsula to the south-west of Loch Torridon in Wester Ross.

Generally speaking, podzolisation proceeds rapidly beneath pine trees particularly if acidophyllous species like the ling are present as an undergrowth. The climate of the Scottish Highlands, with its persistent cloudiness and with rainfall nearly everywhere above forty-five inches, promotes rapid leaching. Well-formed podzols are thus of frequent occurrence and would be even more widespread were it not for the fact that most of the surviving pine and birchwoods are on steep slopes where almost continuous soil creep truncates the profiles and maintains the soils in a state of permanent immaturity. It is very easy to mistake some of these soils for true brown forest soils since it is predominantly the bleached A horizon material which is moved away down-slope and the brownish B horizon material which tends to remain. Chemical analysis usually indicates the true nature of the soil, however, in that it reveals a relatively large percentage of free sesquioxides.

3. THE BLANKET BOG FORMATION

In numerous places in the British Isles, forests seem to be excluded solely by the factor of 'wetness'. In some cases this caused forest recession at the end of the Sub-Boreal Period; in others the forest seems to have been excluded ever since the beginning of Atlantic times [32]. The dominant plants in the resulting bog communities are herbaceous; dwarf shrubs like the ling (*Calluna vulgaris*) are quite subordinate and frequently absent. Since the bog communities extend on to slopes as steep as 15° or more, they obviously cannot be regarded as mere subclimax due to drainage impedance; this vegetation occupies sites which could not possibly support it were it not for high precipitation and moist atmosphere.

Climatic climax blanket bog extends on to undulating lowland areas in the extreme west. It is extensive in western Galway and western Mayo and covers all but the steepest slopes on the outlying islands to the west of Ireland. It also covers much of the Outer Hebrides and some of the western fringes of Sutherland and of Ross and Cromarty as well as many localities in the Inner Hebrides (Fig. 21). On upland areas throughout the British Isles it is widespread, becoming more extensive and occupying lower levels as one passes from east to west. It covers undulating upland surfaces in Ireland, western Scotland, Wales, the western Pennines and on Dartmoor from a height of 800 or 1,000 feet upwards; in the eastern Highlands of Scotland and the eastern Pennines it is only extensive above 1,200 or 1,300 feet while on the Cleveland Hills in north-east Yorkshire there are only very small areas of true climatic climax bog.

Typically this formation is dominated by bog moss (*Sphagnum spp.*) though nearly always other species are co-dominant or plentifully represented. The commonest associate is the cotton grass (*Eriophorum vaginatum*); indeed this species may have been absolutely dominant on some upland blanket bogs even before human interference. In north-western Scotland the deer's hair grass (*Scirpus caespitosus*) takes the place of cotton grass. The extent to which other species were represented under undisturbed conditions is not known; ling (*Calluna vulgaris*), cross-leaved bell-heather (*Erica tetralix*), bog myrtle (*Myrica gale*), rushes (*Juncus spp.*) and sedges (*Carex spp.*) are of frequent occurrence at the present day and must have been present to a certain extent in the past since their remains occur at various depths in the peat.

Most areas of blanket bog are not completely homogeneous; here and there shallow channels are found where an even more hygrophilous flush vegetation can subsist. These tend to become more and more frequent as slope steepens. In these flushes, plants which are adapted to live in flowing water often come to dominance; rushes (*Juncus spp.*), the broad-leaved cotton grass (*Eriophorum angustifolium*) and the purple moor-grass (*Molinia caerulea*) are among the commonest of these.

Blanket bog peat

The soil that occurs within root range beneath climatic climax bog has all been manufactured by the bog plants themselves. This peat varies in thickness. At or above 2,000 feet on the Pennines it has achieved an average depth of twelve to fifteen feet (Plate XXI*b*) but here it has been growing rapidly, generally without interruption, since the beginning of Atlantic times. At 1,300 or 1,400 feet on the eastern side of the Pennines it is usually no more than six feet thick.[1] This lower blanket bog is generally thinner because its growth was retarded [16] or completely interrupted during the Sub-Boreal Period and because its rate of growth at all times has been slower than has been the case in the cooler, wetter areas at higher altitudes.

The thicker peats are often found to be composed predominantly of the remains of bog moss whereas, in the thinner ones, the remains of cotton grass or deer's hair grass preponderate. It is in the thinner peats, moreover, that the most frequent inclusions of the remains of ling, bell-heather and other subordinate species are to be found. All the blanket bog

[1] This is only true of true climatic climax bog which has not had its growth rate augmented by being situated in a basin or hollow (Chapter XII).

A. SCOTTISH PINEWOOD. The basin of Rothiemurchus in Inverness-shire.
(Photograph by John Markham in Fraser Darling, *Natural History in the
Highlands and Islands*, 1947.)

B. ARCTIC-ALPINE VEGETATION. Cushions of thrift (*Armeria maritima*)
and moss campion (*Silene acaulis*) at 3,300 feet above sea level of Ruadh stac
Mor, Beinn Eighe Nature Reserve, Wester Ross, Scotland.
(Photograph by S. R. Eyre.)

PLATE XV

PLATE XVI

A. STAGES IN A HYDROSERE. The deepest water is occupied by a pioneer community of white water-lilies and the peripheral shallow water by a second stage community of reeds while a third stage community of willows has occupied the completely silted zone in the middle distance. Gower, South Wales.
(Photograph by S. R. Eyre.)

B. A STAGE IN A HALOSERE. The dense cover of low-growing plants is composed of the original pioneer species, glasswort, into which the taller plant, sea star-wort, has invaded. This area of salt-marsh on Scolt Head Island, Norfolk, is still covered by salt water every high tide.
(Photograph by S. R. Eyre.)

peats, regardless of thickness, are alike in having a very high acidity however; pH values of just over 4 are normal. This acidity is due to the fact that the high rainfall throughout the year maintains a downward flow in the ground-water. All the water in the upper part of the peat is thus derived directly from precipitation; water which has passed through rock or mineral soil material never reaches the roots of growing vegetation. Organic acids form in the peat and are not neutralised by bases. When once the peat has attained a certain minimum thickness therefore, its subsequent development is quite independent of the nature of the under-lying rocks. In places in western Ireland and the Pennines, blanket peat can even be found overlying limestone. This acidic peat is often referred to as 'oligotrophic peat' as distinct from the 'eutrophic peat' (Chapter XII), with a higher base status, which develops in base-rich ground-water.

At its base the climatic climax blanket bog peat gives way, often with a very sharp junction, to mineral material. This can be regarded as a fossil soil; it is the soil in which the forest was growing when, at the beginning of the Atlantic Period, bog plants smothered the surface, preventing tree regeneration. Quite frequently tree stumps are still in situ in this mineral material—preserved in the acid water percolating downwards from the overlying peat. In these acidic, anaerobic conditions, decay bacteria cannot survive and whatever organic decomposition does take place is extremely slow and due primarily to anaerobic fungi. It was because of this that blanket bog was able to come into existence in the first place.

4. THE HEATH FORMATION

Climatic climax heath would occur under natural conditions in areas where forest communities were unable to establish themselves because of persistently strong winds or, possibly, extreme soil poverty. It is thought that the latter would be capable of excluding tree growth completely, only over extremely small areas, if indeed it were capable of doing so at all; under natural conditions some trees were probably able to maintain themselves on the poorest and most acid of what are now lowland heaths (vide supra). Wind speed is therefore the only significant factor in a consideration of the original distribution of this formation.

There must have been two distinct categories of climatic climax heath—acidic heath on siliceous soils and calcareous heath on limestone soils. Some kind of acidic heath must have dominated above the tree line on mountain slopes composed of grits, slates, gneisses, schists and acid igneous rocks wherever the slope was too steep to permit the formation of blanket bog. At even lower levels it must have dominated over restricted areas along cliff tops or the crests of hills wherever wind speed thinned out or dwarfed the tree growth. Since wind speed varies so enormously over distances of only a few yards, however, and since vegetation, by its very presence, can greatly modify wind conditions, it is extremely difficult to arrive at any conclusions about the exact extent of these original heaths.

The acidic upland heaths would be dominated, in the main, by the plants which formed the undergrowth in the uppermost forests. Ling (*Calluna vulgaris*), fine-leaved bell-heather (*Erica cinerea*) and heath grasses such as bent (*Agrostis spp.*), wavy-hair grass (*Deschampsia flexuosa*) and mat grass (*Nardus stricta*) were probably the most widespread dominants in the formation as a whole (Plate XIXa).

Most of these species are well adapted to living on siliceous materials; they demand very little from the soil and thus perpetuate an impoverished nutrient cycle. Some of the most acid soils are to be found beneath them, pH values of about 3·5 being normal. Soils beneath ling in particular are noted for their extreme poverty in mineral nutrients. In spite of this acidity the peat layer is usually only a shallow one, the roots of the plants normally reaching down into the mineral soil beneath. Even where the slope was insufficient to cause rapid truncation, the peat beneath climatic climax heath would probably rarely be as much as a foot in thickness. Beneath this A_0 horizon the mineral soil would, typically, be strongly podzolised with a relatively deep, bleached A_2 horizon wherever truncation was not too pronounced.

The distribution of climatic climax heath on limestones has been much discussed. In view of the expanses of treeless grassland that extend over the Chalk and the older limestones, many naturalists in the nineteenth century came to the conclusion that these were 'natural grasslands'. There are several good reasons why one cannot accept this view however. Plantations that have been made on some of the most exposed crests of the South Downs and other Chalk ridges have been found to flourish and to grow to maturity. Similar plantations flourish at over 1,000 feet on Carboniferous Limestone in the Pennines and it is quite clear that forest would regenerate naturally up to an altitude of at least 1,200 feet in places if sheep grazing were discontinued (Chapter XIII). The treeless limestone plateaux of the Peak District and Craven must be viewed in the light of more than 4,000 years' intensive usage by human beings. It is probable, therefore, that climatic climax calcareous heath was no more extensive on limestone outcrops than acidic heath was on outcrops of siliceous rocks.

Studies of community successions on the Chalk have indicated that any calcareous heath that may have existed in south-eastern England probably consisted of thickets of scrubby hawthorn (*Crataegus oxyacantha*), juniper (*Juniperus communis*), yew (*Taxus baccata*) and even lower-growing, spiny shrubs with intermixed calcicolous herbs such as brome grass (*Bromus erectus*), oat grass (*Avena pratensis*), bird's foot trefoil (*Lotus corniculatus*), lady's bedstraw (*Galium verum*) and salad burnet (*Poterium sanguisorba*).

Since exposed areas of Carboniferous Limestone in northern England all have heavy rainfall, the climatic climax heaths developed upon them are more complex. Wherever the slope has permitted the accumulation of several inches of soil, the uppermost part of this has become well leached and shallow-rooted acidophyllous plants have been able to invade. Indeed, in places, plants typical of the acid heaths such as ling, mat grass and bilberry can be found growing in association with calcicolous plants, the former being rooted at shallow depth and the latter having roots penetrating to deeper levels. This same kind of mixed association of species can also be found in places on the Chalk plateaux in the south-east.

5. THE ARCTIC-ALPINE FORMATION

It is only above about 3,000 feet in the central Highlands of Scotland that tundra and alpine species become predominant over upland heath species. This is achieved at a somewhat lower level in western Scotland and the change takes place at comparable heights in England and Wales respectively. In some places the transition from heath or moor is marked by a distinct plant community often referred to as 'arctic-alpine grassland'. This occurs particularly on basic igneous rocks or base-rich metamorphic rocks where the slopes between 2,000 and 3,000 feet are at a sufficiently low angle to permit the development of a continuous turf. In spite of the fact that this turf is composed of a mixture of heath species and arctic-alpine species, there seems to be little doubt that it is a true climatic climax community. On Ben Lawers in the central Highlands of Perthshire, bent grass (*Agrostis tenuis*) and sweet-scented vernal grass (*Anthoxanthum odoratum*) from the heaths are blended with alpine sheep's fescue (*Festuca ovina* var. *vivipara*) and alpine lady's mantle (*Alchemilla alpina*).

The communities of the arctic-alpine formation proper often do not form a closed cover over the ground. This is mainly because the surfaces that must be occupied at these heights do not permit complete occupance. Slopes are predominantly steep and the large amounts of material produced from crags by frost shattering build up coarse screes which are mainly in a state of constant disturbance. Vegetation on these screes often occurs in vertical belts or 'stripes' but, at any particular time, the great proportion of the scree surface is almost untenanted. Even the gentler slopes, on benches or at the crests, often have no true soil. The fine-grained mineral fraction of a true soil is mainly produced by the chemical weathering or 'rotting' of rocks but, at these heights, chemical weathering is much reduced. Normally the rocks are disintegrated to quite a considerable depth by the simple mechanical process of frost shattering; consequently a layer of angular boulders, often several feet deep, covers the surface. Though some mosses and crustose lichens do grow on the rock surfaces, the only inhabitable sites for rooting plants are the crannies between the boulders where small amounts of comminuted material and humus are able to accumulate. A superficial glance across one of these areas almost gives the impression of a rock desert; careful examination from above, however, usually reveals quite a rich and well-distributed flora.

There are between one and two hundred distinct species and varieties of flowering plants in the British flora which are regarded as arctic-alpines. They are defined as plants which are confined to mountain tops except for some occurrences at lower levels where sporadic natural or human influences have diminished the competition of more luxuriantly-growing species. Apart from commonly occurring mosses such as woolly fringe-moss (*Rhacomitrium spp.*), the most frequently encountered species are alpine sheep's fescue, alpine lady's mantle, arctic willow (*Salix herbacea*), stiff mountain-sedge (*Carex rigida*), twisted whitlow grass (*Draba incana*), dwarf cudweed (*Gnaphalium supinum*) and purple mountain saxifrage (*Saxifraga oppositifolia*).

All these species have adopted modes of growth well adapted to this exacting environment (Plate XV*b*); the cushion, the rosette, the sub-aerial mat and the sub-surface mat forms are all adopted. All the species hug the rock surface very closely or nestle well within the rocky crannies. This enables them not only to minimise the drying effect of strong winds in drought periods in summer, it also permits complete entombment beneath snow cover during the whole of the winter. This is probably all-important; it ensures an almost isothermal climate at about 32° F. during the whole winter, even though air temperatures may frequently fall to 20° F. below freezing-point; it also ensures that the plant is almost completely protected from transpiration during the period when liquid water is quite unobtainable from the frozen soil. The actual distribution of plants on the uneven surface indicates fairly clearly that any individual plant which does not have the protection of snow cover must perish during its first winter. Since snow lies late in the spring and arrives again early in autumn it is quite clear why plants at this altitude must be adapted to a short growing season. No plant can synthesise food or carry out reproductive processes beneath a blanket of snow.

CHAPTER XII

Seres and Subclimaxes on the Original Landscape

EVEN the original wild landscape was not covered everywhere by climatic climax vegetation (Chapter II). Some of the vegetation which was swept away by the successive waves of Neolithic, Bronze Age, Anglo-Saxon and mediaeval farmers, consisted of priseral and sub-climax communities. Even today, some small areas of these remain almost unaltered but, in great part, like the climatic climax, they have been removed or modified.

The reason why climatic climax had not been achieved in many places was drainage impedence. The retreating ice sheets had left the drainage pattern in an unintegrated and deranged state with widespread lakes and marshes and, although many of these had been drained by the natural processes of river erosion, or filled in by natural silting, many still remained in Sub-Atlantic times. In these the hydrosere must have been observable at all stages of development. Some large, flat areas were so near to sea-level that drainage integration and vertical corrasion made little or no difference to the height of the water-table and the hydrosere was arrested at a fairly late stage as a natural subclimax. Nothing short of a major fall in sea-level with subsequent rejuvenation could have removed the arresting factor in these areas. This was the state of affairs in the Fens of Lincolnshire, Cambridge and Norfolk as well as in the Norfolk Broads, Suffolk Sandlings and Somerset Levels. These conditions also obtained on numerous smaller areas of flood plain along the middle and lower reaches of larger rivers.

There were two other types of area along estuaries and coastlands where the climatic climax had not been achieved. Certain areas where tidal inundation took places at regular intervals were occupied by various stages of haloseres. These are commonly referred to as salt marshes. Other areas were constantly affected by sand blown from the foreshore; around the coast various stages of the psammosere can still be observed on the dunes so formed.

Hydroseres

Although many of the open stretches of water which remained in mediaeval times have now been drained, some still remain. The numerous fresh water lochs of Scotland and some of the meres of Cheshire are examples. These are all in process of being silted up, the rate of the process being dependent upon the nature and extent of the streams which drain into them and the rocks over which these streams flow. Despite these differences, however, all small, shallow water bodies will inevitably silt up relatively quickly though very large ones like Loch Ness and Lake Windermere would take such a long time that earth movements of one sort or another are likely to intervene before the process is complete.

When once the water in a lake is reduced to a certain critical depth by silting, the floor is invaded by rooting plants. They accelerate silting by reducing turbulence and by supplying organic debris. As the depth decreases further, other plants replace the pioneer community and this goes on, stage by stage, until the lake is filled completely and becomes a marsh. Many formerly ill-drained, flat valley floors and level plains are no more or less than silted lakes. Today many of them are covered by a patchwork of cultivated fields (Chapter XIII) but prior to mediaeval times, or even later, they were occupied by late stages of hydroseres.

The nature of British hydroseres can still be studied, however, in those water bodies which remain. The species-composition of the different stages varies very much from place to place because of differences in the chemical nature of the ground-water and in the nature of the deposited material. The water varies from peaty, acid water draining from siliceous, bog-covered uplands to base-rich, alkaline ground-water draining from calcareous rocks. The deposit is coarse-grained and poor in nutrients in places, and fine-grained and rich in organic material and mineral nutrients in others. Also, in some water bodies, deposition has taken place slowly so that the bottom has been stable; in others it has occurred quickly and thus provided a very unstable rooting medium. In view of these great contrasts only very general statements can be made about the nature of hydroseres. Quite often a pioneer community has invaded the bottom when the depth has been reduced to about three feet. Native species which effect this invasion in a variety of conditions are the white water-lily (*Nymphaea alba*), the yellow water-lily (*Nuphar luteum*), pond weeds (*Potamogeton spp.*) and the common arrow-head (*Sagittaria sagittifolia*). As the water becomes more shallow, species like the common reed (*Phragmites communis*) (Plate XVIa) and the great reed-mace (*Typha latifolia*), form a dense community which yields ample organic material. Peat thus builds up and the lake margin is converted into a swamp. Normally the reeds persist and remain dominant after the soil surface has been built to above water-level but they are normally joined by numerous other species among which sedges

(*Carex spp.*), saw-sedge (*Cladium mariscus*), yellow flag (*Iris pseudacorus*) and reed manna-grass (*Glyceria aquatica*) feature frequently.

Haloseres

A great deal of land which today fringes bays or stands at the head of estuaries, was beneath sea water not so very long ago; silting has taken place and the mud, silt, sand and shingle has gradually been stabilised by invading vegetation. The nature of the halosere has varied from place to place even more than has that of the hydrosere. This is mainly because of contrasts in tidal range and differences in exposure to wave-action—phenomena which the pioneer communities of hydroseres usually do not have to contend with. Pioneer communities of haloseres also have to be resistant, not only to a high concentration of salt in the rooting medium but also to periods of complete inundation twice daily. The flowering plant in the British flora which seems to be most resistant in these respects is the grass-wrack (*Zostera marina*) which is only uncovered at low tide. Though this grass forms the pioneer community in some places, its distribution is very sporadic. A much commoner pioneer on estuarine mud is the glasswort (*Salicornia herbacea*) and it is this species which is normally responsible for stabilising estuarine mud and thus preparing the ground for less tolerant communities. Glasswort is adapted to withstand regular inundation even at the neap tides. As the mud builds up and the rooting medium is stabilised, glasswort is gradually ousted by species like sea manna-grass (*Glyceria maritima*) or sea star-wort (*Aster tripolium*) (Plate XVI*b*) which in turn, with decreasing salinity, give way to thrift (*Armeria vulgaris*) and sea blight (*Sueda spp.*). All these species and many others contribute to the invasion of marine mud flats; quite often it is possible to see a more or less regular banding of communities with the pioneer ones farthest seaward. Ultimately the halosere stabilises the surface with a complete cover of vegetation, the nature of which differs according to the physical and chemical nature of the deposit. The surface ultimately rises above the level of the highest tides although, because of the low gradient, the water-table usually remains very close to the surface. Because of this, vegetation development is arrested at a subclimax, the exact nature and extent of which depends upon the nature of the ground-water and the climate.

Fen and carr

All the extensive areas which were covered with this kind of subclimax vegetation in mediaeval times had commenced their development as stretches of open water. In most cases, however, because of small changes in post-glacial sea-level, this development had not been a simple, progressive one. Even as far inland as the Cambridgeshire Fens, marine incursions occurred several times with consequent deposition of silt. This accounts for the alternate layers of peat and silt beneath the Fens [32]; each time the sea encroached, pioneer invasion of shallow, brackish water had to begin all over again. It appears that, by late mediaeval times, a stage had been reached where silting and prisere development had almost filled up the water bodies created by the last marine incursion so that these were now occupied by land whose surface was at or above the mean level of the water-table. In the case of eastern England, when once the sea water had been excluded, peat development and vegetation growth had been predominantly in a base-rich ground-water, the uppermost rocks here being predominantly chalk and chalky boulder clay. Because of this, plants like the saw-sedge (*Cladium mariscus*), reeds (*Phragmites communis*) and flags (*Iris pseudacorus*) had remained dominant after the passing of the swamp stage. These base-demanding plants formed a number of communities which are commonly referred to as 'fen'. From time to time, particularly in winter, flood waters still rose a little above the surface of these fens. Where the sedge and reed peat had grown to a level above the height of the highest of these floods, however, the herbaceous plants were no longer the dominants. On those parts of the Fens which had developed to a higher level, it seems that woody plants, particularly the alder (*Alnus glutinosa*), had reached dominance. These fen woodlands or scrublands are commonly known as 'carr'. It appears that the dominant woody species of carr, though they can flourish with most of their roots in permanently saturated peat, need to have an uppermost layer which is rarely flooded.

The commonest woody associates of alder in the carr woodlands are the grey sallow (*Salix atrocinerea*), common buckthorn (*Rhamnus catharticus*), alder buckthorn (*Frangula alnus*) and guelder rose (*Viburnum opulus*). When once these dominants are established the pre-existing fen dominants die away leaving an almost bare surface which is then colonised by quite a characteristic undergrowth. Bent grass (*Agrostis stolonifera*), stinging nettle (*Urtica dioica*), ferns (*Dryopteris thelypteris*), sedges (*Carex spp.*), marsh marigold (*Caltha palustris*) and meadow sweet (*Filipendula ulmaria*) are all found here but a number of them do not flower during the early stages of carr development when the over-topping shrubs and trees form a particularly dense cover [74].

Areas of gley soil

On a smaller scale, fen and carr communities must originally have been widespread on the silted parts of lakes and meres wherever the ground-water was sufficiently rich in bases to permit a basic peat to accumulate. Rather similar communities were also to be found along the flood plains of rivers and indeed on all kinds of flat or gently-sloping land within generally forested areas. Here the water-table would be high throughout the year but conditions would be different from those in the fens and carrs in that the top soil would be of predominantly inorganic silt or clay.

The dominant plants would be alder and willows (*Salix spp.*) but the associated shrubs and herbs would vary very much according to the soil characteristics and the precipitation as well as to the degree and frequency of waterlogging and inundation.

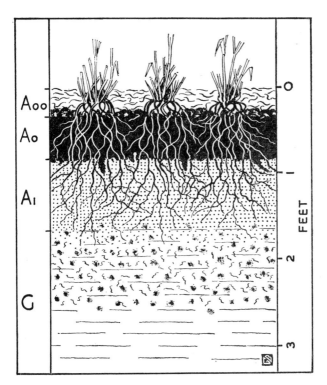

Fig. 22. A gley soil beneath moorland grasses.

A_{00} *Unhumified remains of grasses.*
A_0 *Black, structureless peat.*
A_1 *Podzolised, greyish-brown, peaty sand.*
G *Bluish-grey gley horizon with rusty and yellowish-brown*
 streaks and mottlings above; homogeneous, bluish-grey beneath.

The uppermost horizons of the soils beneath these communities would vary very much. On floodplains they would tend to be homogeneous silts while on heavy parent-material which was not periodically flooded, it would often be a brown forest soil at some stage of podzolisation. These soils were similar, however, in that they possessed a quite characteristic type of horizon at a greater or lesser depth beneath the surface layer. This horizon owed its characteristics to the fact that it was saturated for most of the year. From time to time however, usually during the summer, the water-table sank within it, thus permitting aeration to a certain depth. During these periods some of the iron became oxidised. This involved a change from a ferrous salt such as ferrous carbonate to some form of ferric oxide (Fe_2O_3). The horizon as a whole is referred to as a 'gley' horizon (Fig. 22). Normally it is grey or greyish-green in colour but speckled or streaked with red, brown or yellowish-brown in the partially oxidised layer. The oxidation often takes place around fragments of organic material and quite often rusty filaments develop along the line of former root channels.

Raised bogs and valley bogs

Throughout Ireland and western Britain hydroseres and haloseres proceeded very much as in the east. When once the fen stage was achieved, however, subsequent development was often very different, even with initially base-rich conditions. In the moister climate of the west, when once the fen peat had developed only a few inches above the maximum level of the ground-water, its upper layer began to be leached of nutrients by percolating rain-water. Precipitation was sufficiently frequent and heavy to keep the fen surface saturated and to displace any base-rich ground-water. In these conditions the uppermost peat became sufficiently acid to favour the growth of acidophyllous bog plants such as bog moss (*Sphagnum spp.*) and cotton grass (*Eriophorum vaginatum*). From this stage onwards the surface developed upwards in a very similar way to that of blanket bog. Generally speaking it became too acid and remained too wet to permit the invasion of the common dominants of carr.

In many parts of central Ireland as well as in the lowlands of Lancashire and other western counties in Britain, these bogs came into existence within generally forested areas. Quite often they are founded on silted glacial lakes or on estuarine muds which were originally colonised by haloseres. They are usually referred to as 'raised bogs' because their central regions are at a higher level than their peripheries. This is due to the fact that, after the bog plants had invaded, they continued to flourish and thus to produce more and more peat, accumulation being particularly rapid in the centre. Rainfall was sufficiently heavy and the peat sufficiently retentive to keep the water-table near the surface even in the highest central regions. The former existence of forest on free-draining areas around the Irish raised bogs and the Lancashire lowland mosses indicates clearly that they are subclimax and not climatic climax vegetation.

Similar subclimax bogs are of frequent occurrence in western Scotland and on areas of low gradient on the lower parts of uplands throughout England and Wales. Since the beginning of Atlantic times, the floors of the glens in western Scotland have experienced extremely wet conditions; not only is the precipitation heavy and well-distributed throughout the year but run-off and spring water from the surrounding mountains is also received. This rapidly-moving drainage water, though by no means so base-rich as ground-water in many English lowlands, still contained a certain amount of mineral salts derived from the metamorphic and igneous complexes. In this waterlogged habitat hygrophilous plants like rushes (*Juncus spp.*) and purple moor-grass (*Molinia caerulea*) grew luxuriantly and initiated the formation of peat. This peat ultimately developed above the influence of the ground-water and became populated by species similar to, or identical with, those of the Irish raised bogs. These 'valley bogs' of the Scottish glens often developed so rapidly and to such an extent that they obstructed the 'burn' and pushed it over to one side of the valley floor. Indeed, quite frequently the peat grew up over the top of the burn which subsequently came to flow in a natural 'culvert'.

Although forests extended up the slopes of the Pennines, the North York Moors and the uplands of central Wales to at least 1,500 feet, they were not absolutely universal below this height. In many places, where gentle folds or hollows occurred in the rolling, peneplained surfaces, they were occupied by bogs. This occurred particularly on shale outcrops or where a veneer of boulder clay covered the consolidated rocks. These subclimax 'mosses' obviously began their development in the abundant acidic ground-water which was able to remain at or near the surface for most of the year. As in the case of the Scottish glens, peat formation was begun by plants which tolerate wetness and moderate acidity, and subsequent invasion by bog plants took place. The very fact that these areas were called 'mosses' indicates that bog moss (*Sphagnum spp.*) was once dominant there. This is substantiated by the composition of the deep peat underlying the present surface; it is predominantly composed of the remains of bog moss [16]. 'Mosses' of this type are found on East Moor on the eastern side of the Peak District of Derbyshire at a height of little more than 900 feet (Fig. 23) and where the mean annual precipitation is no more than thirty-five inches. The peat here is, locally, more than twenty feet in depth and must have grown rapidly and continuously before man's interference (Chapter XIII).

The raised bogs of Ireland, the mosses of lowland Lancashire, the valley bogs of Scotland and the mosses at intermediate altitudes in the English and Welsh uplands thus appear to be all of the same generic type. They formed in generally forested areas and therefore must owe their existence to locally impeded drainage. Regardless of original differences in ground-water chemistry and in vegetation, they were all tenanted ultimately by acidophyllous bog plants—predominantly *Sphagnum*—and covered thickly by acid peat.

Cliffs, screes and sand dunes

In the original landscape numerous areas, mainly of small extent, carried seral or subclimax vegetation because of instability of the surface due to gravity or wind. On vertical cliffs no continuous layer of debris

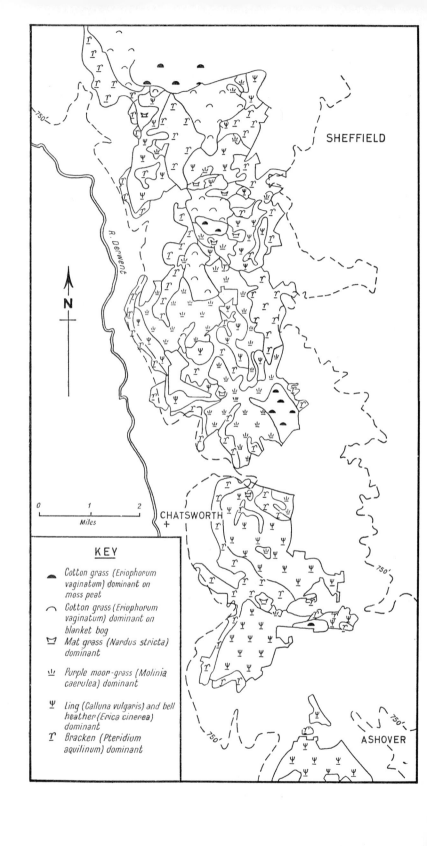

SHEFFIELD

R. Derwent

750'

N

0   1   2
Miles

CHATSWORTH
+

750'

750'

ASHOVER

750'

KEY

- Cotton grass (Eriophorum vaginatum) dominant on moss peat
⌒ Cotton grass (Eriophorum vaginatum) dominant on blanket bog
⊌ Mat grass (Nardus stricta) dominant
⩍ Purple moor-grass (Molinia caerulea) dominant
Ψ Ling (Calluna vulgaris) and bell heather (Erica cinerea) dominant
r Bracken (Pteridium aquilinum) dominant

could accumulate; on very steep slopes coarse debris predominated and even this was extremely unstable. The species found on these cliffs and screes must have varied a great deal from one place to another according to contrasts in the chemical and physical nature of the rock. Apart from those in the vicinity of permanent springs, however, they all had one characteristic in common: they must all have been adapted to withstand a high degree of drought. The xerosere (Chapter II) is arrested at a very early stage on a very steep, unstable scree but at increasingly late stages the more stable the slope; indeed most climatic climax communities are probably able to establish themselves on any slope which is sufficiently stable to permit the formation of a continuous soil cover.

Instability of the environment near the sea coast is due primarily to the fact that the waves are constantly producing sand by the comminution of eroded material. This sand accumulates between the tide marks but frequently becomes dry and incoherent at low tide. It can then be blown inland by only moderate breezes; it can be transported quickly and in huge quantities by gale-force winds. In consequence, wherever wave action concentrates sand from a considerable stretch of coast into a relatively small bay or estuary, there will be a perpetual tendency for the production of landward-migrating dunes. Newborough Warren in Anglesey and Braunton Burrows in north Devon are two such areas. Were it not for the spontaneous invasion of plants on to these migrating dunes, the latter, under quite natural conditions, would have held sway over much larger areas. Large dunes are effective in smothering and killing the most luxuriant types of vegetation; thriving communities are submerged by the leading edge of the dune and a completely devastated surface is left behind it.

Only plants which are markedly xeromorphic and which also possess growth characteristics to withstand this remarkably unstable, mobile surface, can invade migrating dunes. Two distinct types of pioneer community are found. The first is often dominated by sea couchgrass (*Agropyron junceum*). It invades the base of those dunes which are still reached occasionally by the tides and is often associated with other plants which are resistant to both sandy and saline conditions. The succulent sea sandwort (*Arenaria peploides*), sea rocket (*Cakile maritima*) (Plate XVIIa), prickly saltwort (*Salsola kali*) and orache (*Atriplex spp.*) are all common pioneer species in this type of environment. These plants often initiate dune formation since they can intercept in-blowing sand and thus cause it to form 'drifts' over them. They continue their growth, vertically and laterally, through these 'drifts' so that the latter are partially stabilised and continue to enlarge. This can go on until the top of the nascent dune

Fig. 23 (*opposite*). Patchwork of moor and heath communities on the East Moor of north Derbyshire. (Surveyed in 1950 by S. R. Eyre.)

has grown above the highest tide level. It is then that the second type of community is able to invade. This is dominated by the marram grass (*Ammophila arenaria*)—a plant which cannot survive persistent immersion in salt water but which is even better adapted for survival in mobile sand. It throws out *rhyzomes* which are able to maintain rapid vertical and lateral growth and to keep up with dune development. Indeed the rhyzomes often form a close network with such rapidity that they are capable of stabilising the dune and preventing its further migration (Plate XVIIa). When once the mass of the dune has been stabilised in this way, a second stage community which often contains rosette plants such as ragwort (*Senecio jacobaea*), hawkweed (*Hieracium umbellatum*) and thistles (*Circium spp.*), is able to invade. These could not survive, however, were it not for the persistence of the marram grass, holding together the mass of the sand. Usually the communities on the stabilised dune remain open, permitting the surface sand a good deal of mobility until, as a third or fourth stage community, the sand fescue (*Festuca rubra* var. *arenaria*) forms a turf over the surface and ultimately squeezes out the remaining tufts of marram grass (Plate XVIIb).

All stages of these psammoseres were to be found around the coasts of the British Isles when the herdsmen and cultivators first arrived. They interfered here as they did with other plant communities and, since man did not appreciate the nature of this particularly unstable environment, the results of his interference were sometimes even more spectacularly catastrophic than elsewhere.

CHAPTER XIII

The Impact of Man

OF the original pattern of plant communities described in the previous two chapters, only vestiges remain. In places it has been replaced by an orderly, well-tended patchwork of cultivated crops, in others by an exploited and depleted vegetative cover; in a few areas, nothing remains but the bare rock. Man has been an element in the fauna of the British Isles throughout the whole of post-glacial times but for thousands of years, as a hunter and food collector, he modified the vegetation and soils but little. It was only with the arrival of people of the Neolithic Culture, shortly before 2000 B.C., that mankind began to modify the landscape for his own ends. Fire and the axe were used to clear scrub and forest in order to use the land for growing crops [10]; a great deal more forest was prevented from regenerating by the grazing of flocks and herds. For these two reasons, little forest remained by the end of mediaeval times [27]. The effect of these changes on the actual flora must have been profound. Many herbaceous plants are adapted to live either in open sunlight or in the shade of trees and shrubs. Consequently, large numbers of forest plants must have been much reduced in range and in numbers when the thick forests were thinned or removed. The changes were by no means all on the debit side however; a variety of cereals and fruit trees were introduced purposefully to provide food, and along with these came many weeds o cultivation, actually introduced involuntarily from abroad with the seed grain. Native plants which heretofore had been quite rare elements in the flora, having a foothold only on cliffs, screes, foreshores and other disturbed areas, were now able to compete with the crop plants on the deforested arable lands. It is also interesting to reflect that the familiar daisies (*Bellis perennis*), dandelions (*Taraxacum officinale*) and buttercups (*Ranunculus spp.*) of our fields and waysides could not have been anything but rarities in the original primeval forests. Similarly, many other conspicuous flowering plants of the pastures and meadows which are thought of today as being typical of the British landscape, cannot have been common originally. One would be quite surprised today if one came upon a patch of clover (*Trifolium spp.*) or a clump of ox-eye daisies

(*Chrysanthemum leucanthemum*) when walking through an oakwood; how much more unlikely it would be for them to have occurred beneath closed-canopy primeval oak forest! One must not go so far as to over-stress this point however. The exact nature of the primeval forest will never be known, but it must not be forgotten that the natural fauna contained animals such as wild cattle (*Bos primigenius*) and wild deer (*Cervus spp.*, which must have had some effect on forest regeneration. When a large tree died of old age or was brought down by strong winds, light would stream on to the forest floor and stimulate the growth of herbaceous plants. It may well be that grazing animals would congregate in such areas and even enlarge them by preventing tree regeneration. Pasture grasses and other light-loving plants may have persisted for some time in such places so that, at any one time, a scattering of 'natural glades' may have existed within the forest. Though palaeo-botanical evidence indicates that such 'glades' cannot have been of great extent, their possible existence must not be overlooked.

A further way in which immigrant peoples enriched the flora of the British Isles was by the introduction of trees. These islands have a remark-ably impoverished indigenous tree flora; many species which were com-mon in the forests of southern and central Europe failed to return here of their own accord in post-glacial times. Species like the sycamore (*Acer pseudo-platanus*) and the chestnut (*Castanea sativa*) only arrived when introduced by man, possibly in Roman times, while the coniferous larches (*Larix spp.*), spruces (*Picea spp.*) and firs (*Abies spp.*) have been introduced at an even later stage. Along with all these changes in vegetation, there have been changes in soil. Though less obvious, these must be regarded as even more profound in their effect and significance.

The erosional effects of forest clearance

It seems likely that there was a period of exceptionally rapid forest clearance in parts of Britain in the twelfth and thirteenth centuries and although it is only possible to guess at the ecological implications of this, one or two facts are worthy of note. In recent years, soil surveyors have examined a number of hillside soil types in Wales. Superficially these appear to be quite typical brown forest soils but careful chemical analysis of the whole of their mineral profiles has shown them to be abnormally rich in sesquioxides throughout [62]. The implication of this is that the present-day soil may be no more than the lower horizons of a pre-existing and quite heavily podzolised one whose upper horizons were removed with almost catastrophic swiftness immediately after the forest cover was removed. The point has been made that the partially bare forest floor would not be invaded immediately by turf-forming grasses so that, for a number of years, the sloping areas of forest soils would be particularly prone to rapid erosion. In support of this theory it has been pointed out

A. PIONEER COMMUNITIES OF A PSAMMOSERE. Marram grass occupying the seaward-facing dune but sea rocket invading the loose sand at its base which is still saturated with salt water at root level at the highest spring tides. Pendine, Carmarthenshire.
(Photograph by S. R. Eyre.)

B. LATER STAGE IN A PSAMMOSERE. Almost complete turf of sand fescue with a few surviving tussocks of marram grass on stabilised dunes near Margam, Glamorgan.
(Photograph by S. R. Eyre.)

PLATE XVII

A

B

PLATE XVIII

A. A HEDGE BANK. This old-established field boundary runs along a moderate slope at Scackleton on the Howardian Hills in the North Riding of Yorkshire. Throughout its length the height of the bank or 'lynchet' that has formed is about six feet.

(Photograph by S. R. Eyre.)

B. HOLME POST, HUNTINGDONSHIRE. The upper end of this iron post was flush with the peat surface in 1848. Since the land around the post reverted to carr at the beginning of the twentieth century, it seems likely that near-by areas, which have continued under cultivation, will have suffered even further surface lowering.

that it was at the self-same period that many ancient seaports on estuaries in the south and west of Britain were made useless for navigation because of rapid silting. This silt may very well have been partly the product of accelerated soil erosion around the headwaters of the rivers. If this seems rather too cataclysmic a theory, it must be recalled that an exactly similar sequence of events is known to have occurred in the eastern United States in the late eighteenth and early nineteenth centuries. The early tobacco plantations on the Piedmont Plateau of Virginia initiated catastrophic sheet-wash and gullying and, because of silt deposition, the coast of Chesapeake Bay moved seawards to a maximum distance of about seven miles. Similar silting took place in New Zealand during the nineteenth century.

Even if this rather spectacular theory is discounted, there can be no doubt that a great deal of accelerated soil erosion has taken place during the past ten centuries though much of the eroded material has not found its way to the sea or even into the rivers; it has merely been redistributed within relatively small areas. Much of the old-established arable land in many parts of Britain has been cultivated in relatively small patches from mediaeval times onwards. These patches have been separated from each other by balks, hedge-banks or other kinds of permanent boundary. Where these old-established boundaries run across the slope, it is quite common to find an abrupt declivity of several feet between the bottom of the higher field and the top of the lower one (Plate XVIIIa). There is only one possible explanation for this kind of feature; it has come into existence since the boundary was established because of accumulation above the barrier with concurrent loss of soil directly below it. The inescapable corollary is that changes in soil thickness over the fields involved, must have been very considerable, the situation today being very different from that at the time when the division of the land was first made. Farmers at the present day regard it as almost axiomatic that the soil at the top of a sloping field should be thinner than the soil at the bottom, but where a field is in the middle of a hill slope this is quite an illogical view; if the slope is an even one, it is likely that the soil thickness was everywhere similar when the field boundaries were first established. The change that has occurred is inevitably one for the worse since the resulting thin soils at the tops of the fields are very prone to drought in dry weather and have reduced crop yields in consequence. Indeed, they have often been so reduced in thickness that the rock is not far beneath the surface and thus causes difficulties when ploughing is in progress. On the other hand, large amounts of formerly good soil have been completely entombed at the bottom of the field and now lie sterile and useless beneath the material which has come down from above. The condition of many of these fields is being continuously worsened. It is much easier to plough and to sow root crops up-and-down a moderate slope than across it, and, at the

Nvs

bottom of any field that has been cultivated vertically, large amounts of silt and sand can be found after any heavy downpour of summer rain. There can thus be little doubt that removal of material from the top of the field is proceeding much faster than new soil is being produced by natural pedogenic processes.

Constructive soil modifications

With the removal of the climatic climax forests, the farmers of the British Isles must have begun to appreciate how favourable the climate was for the growth of grass. Full realisation of these potentialities only came after many centuries however. Throughout Anglo-Saxon and mediaeval times, land which was improved was kept mainly as arable land. Throughout spring, summer and autumn, the cattle, sheep and goats obtained much of their food by foraging on extensive, unimproved common pastures; there were only relatively small areas of meadow which were set aside to be cut for hay so that some fodder would be available to keep the animals alive through the winter [37]. Most of these meadows were areas along stream and river sides where natural flooding maintained a high base-status. These early farmers do not seem to have visualised the possibility of having improved meadows and pastures over the landscape as a whole. Even if they had thought of collecting the seeds of good meadow grasses such as ryegrass (*Lolium perenne*), cocksfoot (*Dactylis glomerata*) and timothy (*Phleum pratense*), it is probable that social structure and preconceived notions would not have permitted them to create and manage improved grasslands either as commons or as privately-held fields.

The mediaeval and Tudor grasslands were, therefore, mainly rough commons. Many of them were over-grazed and woefully neglected in many ways but the soils of some parts of some commons must have benefited inevitably. The heavy manuring by the animals and the elimination of woody plants must have caused the original forest soil to be transformed into one which was very similar to a prairie soil (Chapter IX). The change would be one in which soil, vegetation and animals mutually benefited one another. With manuring, the base-status would be raised and more demanding grasses would invade. The roots of these would ramify through the upper horizons thus bringing into existence the fine crumb-structure, typical of grassland soils. The grasses would also keep the nutrients, derived from the manure, within the nutrient cycle so that the richness of the turf would be further improved. This would mean that the carrying-capacity of that particular area of common would be increased and more manure would be available. The difficulty of preventing over-grazing on the one hand and avoiding under-grazing on the other must have been very great on this communal land but, by sheer chance, good grassland must have come into existence from time to time. Thoughtful farmers would observe this and must have begun to speculate on the advantages of purposefully creating grasslands in private fields, where grazing could be properly controlled.

The desire for improved grassland was, almost certainly, one of the main factors which brought about the so-called 'agricultural revolution' in the eighteenth and nineteenth centuries. This was merely the culmination of developments which had been taking place for several centuries. The result was that, by the beginning of the twentieth century, not only were grass crops being rotated with cereals and root crops on arable land, but very large areas had become permanent pasture and meadow. There was a good deal of interchange between arable and permanent grass but, early in the present century, there were many fields in most of the counties of Great Britain and Ireland which had been continuously under grass for a century or, indeed, very much longer. Many still remain as such but, during the past thirty years, much permanent grass has been ploughed and brought into a long-ley system.

Vegetation in the permanent meadows and pastures is in equilibrium not only with climate and soil conditions but also with annual mowing or perennial grazing. It is true plagioclimax vegetation. There are differences in species-composition from one type of soil-parent-material to another but, generally speaking, the same set of dominants are characteristic of permanent meadow almost wherever it occurs; permanent pastures everywhere, also have the same dominant species. Whereas crested dogstail (*Cynosurus cristatus*) along with perennial ryegrass and cocksfoot are the main dominants in pastures, it is the meadow foxtail (*Alopecurus pratensis*) and timothy (*Phleum pratense*) which tend to be associated with the two main dominants in meadows [74].

The great stability and tolerance of these grassland communities in the British climate is emphasised by the fact that the permanent pastures and meadows in which they are found originally came into existence in very diverse ways. Some permanent pastures, particularly on steep slopes, have almost certainly never been ploughed. One is bound to assume that the original woody vegetation was reduced by firing, cutting and grazing and that the pasture grasses then invaded spontaneously and have maintained themselves under grazing ever since. A second type of permanent pasture is widespread in lowland Britain, being particularly frequent on heavier land in Buckinghamshire, Northamptonshire and Leicestershire. This land is in ridge-and-furrow and, quite obviously, was once ploughed. Documentary evidence and the shape of these fields show quite clearly that, in many cases, they were once part of open fields [26]; the tenants were evicted from these at one fell swoop in Tudor or Stuart times, in order to make way for large sheep farms. Though there is little evidence about the details of farming operations carried out immediately after these evictions, it seems unlikely that the land was actually sown with grasses.

It is much more likely that the weeds of cultivation (Chapter II) were rapidly eliminated by competition from invading grasses which ultimately came into equilibrium with the grazing animals. Thirdly, much permanent grassland must have been sown in the nineteenth century. In many cases, doubtless, exotic species such as Italian ryegrass (*Lolium italicum*) were introduced, but these were soon ousted by the hardier and more aggressive native species. These three types of improved grassland, with such diverse histories, are today occupied by plant communities which are indistinguishable. Together they are classed as 'neutral grassland'. This title is somewhat misleading since most of the soils beneath them are acid in reaction. They are by no means so acid as they would be were it not for the constant manuring however; indeed, both in structure and chemistry, they have a good deal more in common with true prairie soils than with the forest soils from which they are derived [64].

Plagioclimax acidic heath

Many areas from which the forests were removed have never been improved. For centuries they lay as rough commons and, ultimately, many of them have been bought by local authorities as water catchment grounds or for their amenity value. This applies to many upland areas below about 1,500 feet as well as to many patches of sandy lowland. When upland oakwoods or oak-birchwoods on well-drained areas of grit and sandstone were swept away, the plants of the upland siliceous heaths usually invaded and became dominant. This happened particularly where the mean annual precipitation did not exceed forty inches. It will be remembered (Chapter XI) that plants like the ling (*Calluna vulgaris*), fine-leaved bell-heather (*Erica cinerea*), wavy-hair grass (*Deschampsia flexuosa*) and bent grass (*Agrostis spp.*), apart from composing climatic climax heath, would also be present as the undergrowth in the uppermost, more open forests. When the competition of the trees was removed, these acidophyllous plants were able to occupy the ground completely and even to expand downhill into areas where the forest had been denser. These heath plants were resistant to periodic firing and to almost continuous grazing and thus persisted, from then onwards, as plagioclimax dominants. The ling and the bell heather tended to become predominant on very poor soil-parent-material such as gritstone, while bent and other grasses tended to prevail on richer materials such as ferruginous sandstones, shales and slates [28]. This correlation is far from complete however; grazing intensity was also important. Extremely heavy grazing of ling can impair its growth sufficiently to allow heath grasses to take its place even on the poorest types of parent-material. On the other hand, ling can invade grass heath and displace it, where grazing is only light, even on much better soils. Different intensities of burning also have different effects on vegetation development even on the same parent-material. Very heavy burning of ling can penetrate into the uppermost, peaty horizons of the soil and thus destroy the roots; this often permits the heath grasses to invade.

Very similar plant communities occupied sandy areas in the lowlands of England when the forests were cleared. On Bagshot Heath and in the Breckland of Norfolk the ling is a common dominant, as are the typical grasses of the upland grass heaths. A further plant which is often associated with dominants of lowland heath is gorse (*Ulex spp.*). Until recently the heath grasses were absolutely dominant over wide areas of the lowland heaths, the sheep's fescue (*Festuca ovina*) along with bent grass and wavy-hair grass forming a fine, dense turf. It was only with the advent of myxamatosis and the near-elimination of the rabbit that one was made to realise the extent to which the fine, grassy turf had owed its existence to this animal. The rabbit had been present in its thousands in these sandy areas and, since it is a very close grazer, the invasion of shrubs and even of taller grasses had been prohibited; it was only the fescue that could maintain a closed turf under these conditions. When the rabbits almost disappeared, the rapidity with which this turf was invaded by the coarser heath plants was truly remarkable. It is obvious that, here again, differences in firing and grazing had brought into existence an intricate patchwork of heath communities.

The extent to which the soils beneath upland and lowland heath are different from the pre-existing forest soils is difficult to determine but it is thought that profound changes must have taken place, particularly beneath the ling community in the upland heaths.[1] The original trees rooted deeply and abstracted nutrient minerals from weathering parent-material. These were then returned to the soil in a loose mulch of dead leaves and, although the combination of rapid leaching and sterile parent-material probably induced the growth of some heath plants in the undergrowth, it seems likely that other, more demanding woodland plants would have been able to compete successfully in this environment. When the trees were removed however, the source of most of the mild humus disappeared and an acid mor, mainly derived from the remains of ling and acidophyllous grasses, took its place. Some brown forest soils may have been converted into podzols during the centuries following this vegetation change. Woodland plants would disappear and whatever earthworms may have been present would rapidly die out. Even if the original soil was more like a podzol than a brown forest soil, podzol characteristics must inevitably have become more pronounced. Such a change must have caused a more intense bleaching and an expansion of

[1] These are usually referred to as 'heather moors' in the north of England but ecological study has shown that most of them are so similar to lowland heaths that the term 'upland heath' is probably the most suitable name for them.

the A_2 horizon and, more significant, more rapid deposition of humic and ferric material in the B_1 horizon (Chapter IV). In some cases there can be little doubt that the B_1 horizon became an indurated, rock-like pan which the roots of more deeply-rooting plants would find difficult or impossible to penetrate. Under certain conditions this pan became almost impervious so that downward percolation was restricted and a gley horizon began to develop on top of it, in spite of the original perviousness of the parent-material. Such soils are known as 'gley podzols' [63] and the areas they occupy have been called 'wet heaths' [74]. The plant communities which subsequently developed on the 'wet heath' are very similar to those moorland communities which became widespread on more argillaceous parent-material (*vide infra*). The term 'wet heath' is, in fact, rather an undesirable one since the increased wetness caused the A_0 horizon to develop into a layer of peat which, even in lowland areas, may be a foot or more in depth. Beneath true, freely-drained heath, both in uplands and lowlands, the peat rarely develops to a foot in depth; beneath grass heath it is usually no more than a two- or three-inch turf.

Subsere invasion of plagioclimax heath

By persistent grazing and burning from mediaeval times onwards the re-invasion of forest on to the plagioclimax heaths was prevented. It was not until the late nineteenth centuries that economic changes caused a widespread relaxation in the intensity of these deflecting factors. By the middle of the nineteenth century many of the former commons, upland and lowland, were technically 'enclosed'[1] though very large parts of them remained wild and unimproved [27]. They fell into the hands of large landowners who used them predominantly as grouse moors or, in the case of the Scottish Highlands, as 'deer forests'. Since burning and grazing still took place, however, this would probably not have caused any important vegetation changes had it not been for subsequent developments. The growth of large industrial towns in the nineteenth century created a need for large quantities of piped water so that a high percentage of upland heath and moor, particularly in England, was bought or taken on lease by local authorities for the purpose of water catchment. Then, in the late nineteenth and early twentieth centuries, the import of cheap frozen and chilled meat caused acute agricultural depression in upland areas where animal breeding and rearing had been the mainstay of the farm economy. Because of this and restrictions on the number of grazing animals imposed by some water boards, animal numbers on the heaths were greatly reduced and, in places, subsere development began. Old inhabitants in many upland areas can describe how they have seen the bracken spread across the hillsides like wildfire during the past sixty or seventy years. This plant,

[1] i.e. They had become private property and were no longer subject to common rights.

with its luxuriant growth and rapid, aggressive rhyzome development, has completely swamped extensive areas of former ling and grass heath. Both ling and grass are far over-topped by this plant; they are thus subjected to heavy shading and the copious amounts of dead bracken fronds produced each year rapidly smother them.

Bracken usually dominates the first stage of subsere development on heath. Within a very short time the ground beneath it becomes almost bare of other herbaceous plants; only plants like the bluebell (*Endymion non-scripta*), which almost complete their annual cycle of growth before the bracken has attained its full height, are able to survive, and even they are much restricted by a dense mass of bracken litter. Within a few years, any former turf disintegrates and is replaced by a thick A_0 horizon of relatively mild, dark-brown humus derived from the bracken remains. When once established, this plant is most difficult to dislodge; few if any wild competitors seem to be able to survive the seedling and sapling stages if their seeds germinate in a patch of dense, healthy bracken. Given sufficient time, however, there is evidence to suggest that conditions might arise which would cause a weakening of bracken growth [80] so that successful invasion by trees such as birch (*Betula spp.*) and rowan (*Sorbus aucuparia*) could occur and the bracken patch thus be over-topped. Furthermore, contiguous stands of birch are able to shade out the bracken peripherally so that, very gradually, successive generations of birch can encroach and ultimately invade the bracken patch completely. It is clear, therefore, that the bracken stage of the subsere is normally a protracted one but that, in time, trees like the birch will succeed and, if parent trees are near at hand, even oaks can become established.

In spite of the widespread reduction of grazing, a great deal of plagio-climax upland and lowland heath still remains as such. This is mainly because fires are still frequent on most heaths. Even where gamekeepers are not present to burn off the old ling under controlled conditions, the weekend picnicker and his carelessly discarded cigarette-end is today such a common phenomenon that, by sheer chance, only a few areas escape fire in the course of a few years. During the past few decades, very large areas of upland heath have been so devastated by fire that a very pro-tracted sere would now be necessary before trees could re-establish themselves. During the very dry summer and autumn of 1959, accidental fires devastated thousands of acres of heath (and some moor) in upland Britain. Many of these fires burned so fiercely that, not only was the vegetation completely destroyed, but the whole of the A_0 horizon was completely consumed, leaving nothing but the inorganic fraction of the soil overlying the rock. Subsequently, strong winds drifted most of this unconsolidated dust and sand into hollows or against stone walls so that nothing but a bare, rocky surface remained (Plate XXII*a*). Five years after-wards hardly any plants had established themselves on some of these areas.

Cultivated fields, supporting good grass and even crops, are to be found at just the same height and on just the same kind of parent-material as some of these devastated areas; both carried upland oak forest originally. Because of sheer historical accident, the former have been able to maintain a relatively rich (though transformed) nutrient cycle while the latter have been impoverished, first as commons and latterly as catchment areas and public playgrounds. Those upland heaths which still carry some soil would doubtless be invaded quite quickly by trees if firing could be stopped; those which have lost their soil would now require the full course of a xerosere before true climatic climax forest, with its associated soil, could become established.

Plagioclimax moor

Heath communities did not occupy all those upland areas which remained wild after deforestation had taken place. On outcrops of shale and slate, where the surface was level or only slightly undulating, the more impeded drainage favoured the encroachment of bog and moor species even where the mean annual precipitation was below forty inches. Where precipitation was above forty inches, this encroachment took place on to steeper slopes and on to more pervious rocks.

Under natural conditions the junction between the forest communities on the one hand and climatic climax bog and heath on the other would often not be a sharp one; the trees would thin out and there would be a zone of ecotone in which scattered trees were interspersed with bog or heath species. Where forest gave way to bog, the commonest dominants in the latter would probably be cotton grass (*Eriophorum vaginatum*) or, in the case of north-west Scotland, deer's-hair grass (*Scirpus caespitosus*). When the uppermost outposts of forest were removed by man, these representatives of the bog communities merely extended a little and occupied the whole of the territory of the former ecotone and, in many places, they doubtless extended even further downhill on to areas of former closed forest. On many areas of steeper slope and lower precipitation, however, other plants were better adapted to establish dominance. Two of these, the mat grass (*Nardus stricta*) and the purple moor-grass (*Molinia caerulea*), became particularly important. Both these plants had probably been relatively unimportant before forest clearance, the former as a minor plant on steep, wet slopes or even in climatic climax heath, the latter along stream sides or in flushes in climatic climax bog. The mat grass extended its range enormously on to moderate slopes particularly between 1,200 and 1,700 feet where the mean annual precipitation was between forty and fifty-five inches. The purple moor-grass became dominant on rolling outcrops of shale and slate, particularly where frequent springs provided a copious and reliable supply of ground-water but where this water was permitted to seep laterally with freedom. Neither of these grasses can tolerate the

A. UPLAND HEATH. A community dominated by ling (*Calluna vulgaris*) with invading tussocks of mat grass (*Nardis stricta*) near Hathersage, Derbyshire.
(Photograph by S. R. Eyre.)

B. PLAGIOCLIMAX MOOR. Large tussocks of purple moor-grass (*Molinia caerulea*) on relatively thin peat with some sporadic invasion by birch (*Betula sp.*) and willow (*Salix capraea*). Leash Fen, Barlow, Derbyshire.
(Photograph by S. R. Eyre.)

PLATE XIX

A. SUBSERE BIRCHWOOD. Birch of different ages advancing on an area dominated by purple moor-grass (*Molinia caerulea*), ling (*Calluna vulgaris*) and bracken (Pteridium aquilium) which was subject to common grazing until 1820. Ramsley Moor, Holmesfield, Derbyshire.
(Photograph by S. R. Eyre.)

B. INVADING ASHWOOD. Young ash trees rooted in the grykes of a limestone pavement. A Mat of mixed calcifuge and calcicolous plants already covers part of the clint surfaces. Chapel-le-Dale, north-west Yorkshire.
(Photograph by S. R. Eyre.)

PLATE XX

A. COTTON-GRASS BOG. Cotton-grass (*Eriophorum vaginatum*) growing on deep *Sphagnum* peat on Ringinglow Bog, Dore, Sheffield. (Photograph by S. R. Eyre.)

B. DISSECTED BLANKET BOG. Only a small proportion of the original surface remains on this heavily dissected peat on Kinder Scout, north Derbyshire. (Photograph by J. Radley.)

PLATE XXI

A. BURNED MOOR. A desert of rock and sand dunes created by the burning of
vegetation and peat. This photograph was taken in the summer of 1964, almost
exactly five years after the original destruction by fire in 1959.
(Photograph by S. R. Eyre.)

B. ELFIN WOODLAND. Engelmann spruce (*Picea engelmanni*) at the tree limit in
New Mexico.
(Photograph from Henry A. Gleason and Arthur Cronquist, *The Natural Geography
of Plants*, 1964, Columbia University Press, New York.)

PLATE XXII

stagnant wetness of the climax bog and the purple moor-grass in particular requires ground-water of moderate base-status.

Though mat grass is tough, unpalatable and lacking in nutritional value, the purple moor-grass, for centuries, was regarded as one of the main assets of the moorland commons. The ancient practice of trans-humance was almost universal on the uplands of England and Wales and it is significant that the usual date when the livestock were moved up to the common hill pastures was around the middle of May. This is the time when the first bright-green flush of purple moor-grass is just making its appearance and there can be little doubt that the movement was timed so that the livestock, particularly the cattle, could make the best use of it. If this grass is allowed to grow freely, it forms large tussocks (Plate XIX*b*); furthermore, if it is ungrazed throughout the early stages of growth, it soon loses nutritional value. If grazed persistently, year after year, from the time of the first flush onwards, it forms a level sward which is of nutritional value throughout the whole growing season.

The surface of these plagioclimax moors is often so very wet, even in summer, that it is difficult to imagine how the former oak forest regenerated upon it. It is necessary to realise that the pattern of water movement was completely changed by the very process of forest removal. A large oak tree loses several gallons of water by transpiration from its foliage on an average day in the growing season. All this water has to be replaced by water taken in by the tree's rooting system. These roots ramify through a considerable range of depth and the deepest of them normally penetrate into the joints in the rock beneath the sub-soil. The rooting system of an oak tree is, in consequence, one of the most efficient kinds of sub-surface drainage system that can possibly be imagined. Although it does not function in winter when the tree is leafless, it causes the sub-soil to be well-aerated throughout the summer and leaves a large, vacant reservoir in the soil and rock interstices which has to be filled by autumn and winter precipitation before the soil itself can become thoroughly waterlogged.

A system like this existed beneath the original upland oakwoods; it was completely disrupted when the trees were removed. From then on-wards, water was released to the atmosphere by plants with only shal-low roots. The only way in which water could be lost from the sub-soil was by lateral seepage—a very slow process indeed in fine-grained material. The result was that the sub-soil became permanently water-logged even though, at times during the summer, the upper soil horizons were dried out. Without the reserve water-holding capacity which was formerly present in the sub-soil, a single rain storm, even in summer, would be sufficient to waterlog the whole system. This is the situation today beneath plagioclimax moor.

Apart from the increased wetness, a complete change in humus type also occurred when the products of leaf-fall were replaced by the humified remains of acidophyllous grasses. With the increased wetness and acidity, peat developed on the surface at the same time as a gley horizon developed in the mineral soil beneath (Fig. 22). It is rare to find undisturbed peat of more than a foot in depth beneath purple moor-grass and mat grass so that these plants, like plants of the upland heaths, usually have their deepest roots in the mineral soil horizons. Under most conditions it seems likely that cotton grass automatically displaces these grasses as the main dominant if the peat develops beyond a depth where this is possible.

It requires an effort of the imagination to visualise the former forests on these extensive plagioclimax moors. Large areas are almost treeless today. Where grazing activity has been relaxed, however, particularly where near-by trees can supply seeds, birch, rowan, goat willow (*Salix capraea*) and alder (*Alnus glutinosa*) are all capable of invading the moor surface. Alder, willow and rowan normally only invade where the peat is thin and the ground water contains appreciable quantities of mineral nutrients but the birch appears to be able to regenerate spontaneously on quite deep peat. It appears possible, however, that man's interference, in places, may have initiated changes which are irreversible. It is possible that, unless they are drained artificially, some of the deeper peats beneath cotton grass moor are sufficiently wet and stagnant to repel even birch. In many marginal cases, however, it has not yet been ascertained which are areas of true, climatic climax blanket bog and which are areas of former forest or ecotone. On much of the Lewisian gneiss in northern Scotland, for instance, it is not known how much of the blanket bog is climatic climax and how much was initiated by forest clearance by Scandinavians at the end of the Dark Ages. On the other hand, in north Derbyshire, pollen analysis has demonstrated quite clearly that some thin peat has only been forming since the 12th century and must therefore have been initiated by mediaeval forest clearance [28]. Furthermore, large areas of mat grass and purple moor-grass would clearly revert to forest given a sufficient length of time without human interference. In many places, young birch trees can be found growing in peat almost a foot in depth and the changes taking place beneath their shade are already apparent (Plate XX*a*). The moor grasses have been shaded out, a mulch of milder humus is accumulating on top of the peat and, as the tree roots ramify, there can be little doubt that the water-table will fall, the peat will disintegrate quickly and a kind of brown forest soil, though probably with gleying beneath a certain depth, will again come into being. This subsere development is inevitably slower than that on heath, mainly because bracken cannot tolerate a waterlogged soil.

Upland complexes

The plagioclimax heaths and moors of the British uplands are not quite so separate and homogeneous as the foregoing sections might seem to imply. Particularly on uplands where outcrops of arenaceous and argillaceous rocks alternate, moor and heath communities are found in an intricate patchwork (Fig. 23). This is made even more difficult to analyse by the fact that transition communities occur wherever the two types are in contact. Furthermore the dominants of the heath communities frequently occur as subordinate plants on moors; they may even dominate on moorland peat if artificial drainage operations lower the average height of the water-table. On the other hand mat grass and purple moor-grass are frequently associated with ling and heath grasses (Plate XIX*a*), particularly where hard pan has reached a certain stage of development in an originally pervious material. Though the concepts of 'heath' and 'moor' should be kept quite distinct, the one being leached and freely-drained and the other wet and impeded, it must be realised that all gradations between the two exist and that many of the plants occurring in each have wide tolerances of habitat conditions.

Plagioclimax communities on limestone

Because of their naturally free drainage, the limestone hills of Britain were among the first areas to be cleared extensively for cultivation and grazing. The forests of the Peak District and the Craven Highlands were probably much depleted in Neolithic times and the presence of widespread human activity on the Chalk in the same period leaves one in some doubt about the original vegetation on the North and South Downs and the Chilterns. Although the beech must be regarded as the theoretical climatic climax dominant at the present day, it is almost certain that this tree was not present in any quantity in the original forests of these chalklands. As a common forest tree the beech, like the ash, arrived relatively late in Britain (Chapter XI) and was almost certainly preceded by Neolithic Man. As in the case of the harder limestones therefore, elm may well have been the dominant tree that was cleared from the Chalk. Since the Sub-Boreal climate was considerably drier than that at the present day, it is also possible that a low-growing scrub of yew and thorny shrubs may have covered the higher and more exposed areas.

Centuries of persistent grazing on the Chalk have induced the development of a widespread and characteristic calcareous grassland. Where this grazing has been particularly intense, one species, the sheep's fescue (*Festuca ovina*), has excluded most other plant species and has formed a dense, fine turf. As in the case of the lowland grass-heaths, this turf was maintained in a particularly fine condition by a heavy rabbit population; now this has almost disappeared, other taller-growing grasses have become more important in many places. The brome grass (*Bromus erectus*) and perennial oat grass (*Avena pratensis*) are particularly common. These taller grasses, along with numerous lime-tolerant herbs, replace the fescue completely wherever the Chalk grassland is used as meadow.

The activities of man have produced a greater variety of results on the harder limestones. On steep slopes, fescue grasslands similar to those on the Chalk have come into existence but here, more acidophyllous grasses such as bent (*Agrostis spp.*), are usually associated. On flatter plateau surfaces, where there was a greater depth of soil when the original vegetation was reduced, even more acid conditions have been achieved. The greater precipitation on these more northerly uplands ensures more rapid leaching so that the uppermost horizons become base-desaturated and, in extreme cases, really acidophyllous plants like ling and mat grass are able to invade. Indeed, on Carboniferous Limestone in western Ireland and on the Durness Limestone of Cambrian age in north-west Scotland, even acid blanket bog has developed. In the leached inorganic material beneath these plagioclimax heath and moor communities, quite pronounced podzols have developed.

Probably one of the most interesting problems concerning vegetation regeneration on limestone is to be found on limestone pavements such as those in the Malham and Ingleborough districts of north-west Yorkshire. A number of these—as at Malham Cove—appear at first glance to be quite bereft of soil and rooted vegetation; one has to search deep within the grykes in order to find pockets of soil and growing plants. A careful examination of this almost 'troglodyte' vegetation however, reveals that it consists of the saplings of trees such as the ash and sycamore along with a variety of herbaceous plants which are normal components of the undergrowth of upland ash woods. Other pavements have quite a different appearance; on the lower slopes of Ingleborough near Chapel-le-dale, for instance, many ash trees and some sycamores are found growing up out of the grykes and there is a considerable depth of soil, almost completely composed of peaty material, covering a large percentage of many of the clints (Plate XX*b*). It is noteworthy, that almost all the trees growing here are no more than 20 to 30 years old so that, only a few decades ago, this pavement must have been almost as treeless as the one at Malham Cove. Furthermore, the turf and soil cover on the clints appears to be advancing rather than the reverse. Since at the present time the pavement at Chapel-le-dale is fenced so as to prevent sheep from straying on to it, while the one at Malham Cove is open to grazing animals, there is much evidence to support the view that, in the past, grazing has been responsible for preventing tree regeneration. The extent to which a soil of some kind will be able to develop beneath the trees is an obvious subject for speculation but there seems to be little doubt that, if the trees increase in size and numbers, their massive rooting systems will invest the grykes more intensively and will further restrict the loss of material by downwash. Whatever the exact nature of the original vegetation and soil here, there seems to be every reason for postulating a much more substantial cover than that which exists today.

Modifications of acid bog

Although the upland and lowland areas of acidic peat were amongst the least useful from the point of view of agriculture, even they have been much altered. The raised bogs and blanket bog of Ireland have been used extensively for fuel, and deep, acid peat has been utilised similarly in upland areas throughout Great Britain. Locally quite large areas of the original surface have been removed but, even where this has not occurred, the deep peat cuttings have lowered the water-table sufficiently to affect the vegetation over many surrounding acres of untouched bog. The widespread effect of these peat cuttings was doubtless noted by farmers and, since the induced vegetation was more valuable for grazing than the original bog plants, they began to drive deep ditches across the bogs as an organised agricultural policy. This took place increasingly from the beginning of the nineteenth century onwards.

Much of the deep peat is today covered by a plant community dominated by cotton grass (Plate XXIa) though this is often associated with a good deal of ling. This is true of many of those places on the Pennines which are underlain by peat exceeding fifteen feet in thickness and which are locally referred to as 'mosses'. In spite of this name it is often very difficult to find any *moss* in the present vegetation. If one digs to a depth of only a few inches into many of these bogs, however, one discovers peat which is composed almost entirely of the remains of bog moss and it is clear that up to about a century ago this plant was dominant [16]. Several theories have been suggested to account for this profound change. One possible reason for the recession of bog moss is that, since the mosses in general are very sensitive to atmospheric pollution, the onset of the industrial revolution along the Pennine flanks might have reduced them very rapidly. However, since bog moss can still be found growing luxuriantly in very wet localities very close to sources of most intense pollution, this cannot be accepted as an explanation. It seems much more likely that intensive drainage and grazing, associated with the agricultural revolution, were the main factors which initiated the change.

Even more striking changes have taken place in large areas of blanket bog on the high plateau surfaces, particularly around 2,000 feet on the Pennines. Here bog moss peat is being dissected and removed at a very rapid rate. It appears that other plants were unable to establish themselves here as the bog moss receded; gullies invaded many acres so that today, very little of the original surface remains; the whole of the peat layer is at an advanced stage of dissection. Hill-walkers from industrial areas in the

north of England will be familiar with the peculiar, desertic landscapes
on the flat tops of Kinder Scout, Bleaklow and other Millstone Grit hills
(Plate XXI*b*). It seems likely that a certain amount of peat dissection
must have occurred as a purely natural phenomenon around the edges of
these upland peat bogs, particularly where flat plateau surfaces were
bounded by steep escarpments or valley heads. Rapidly deepening peat
must often have been undermined by ground-water in unstable situations
such as these. The present widespread dissection will, very shortly, almost
completely remove what were once continuous areas of peat cover
however. This peat had been developing, without interruption, for a
period of about 7,000 years; it is now being removed over a period of only
a century or two. Since the last century and a half has been a period of
unprecedented draining, grazing and burning, it would be a remarkable
coincidence if this were not the basic reason for the wholesale dissection.
On the other hand, a certain amount of blanket peat dissection had
occurred at lower levels before the industrial era. Quite a number of
gently-sloping moorland surfaces at between 1,000 and 1,200 feet on the
Pennines are today covered with plagioclimax communities of purple
moor-grass and mat grass which were formerly occupied by climatic
climax blanket bog. The present soil beneath the moor grasses consists of
a mere inch or two of peat underlain by shallow, predominantly mineral
horizons but here and there, usually no more than a few yards in diameter,
there are peat 'hags' with a depth of two or three feet. Quite clearly these
are relics of a once continuous cover. Since these areas are nearer to
centres of settlement than are the highest plateaux, it is reasonable to
suppose that peat-cutting, ditching, grazing and sporadic burning
caused peat removal at an earlier date in the post-mediaeval period. Very
destructive exploitation must have taken place on some of the more
accessible upland commons since the commoners had the right to remove
wood, peat and stone as well as the right to graze their animals.

The drainage of fen and carr

From Roman times onwards, sporadic and small-scale attempts were
made to lower the water-table in the fen and carr peats of the lowlands.
The deep, basic peats had a high potential fertility but, in their original
state, were quite useless for cultivation. Here and there the edge of fenlands
had been embanked, drained and cultivated by the end of mediaeval times
but the bulk of the Fens, the east Norfolk fenlands, the Somerset Levels
and other extensive areas of lowland fen peat, remained almost in their
original state until the seventeenth century. An accumulation of capital
and a degree of engineering knowledge were necessary before such
extensive areas, so near to mean sea-level, could be improved. It was not
until 1631 that the Duke of Bedford with some associates, advised by the
Dutch engineer Vermuyden, put a scheme in motion by which it was

intended to drain the fenlands of most of north Cambridgeshire and north-west Norfolk [33]. The 'Old Bedford River' was dug across the area for a distance of twenty-one miles; this shortened the course of the River Ouse and was intended to permit the more rapid drainage of flood waters. The original installations were not fully effective but subsequent alterations were completely successful and a large range of crops were being grown on the peat before the beginning of the eighteenth century. A series of other drains were cut soon afterwards with similar success [76] although it was not until the next century that the whole of the Lincolnshire Fens was brought under cultivation.

Unfortunately the problems of these areas were not finally solved by drainage. Apart from thin beds of marine silt, the deep deposits here were almost entirely organic; this organic material persisted almost unchanged under the waterlogged conditions in which it had accumulated, but, as soon as the water-table was lowered, decay bacteria were activated in the aerated upper layer. Very little insoluble material remains when peat decays so that, inch by inch, the soil in the drained, cultivated fields has fallen below its original level. It is said that in the sixteenth century the level of the peat was, on the average, about six feet above the silty marshlands around the Wash; generally speaking it is now about ten feet below that level. Indeed it is quite clear that in places the height of the surface of the peat has fallen about twenty feet. At one point near Peterborough a long, iron post was driven into the peat so that its upper end was flush with the surface; after little more than 100 years this pole now projects to a height of fourteen feet (Plate XVIII*b*). The cultivated Fens are thus well below sea-level at every high tide and are constantly getting lower as more and more peat disintegrates. The problem is exacerbated by the fact that peat along the rivers and drains remains waterlogged and unaffected, so that all the main water-courses now flow well above the general level of the cultivated land; this increases the danger in times of flood and also necessitates all the water from the field drains being pumped up to a higher level. Diesel engines now work the pumps day and night; a complicated system of locks and sluices keep back the sea at high tide and permit outward drainage at low tide.

In fen peat we thus have an example of a natural soil which cannot possibly survive improvement and cultivation. Its innate fertility has led farming communities and larger organisations to devote vast resources to its initial drainage but, when once this has been achieved, its complete destruction is ultimately assured.

Misuse of sandy lands

Very light soils composed predominantly of sand grains are of frequent occurrence in the British Isles. Some of them, such as the soils on the

Bunter Sandstone in and around Sherwood Forest in Nottinghamshire [42], have developed from the solid rock beneath them; others, like those in the southern part of the Vale of York, have developed in glacial outwash deposits [56]; others, of frequent occurrence near the coasts, have developed on dunes during the course of psammoseres. All these soils require very careful treatment when the wild vegetation is cleared since strong winds over a bare, dry soil can soon remove sufficient material to fill ditches, block roads and cause all kinds of local disruption. At times, migrating dunes have been initiated even in inland areas like the Breckland of East Anglia; whole villages have been known to be engulfed by sand [79]. In the humid climate of Britain, however, such occurrences are rare.

The soils developed on consolidated dunes near the coast are, fortunately, so sterile and prone to drought that they discourage cultivation. Nevertheless it appears that human interference in places may have caused a fairly stable situation to get out of hand from time to time during the past. A notable example is to be found at Newborough in the south of Anglesey where a large proportion of the arable land was engulfed by dunes in late mediaeval times. It seems clear from contemporary documents for the neighbouring township of Aberffraw, that some of the inhabitants of this part of Anglesey supplemented their incomes by weaving basketwork from marram grass which they obtained from the near-by dunes [41]. It is reasonable to suppose that, the more impoverished they became because of loss of their land, the more would they concentrate on this supplementary work. They were probably quite unconscious of the fact that, by harvesting the marram grass, they were weakening the growth of the only organism which could provide any long-term protection for their threatened lands. This is but another example of the ways in which man, though a reasoning animal, because of his lack of understanding of ecological phenomena, has over-exploited natural assets upon which he is dependent for survival. Though one can find much more blatant and terrible examples of this over-exploitation in other parts of the world, these islands furnish a sufficient variety to provide food for thought. It is fortunate that the destructive changes wrought by man on the soil and vegetation of this country are, to a certain extent, balanced by beneficial and constructive ones.

B

A

PLATE XXIII

A. TROPICAL RAIN FOREST. Virgin forest at about 2,500 feet above sea-level in the Arfak Mountains, New Guinea. (From Richards, 1952.)

B. A CAULIFLOROUS TREE. Fruit pods on *Parmentiera cereifera* in Ceylon (From Schimper, 1903.)

A. PLANK BUTTRESSES. The buttressed bole of a specimen of *Mora excelsa* at Moraballi Creek, Guyana.
(From Richards, 1952.)

B. STILT ROOTS. Mangroves (*Rhizophora mucronata*) in coastal Tanzania.
(From Shantz and Marbut, 1923.)

PLATE XXIV

PART FOUR

TROPICAL REGIONS

CHAPTER XIV

Introduction to Tropical Vegetation and Soils

In the English language the expression 'in the tropics', for some time, has been used to refer to the area which is, quite literally, in between the Tropics of Cancer and Capricorn. The adjective 'tropical' has been applied to phenomena which are characteristic of the whole or parts of this inter-tropical belt. Botanists and foresters thus refer to 'tropical rain forest' and agriculturalists and pedologists to 'tropical soils'. Confusion about the word 'tropical' has arisen mainly because some systematists, particularly climatologists, have used it to refer to areas *in the vicinity of the geographical tropics* as distinct from areas around the Equator which they have referred to as 'equatorial'. If this latter usage had achieved any degree of precision, there would be a good reason for retaining it, but since this is not the case, the wider and more general sense will be adhered to here. It must be emphasised, however, that complete avoidance of ambiguity is impossible since biological phenomena do not conform exactly to terrestrial geometry; numerous plant communities and soil types overlap the Tropics of Cancer and Capricorn so that the distributions of so-called 'tropical' phenomena are, to a certain extent, 'sub-tropical'.

Although treatment of tropical vegetation and soils has here been relegated to last place, this is rather misleading from the standpoint of plant evolution. Many of the genera and species which compose the plant communities of extra-tropical latitudes belong to families which are predominantly tropical. The general world trend appears to have been the adaptation of tropical plants to cooler regions rather than the reverse. This might well be expected in view of the fact that, throughout most of Tertiary times, the whole of middle latitudes was subjected to much higher temperatures than those of the present day and, in consequence, supported plant communities very similar to those now occupying tropical regions (Chapter VII). Even the conifers, which today are only poorly represented in the tropics, probably originated in tropical climatic conditions during Upper Palaeozoic times [58]; they have adapted themselves to the marked seasonal rhythm of higher latitudes but have retreated in face of competition from angiosperms in tropical lowlands.

The variety of plant communities is just as great within the tropics as it is in higher latitudes. Although plants resistant to low air temperatures are found only on high mountains, every degree of moisture availability is found, from the very wettest type of area on the one hand to absolute deserts, where rain has never been observed, on the other. A series of plant formation-types adapted to these different degrees of wetness is to be found on all the main continental areas which extend into the tropics. Rain forests, seasonal forests and scrublands are all represented in the lowlands and contrasted vegetation zones, adapted to different temperature conditions, are to be found on tropical mountains. So far as the lowlands are concerned, however, there is a basic environmental difference between tropical and extra-tropical regions: low temperatures and the occurrence of frost have little relevance to plant growth in the tropics; wherever periodicity or seasonality is encountered, it is due to periodic water deficiency and not to temperature fluctuation. There are other general climatic differences which are important. Firstly, even in those areas with a heavy mean annual rainfall, the intensity of insolation and the number of hours of sunshine are both great. This is because the rainfall is concentrated into relatively short, heavy storms; with few exceptions, it is only on tropical mountains that cloud and rain can persist, unbroken, for days at a time. Secondly, apart from areas where tropical cyclones are of frequent occurrence, high winds are shortlived, and even where tropical cyclones do occur, the *average* wind speed is usually by no means as strong as it is in most parts of middle and high latitudes. Vegetation does not have to withstand the high transpiration rates due predominantly to wind speed as it does, for instance, along western coastal fringes between 40° and 60° in both the northern and southern hemispheres. If tropical vegetation did have to withstand such winds along with the persistently higher temperatures, there can be little doubt that the thousands of miles of tropical shore, now fringed with coconut palm (*Cocos nucifera*) and mangrove swamp (*Nipa, Pandanus, Rhizophora, inter alia*) would have a rather different appearance.

As compared to the soils of higher latitudes, the soils of tropical regions are little understood. Apart from the few broad generalisations which have been made with great frequency, little can be said about the characteristic soil-forming processes within the tropics. This lack of information sometimes gives rise to the impression that the pattern of soil distribution is much more uniform here than it is in cooler regions; in all probability this is not the case; there is good reason to suspect that the pattern of soil distribution may be just as varied in tropical regions as elsewhere. The term 'laterite' readily springs to mind when the subject of tropical soils is raised; it must be emphasised at the outset, however, that tropical soils are very far from being all 'laterites'; indeed, many are not even 'lateritic soils'.

The inter-dependence of vegetation and soil in middle and high latitudes has already been stressed. In the tropics, where rainfall tends to be more torrential and the sun's heat even more intense, this relationship is even more intimate. As will be seen in the following chapters, the price to be paid for careless disturbance of natural equilibrium may here be very much higher.

CHAPTER XV

Tropical Rain Forest

THROUGHOUT those parts of tropical regions where a heavy and reliable rainfall is experienced throughout most of the year, a plant formation-type of remarkably constant structure is found under undisturbed conditions. It has often been referred to as 'equatorial forest' but since almost identical communities extend into regions which are far away from the equator, this term is obviously undesirable. The term 'tropical rain forest' is much more appropriate.

The extent of tropical rain forest

The tropical rain forest formation-type comprises three distinct formations, usually referred to as the American, African and Indo-Malaysian tropical rain forests. The American formation has its most extensive development in the Amazon Basin which it occupies almost completely. It also extends along river valleys southwards on to the Mato Grosso, south-westwards in a narrow belt along the sub-Andean region of central Bolivia and northern Argentina, north-westwards along the foothills of the Andes in north-central Colombia and south-west Venezuela and, in a wider belt, north-eastwards to occupy almost the whole of the Guianas and eastern Venezuela. A southern outlier occupied most of the coastal lowlands of south-eastern Brazil from Bahia southwards to Santa Catarina Island; a quite separate northern outlier extended from the lowlands of northern Ecuador through coastal Colombia, eastern Panama, eastern Nicaragua, eastern Honduras and northern Guatemala to occupy most of southern Yucatan and other eastern parts of Mexico as far north as Tampico. Most of the eastern parts of Cuba, Hispaniola and Jamaica and the northern parts of Puerto Rico were also covered with tropical rain forest (Maps 5 and 6). Apart from the Amazonian forests, much of this American formation has now been either removed or modified, but most of these changes have occurred sufficiently recently for its former extent to be outlined with a high degree of confidence.

Large areas of the African formation are still to be found in the Congo Basin and Cameron but there can be little doubt that its former extent was much greater. Even during the past fifty years, vast areas of forest have been cleared in Nigeria, Ghana and other West African countries as far west as Sierra Leone; until relatively recently a broad, if not quite continuous, belt of forest extended across southern West Africa. The tropical rain forest of eastern Madagascar, Mauritius and the main river valleys of south-eastern Tanzania has also been almost completely destroyed. Nevertheless, there can be little doubt that, in historical times the African formation has never been so extensive as the American one. Apart from the areas already mentioned, it is doubtful if true tropical rain forest ever extended further east than south-western Uganda, Burundi and north-eastern Katanga; it also seems certain that the northern and southern parts of the Congo Basin itself were occupied, under natural conditions, by other types of vegetation. Human interference has been so extensive and prolonged in Africa, however, that it is now extremely difficult to infer what the distribution of the different climatic climax formations would be if the interference were to cease (Chapter XVIII).

The Indo-Malaysian formation is the most widely-distributed formation of the tropical rain forest. It extends outside the Tropic of Cancer in Assam and outside the Tropic of Capricorn in south-eastern Queensland; it occurs as far west as the Western Ghats in India and as far east as the Fiji Islands in the southern Pacific. It covered most of Malaya, Sumatra, Borneo and New Guinea and can still be seen here at its most typical; but there are also extensive areas in East Pakistan, Burma, Viet Nam and the Philippines. Even in the smaller outliers in south-west Ceylon, Hainan, Formosa, Java and eastern Queensland, the Indo-Malaysian formation possessed the typical structure and composition of true tropical rain forest. Although vast areas have now been cleared or modified, recognisable tropical rain forest can still be found over most of its former range; it is only in Ceylon, Java and some of the Pacific islands that almost complete obliteration has occurred.

Plant forms

Although the three formations of tropical rain forest each have their own distinct assemblages of plant families and genera, they are remarkably similar in structure and general appearance. The same types of structure are found in hundreds of species in all three formations and in a great range of families which are only distantly related to each other. The struggle for survival in these conditions of high temperature and frequent precipitation is so fierce that it appears as though each species, in order to survive, has had to evolve in one of a limited number of directions. A great number of life-forms which have been permitted to survive in other formation-types have been eliminated here. This is illustrated particularly well by the fact that, of all the many thousands of species of woody plant composing these forests, nearly all are evergreen in habit. The vast majority cast their old leaves and grow new ones continuously and simultaneously and, in consequence, are rarely if ever leafless. A minority sometimes shed all their leaves for a short period but do so at odd or irregular intervals which bear no relationship to any annual climatic régime. It appears that a tree of regular deciduous habit would have no chance of survival in this environment where temperature and precipitation permit continuous assimilation and growth throughout the year; such a tree could not compete with the constantly-growing evergreen species.

Many other plant forms recur with such remarkable frequency as to indicate that they, too, bestow some competitive advantage in the struggle for survival. Very often, however, it has been found impossible to diagnose the exact nature of this advantage. A case in point is the remarkable monotony in size, shape and texture of the leaves of the large trees forming the forest canopy. It has been remarked that the layman ' . . . might easily be excused from supposing that the forest was predominantly composed of species of laurel' [59]. The majority of the trees, though very diverse *taxonomically*, have leathery, dark-green leaves remarkably similar in shape and size to the ones which are typical in the laurel family (*Lauraceae*). Large-leaved trees such as the palms (*Palmae*), though of common occurrence, are by no means so numerous in tropical rain forest as one is sometimes led to expect; palms in particular are more typically components in tropical prisere communities—in swamps, on mud flats and on sand-dunes. The reason for the heavy cutinisation or 'leatheriness' of the leaves in the forest canopy is quite apparent; the intense insolation and high temperatures at midday would cause less well-protected leaves to collapse and become useless for photosynthesis. The reason for the apparent optimum size and shape are far from obvious however; it is a remarkable phenomenon in the light of the great variations in size and shape in the dozen or so forest trees encountered in the British Isles.

A similar problem exists with regard to the shape of the leaves of many species which compose the lower storeys of the tropical rain forest. These leaves are remarkable for the exaggerated development into what has usually been called a 'drip-tip' (Fig. 24). Although the view has been challenged, a number of authorities have maintained that this elongated tip permits the leaf to shed water easily and thus to dry more quickly after rain. This may be important for two reasons. In the first place epiphytic mosses and other small plants are more likely to infest a moist leaf than a dry one and thus to impair its ability to photosynthesise; in the second place, a wet leaf will not transpire so quickly as a dry one. In the low light intensities that obtain inside this type of forest, photosynthetic efficiency is obviously important; furthermore, because of the prevailing low evaporation rates in this almost windless environment, efficient transpiration is equally important; unless water is lost from the leaves, there will be no mechanism by which water can be brought up from the roots, carrying with it the essential, dissolved mineral nutrients. Which of these two functions is the more important remains to be discovered.

There are numerous other forms which are characteristic of tropical rain forests but which are rare or absent in other types of forest. Of

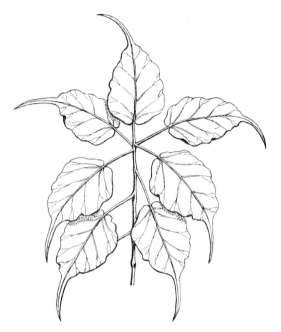

Fig. 24. Leaves with drip-tips; a shoot of *Ficus religiosa*. (Drawn from a specimen kindly supplied by Dr. M. E. Marston, Department of Horticulture, Sutton Bonington.)

particular interest are the features which are often developed by the boles and major branches of the trees. Many of the largest species develop large 'plank buttresses' from the lowest few feet of the trunk (Plate XXIV*a*). These great flanges of woody material join the upper sides of main lateral roots to the trunk. It was originally surmised that, since buttresses may be required by a building, so these large trees had evolved natural 'buttresses' as an added means of support; competition for light had involved greater and greater height and thus greater susceptibility to 'wind-rock'. Recent observations have provided good reasons for doubting this relatively simple explanation however; in particular it has been noticed that trees with buttresses appear to be just as prone to uprooting and damage in high winds as trees of similar height without buttresses [59]. The real reasons for the evolution of this type of feature are therefore, still obscure. Similarly the 'stilt roots' which grow from the lower parts of the trunks of many smaller trees have not been fully accounted for (Plate XXIV*b*). Many of the species growing in swampy areas possess these features and, quite obviously, obtain extra stability thereby; many others, however, grow on quite stable ground where trees without this characteristic appear to compete on equal terms. Finally, the trunks of many tropical rain forest trees present a most striking appearance (Plate XXIII*b*) in that they bear flowers and fruit. It will be recollected that the pods of cocoa beans (*Theobroma cacao*) are harvested from the trunk and lower branches. This habit of 'cauliflory' is possessed particularly by many of the smaller trees, but any competitive advantage which it may provide, remains unidentified.

It is possible that cauliflory, along with the propensity to produce other types of outgrowth from the trunk, need not of necessity be adaptations born of the struggle for survival. It is probably significant that the bark of most of these trees is remarkably thin as compared to that of trees in higher latitudes; constant high humidity beneath the canopy of tropical rain forest makes it unnecessary to protect the tree by a thick, corky bark. A bark thickness of only one or two millimetres is therefore quite common, indeed it may be so thin as to be almost transparent. It is thus possible that external stimuli of a variety of kinds may stimulate direct outgrowths from the *cambial* layers in the trunk and that these outgrowths, whether functional or not, may be neither beneficial nor detrimental to the tree's chances of survival.

Structure

These unusual organs and developments in the species of tropical rain forest have arisen, in part, from the fact that this forest type is so tall and dense. The density of the shade cast by it does not arise merely by virtue of its height however. Although the height of the tallest trees is usually more than 100 feet, it is the complexity of the structure beneath the canopy which is, in great part, responsible for the deep shade on the forest floor. Quite often the so-called 'trunk space', trapped beneath the canopy, is fairly well filled by foliage down to a height of twenty or twenty-five feet from the ground (Plate XXIII*a*). This is not due to a large number of immature individuals at intermediate stages of growth but primarily to the fact that a very large number of species found in tropical rain forest are actually adapted to live in the lower light intensities beneath the shade of the forest dominants. If a census of all the mature individuals is taken in a sample area, it is usually found that, on the basis of height, they fall, quite distinctly, into three categories. These are usually referred to as the A-layer, B-layer and C-layer trees. This stratification is never obvious to an observer standing in undisturbed forest; even within the tree 'walls' of a newly-made clearing, all that is usually apparent is a seemingly chaotic structure with individuals at all heights. Only a complete statistical examination reveals that there are three distinct maxima when the heights of all mature trees in a sample area are plotted on a height-frequency graph. Near the coast in Guyana [59] a peak-frequency in tree heights was found at about 110 feet, though numerous individuals attained a greater height than this. A second maximum was discovered at about sixty-five feet and a third at between forty-five and fifty feet. The equivalent heights of the three strata in tropical rain forest on Mount Dulit in Borneo were approximately 110, sixty and thirty feet. Analysis of tropical rain forest in other places in Africa, Asia and America, have shown a similar structure though with various height-frequency maxima. In most of the associations the A-layer is not completely continuous so that each individual tends to have more room for the development of its crown; these tallest trees are thus characterised by an umbrella-shaped crown. The B- and C-layer trees on the other hand tend to be more closely-packed and to develop more conical crowns. Regardless of crown-shape, however, all the trees have straight trunks almost devoid of twigs and branches beneath the level of the recognisable crown (Plate XXIIIa).

Although the density of average tropical rain forest appears to have been exaggerated by some earlier writers, the light intensity beneath the three tree layers is very much reduced. Direct sunlight only reaches the forest floor in small, local flecks and, even then, only during the middle four or five hours of the day. Away from the sun flecks, the light intensity is usually less than 1% of that just above the forest canopy. It is almost certainly mainly due to this low light intensity that low-growing plants are relatively unimportant. The shrub layer, which is so important in some associations of deciduous summer forest, is very poorly represented in tropical rain forest and herbaceous plants are also quite unimportant. The ground, with its litter of fallen leaves and wood, is almost bare of low-growing plants (Plate XXIVa). The main exceptions to this are the seedlings of the trees themselves. These are often very numerous as compared to the number of trees at intermediate stages of growth, and it is here again that light intensity seems to be a most significant factor. It appears that, in the deep gloom of the forest, most of the seedlings of the dominant trees have the facility to remain dormant for long periods, growing perhaps at the rate of only an inch or two each year. It is possible that some species can survive this suppressed growth for decades while achieving a height of only a few feet; indeed it is doubtful if many species could survive if their saplings were not capable of this. It has been said, for instance, that most species of *Eucalyptus* in Australia are incapable of invading tropical rain forest for this reason. Rapid growth by the saplings can only be achieved when a major forest tree dies and falls, allowing the light to stream down to the forest floor. When this occurs, growth is enormously accelerated and the gap is very soon filled (*vide infra*). During this period of rapid growth, the sapling either succeeds in growing to its full, mature height in the canopy, or succumbs in the struggle for space and light above the soil, and for nutrients beneath it.

Through the millions of years in which the plants in this formation-type have been competing for light and mineral nutrients, large numbers of them have adopted abnormal growth-forms. These forms fit into the basic structure furnished by the forest trees and are instrumental in making the total structure even more complex and closely-integrated. One of the most conspicuous of these life-forms is that of the *liane* or woody climbing plant. Like the trees, the lianes are adapted to commence their growth in deep shade but require high light intensities when mature. In spite of the fact that they rarely achieve a thickness greater than that of a man's thigh, lianes have overcome their deficiencies in strength of stem by adopting a twining habit (Plate XXV*a*). They frequently pass from tree to tree, achieving a length of some 200 feet, and often succeed in penetrating or even over-topping the crowns of the tallest trees. The lianes thus bind together the forest structure and do it so effectively that trees which break off at the base frequently remain standing until they decay and literally disintegrate.

Another category of plants known as *epiphytes* have solved the problem of survival in tropical rain forest in an entirely different way. Though they are only small shrubs and herbs they have escaped from the gloom of the forest floor by germinating and spending the whole of their lives on the boughs and twigs of the A- and B-layer trees (Plate XXV*b*). Many of them develop a tangled mass of aerial roots which acts like a net and catches great quantities of leaves and other debris which fall from the trees. This material not only supplies mineral nutrients as it decays; it also acts like a sponge and is able to collect and hold a water supply adequate for the plant's requirements. The canopy of one A-layer tree may support hundreds of individual epiphytes belonging to scores of different species. Less numerous, but occupying a similar situation in the forest structure, are the *semi-parasites*. These plants not only lodge on the branches of the taller trees; they also penetrate the living tissue of their hosts with specialised, sucker-like growths and extract water and mineral nutrients from it. Because of this they have no need to develop the complicated structures for obtaining food and water as have the epiphytes. Another group of plants which begin their lives epiphytically, but which develop into very much larger inhabitants of the forest community, are known as the *stranglers*. These plants evade the risks and rigours of early suppressed development in the dimness of the forest floor by germinating high up on tree branches. When once they have built up a sufficient store of food by this mode of existence, however, they put out long, pendant roots which ultimately reach the soil beneath the tree. From then onwards the stranglers obtain water and nutrients like normal plants; their roots thicken and increase in number and ultimately form dense 'umbrellas' which envelop the hosts and strangle and kill them. The vast bulk of mature stranglers forms an important element in many associations of tropical rain forest.

Apart from the sparse population of apparently normal herbaceous plants which populate the better-illuminated areas of the forest floor, other low-growing plants which are condemned to this environment have adopted a novel way of thriving in it. These are the *saprophytes*. They have solved the problem of light-deficiency by divesting themselves completely of all necessity for photosynthesis. They are completely devoid of chlorophyll and thus have to obtain their manufactured organic materials from the humus in the soil. They are only able to do this by entering into intimate conjunction with certain fungi which invade the root tissues of the saprophytes to form *symbiotic mycorrhizae*. The fungi break down the humus chemically and produce an excess of relatively simple organic materials which are taken up by the saprophytic hosts.

Each one of these categories of plant, including the trees comprising the different layers of the forest, has adopted some specialised 'gimmick' for survival in this complex community. From the uppermost part of the tree canopy down to the forest floor there is a range of microclimates in which light intensity, humidity, temperature and wind speed differ enormously. To say that the plants of tropical rain forest, as a whole, are adapted to *one* particular type of climate is unrealistic and misleading; the leaves of the A-layer trees have to function in conditions which are utterly different from those in which the leaves of C-layer trees find themselves. To make the situation even more complex, the conditions experienced by the mature individual are frequently entirely different from those which must be survived by the same individual during the early phases of its development. Each category of plant is therefore adapted to its own particular niche within the structure and microclimatic complex of the forest and is thus referred to as a *'synusia'* [59].

Flora

Apart from its physical structure, tropical rain forest is also remarkably different from the forests of higher latitudes in its floristic richness. Even in eastern U.S.A., where the deciduous summer forest is abnormally rich in species, one can rarely find more than twenty-five species of tree on an acre of ground and in many northern forests one can travel for miles and see no more than four or five species. In tropical rain forest there are seldom less than twenty species of tall tree to the acre and sometimes more than fifty. Indeed, if counts were made over larger sample areas, the contrast would be found to be much greater. Whereas in the undisturbed deciduous summer forest of north-western Europe there can have been no more than a dozen species of tall tree, in the Malay Peninsula alone there are about 2,500 such species. In the Amazon Basin also it is calculated that there are at least 2,500 species of 'large tree'—1,000 in the State of Para alone [59]. It is in great part due to this mixture of species that much of tropical rain forest has not been utilised more intensively for its timber; in order to obtain any considerable quantity from one species, it may be necessary to exploit many square miles—an altogether uneconomic proposition. On the other hand, this diversity in species must not be over-emphasized; many associations of tropical rain forest are definitely characterised by the dominance of one or two species which may make up more than half of the total stand. Even in such associations, however, there are usually dozens of other species of tall tree present in a single hectare of forest.

The richness of the tropical rain forest flora is not limited to its trees; the liane and epiphyte synusiae comprise many thousands of species belonging to hundreds of genera. It should be noted, however, that the flora of the African formation is relatively impoverished as compared to the Indo-Malaysian and American ones. It has been estimated that there are more species of flowering plant in the island of Borneo than in the whole of the Congo Basin in spite of the fact that the latter is many times larger and contains formations other than tropical rain forest. The reasons for this relative poverty of the African formation are not yet understood.

It would be unhelpful to give a list even of the more important families comprising the synusiae of the three formations particularly since large numbers of them are quite unrepresented in the better known floras of higher latitudes and will, in consequence, be quite unfamiliar to most readers. It is sufficient to note that, among the numerous species of these unfamiliar families, there are many others belonging to families like the more familiar *Rosaceae*, *Compositae* and *Leguminosae*. There are also a number of species, from a variety of families, which have become well known because they have been exploited or cultivated. A-layer trees like the india rubber tree (*Hevea braziliensis*) and lower-growing trees like the bananas (*Musa spp.*) and the cocoa (*Theobroma cacao*) are in this category. There are also several well known groups of plants which have been particularly successful in one or another of the various synusiae. A particularly good example is furnished by the orchids (*Orchidaceae*). The majority of this family are epiphytic plants, nearly all of which grow on the branches of trees in the Indo-Malaysian and American formations. Another good example is provided by the figs (*Ficus spp.*) many of which have adopted the strangler habit in the Indo-Malaysian formation. Yet another example of specialisation within a group is shown by the mistletoe family (*Loranthaceae*) the majority of which, like the common mistletoe (*Viscum album*) of north-western Europe, are semi-parasitic on the branches of trees; they are most numerous in the tropical rain forests of Asia and Africa. It must not be assumed, however, that most of the families in these forests have developed predominantly in one particular synusia; in many cases very close relatives have developed very different modes of growth, adapted to very different microclimatic conditions. An extreme case of this is to be found in a group of epiphytic cacti (*Epiphyllum spp.*) in western Amazonia. These are obviously related to the semi-desert cacti of South America but were able to survive in dense, wet forest by becoming epiphytic. Within these very ancient and complex formations, there must be many groups and many species which have survived only because they have evolved in a very specialised way in a particular synusia.

Coastal and riverine hydroseres

Many early descriptions of areas of tropical rain forest are misleading. This is mainly because the travellers responsible for them gained most of their impressions as they sailed up rivers or along the shores. Their observations were therefore unrepresentative for two reasons. Firstly, even when they were actually looking at true climatic climax forest, they were only seeing the margin of it where sunlight impinges upon the forest laterally as well as vertically. Along such margins, the forest nearly always appears very much more dense throughout its whole depth than is really the case. It was primarily because of this that tropical rain forest was so frequently referred to as 'dense, impenetrable jungle'. Secondly, the vegetation beside rivers and shore lines was, quite frequently, not climatic climax vegetation at all, even before it was disturbed by man. Large rivers are normally flanked by wide flood plains and many coastlands are estuarine marshes, salt flats, coral reefs or areas of sand-dunes. In all these areas the travellers would see subclimax or merely seral vegetation.

Along the flood plains of a river in an area of tropical rain forest, wherever the current in times of flood is insufficient to sweep away all vegetation, all stages in a hydrosere can be observed. The earlier stages are often structurally similar to those in higher latitudes with water lilies and tall, reed-like grasses featuring very widely. In most places, however, species of mangrove invade at a relatively early stage. The term 'mangrove' has been applied to a large number of shrubs and trees which produce stilt roots prolifically and are thus able to survive in unstable, submerged mud and silt (Plate XXIV*b*). A number of them produce viviparous seeds which germinate while still attached to the tree and whose *radicles* and *plumules* may have achieved a length of over a foot before they separate from the parent. Ultimately they break away and fall like darts to the mud below. The radicle sticks upright in the mud so that a young, 'ready-made' plant takes up a position beside its parent (Fig. 25). Ungerminated seeds would be in much greater danger of being overwhelmed by mud and water.

When once the mangroves have established themselves on submerged mud, the rate of silting is greatly accelerated and the surface ultimately builds up to the level of the highest floods. When this has occurred, later stages in the hydrosere can follow. Along American rivers, palms such as *Euterpe oleracea* and *Mauritia flexuosa* invade and rapidly overtop and shade out the mangroves (*Rhizophora racemosa, inter alia*) [59]. Similar successions of mangroves and palms occur in the saline waters of silting estuaries and deltas. Although there are a number of tree palms which are normal constituents in climatic climax tropical rain forest, it is apparent that the majority in these areas compete successfully in seral communities only.

Fig. 25. Germinating mangroves near Goa. (From Marilaun and Oliver, 1904.)

There are certainly many hundreds of species of palm in the wet tropics; some, like the coconut palm (*Cocos nucifera*), are extremely widespread. Like this latter species, however, the majority appear to be almost universal merely because they are peripheral to the main blocks of forest. Even the West African oil palm (*Elaeis guineensis*), is not a climatic climax species.

A. LIANES. In the Shasha Forest Reserve, Nigeria.
(From Richards, 1952.)

B. EPIPHYTES. Various herbaceous and shrubby species on a single tree branch.
(From a photograph by Schenck in Schimper, 1903.)

PLATE XXV

A. AFRICAN DRY FOREST. Grassy, open forest with some tree regeneration near Elizabethville, Katanga Province, Congo. The photograph was taken during the wet season. (From Shantz and Marbut, 1923.)

B. THE VENEZUELAN LLANOS. The grassland is shown to be sprinkled with palms (*Capernicia tectorum*), some of which are infested with a strangler species of *Ficus*. (After Carl Sachs in Schimper, 1903.)

PLATE XXVI

Soils

Many European natural historians who saw the tropical rain forest in its undisturbed state concluded that such luxuriant vegetation must be underlain by soils of great natural fertility. At a later stage, plantation farmers who tried to wrest a profit from this forest soil often came to exactly the opposite opinion. Unless the statements are qualified, it is almost certainly misleading to say that these soils are either 'fertile' or 'barren'; in any one place either term may be applicable according to circumstances.

Because they are subjected to even more vigorous leaching than are most forest soils in higher latitudes, these soils might be expected to be poor in mineral bases and quite acid in reaction. The leaching process is offset by the persistently high temperature however; this ensures that the copious supplies of dead plant debris which reach the forest floor are very rapidly decomposed so that any bases they contain are released quickly. This constant supply of bases maintains the base-status of the soil at a fairly high level so that normally the soil reaction does not fall far to the acid side of neutrality. The nutrient cycle is thus a very rich and very rapid one. Since the rate of decay is so rapid there is, generally speaking, very little tendency for organic material to accumulate; normally there is an A_{00} horizon of several inches of plant debris, this being underlain by a negligible A_0 horizon.

Although these soils do not become very acid, the high temperatures throughout the whole year do cause the clay minerals to decompose quite rapidly. Because of this the soils are predominantly red, due to the presence of sesquioxides. In these conditions of high temperature and only moderate acidity, however, it is the silica fraction of the clay (Chapter III) which becomes most mobile. This silica is leached downwards in the soil either to be re-deposited in the weathering material beneath [53] or to be removed altogether and lost to the ground-water. The clay fraction of the upper eluviated horizons thus becomes relatively enriched in sesquioxides. The net result is, therefore, the exact opposite of that achieved by the podzolisation process. These silica-leached, sesquioxide-rich soils are referred to as 'lateritic soils' and the process that gives rise to them as 'laterisation'. This process is going on beneath much tropical rain forest and the deep, red soils so produced are often called *tropical red soils*.

It will be obvious that if the laterisation process continues in the same material for a sufficiently long period, the clay minerals will ultimately lose all their silica and be converted into a substance composed entirely of the two sesquioxides. This is the substance known as 'laterite'. The really absorbing problem which faces soil scientists in the tropics is that, although nodules of laterite are found quite frequently in the tropical red soils, no case has been described where the upper horizons of a clay soil beneath tropical rain forest have become a continuous layer of pure laterite. One might suppose that, throughout the millions of years in which some areas must have been occupied by this type of vegetation, there should have been sufficient time for the laterisation process to be completed. There are rather baffling complications however. Firstly, since pure laterite, near the surface, often forms into a hard, rock-like substance, it is fairly clear that the forest could not survive upon it if a complete crust of such material were to form. Secondly, it is quite impossible for the same soil to rest, undisturbed, on anything but an entirely flat surface if it is to have a sufficiently long period for the laterisation process to be completed; erosion would persistently remove the partially laterised layer even on gentle slopes and beneath thick, undisturbed forest. In consequence, the only areas upon which the process could have been completed are completely flat, old peneplains. Some pedologists [53] claim that, where certain types of basic igneous rock are the soil-parent-material, complete laterisation has taken place; a crust or 'cuirass' of laterite now covers the surface and a postulated former cover of tropical rain forest has deteriorated and given way to a poor type of scrub or heath forest. As yet there is great uncertainty or complete ignorance about the processes operating in nearly all types of parent-material, however, and we are left with the one simple fact that a partially-laterised red soil underlies much of tropical rain forest.

It has become more and more apparent during the past thirty years, however, that the laterisation process does not prevail everywhere in free-draining soils beneath rain forest in tropical lowlands; indeed it is now doubtful if laterisation occurs at all where the soil-parent-material is a coarse sandstone, a grit or a quartzite. On base-poor materials such as these, where the clay content is small and percolation is rapid, it appears that strong acidity is achieved shortly after weathering begins. Because of this, quite deep A_0 horizons of acid peat, sometimes over a foot in thickness, are able to develop, and typical podzol profiles form beneath them. Tropical podzols such as these are widespread in the catchment area of the Rio Negro in Amazonia, in Guyana and in numerous places in Malaya, Borneo and Thailand. The vegetation on these podzols is usually of a quite distinct type; of necessity it is composed of undemanding plants and, quite frequently, these are related to the heaths (*Ericaceae*) of higher latitudes [59]. On the island of Mafia off the coast of Tanzania an area of tropical podzols is dominated by an ericaceous shrub (*Philippia mafiensis*) and the so-called 'heath forest' or 'padang' of south-east Asia is all underlain by tropical podzols. It is quite clear that extreme kinds of parent-material have permitted the evolution of acidophyllous associations of tropical rain forest which, in turn, have accelerated the development of true podzols; a base-poor environment can permit the accumulation of acid peat in spite of high temperatures. Today there is increasing doubt as to whether the laterisation process can proceed normally even on acid igneous rocks like granite. The widespread brown forest soils on granite in the southern part of West Africa indicate that, here again, parent-material may cause deviations from the norm. Finally, a great range of soils are found in areas with impeded drainage beneath late stages of hydroseres within tropical rain forest areas. In spite of the prevailing wetness, however, these soils are not typically characterised by peat accumulation; it is only where there is acid ground-water percolating laterally from siliceous rocks, in conditions of heavy reliable rainfall, that peat builds up. Great areas of thick peat, sometimes attaining a depth of over twenty feet, have been described from the coastal plains of Java and Sumatra [53]. An interesting difference between this peat and that of analogous areas in higher latitudes is that the former is composed almost entirely of the remains of trees and other woody plants of the swamp forests.

Forest regeneration

It seems likely that the soils beneath the shade of tropical rain forest have been developing over large areas, without interruption, for millions of years. During this unimaginably long period, although the surface, in most cases, will have been gradually lowered by natural erosion, at no time will it have been open to the desiccating effects of full sun and wind. At any one point it will only have experienced bright sunshine for a few months once every few score years. The only time when bright light has streamed in will have been when one of the large A-layer trees has died from old age and has fallen, crushing beneath it other smaller trees. Immediately this has happened, however, the saplings which have survived suppressed growth in the undergrowth will have been activated (*vide supra*) and, within only a year or two, the gap will have been completely filled. It is probably highly significant that the species which grow most rapidly and fill these gaps in the initial stages of regeneration, are not the true dominants of the primary forest. There appear to be many species of tree scattered throughout the forest which, though incapable of competing with the A- and B-layer dominants, are usually present as seedlings or suppressed saplings when a gap in the canopy occurs. Mainly they are species which produce vast numbers of fruits, the seeds from which are most efficiently distributed by birds and fruit bats. The trumpet trees (*Cecropia spp.*) of America, *Musanga cecropioides* (*Moraceae*) of Africa and species of *Macaranga* (*Euphorbiaceae*) in Malaya are all good examples of this type of tree [59]. When a gap in the canopy appears, they are able to grow to maturity at almost lightning speed only to be shaded out very shortly after they have produced a few crops of seeds. It appears therefore as though a condensed subsere is initiated every time a major forest tree falls. The true dominants of the primary forest do not immediately re-establish themselves; a mass of bushy, lower-growing trees first fills the gap; saplings of the climatic climax dominants push their way through this and, within ten or twenty years, overtop it.

Human interference and secondary forest

Many of the first cultivators to use the soils beneath tropical rain forest did little more than accentuate the regeneration process just referred to. They made tiny clearings in the forest, planted their seeds or tubers in the sunlit soil, and, after a very short time, moved to another spot. Though an area of soil was actually disturbed in one of these clearings, it would usually be sufficiently small to receive a certain amount of natural leaf-fall from the surrounding forest and, when the cultivators moved on, the normal regeneration species invaded almost immediately only to be overtopped themselves by climatic climax dominants shortly afterwards. This primitive type of small-scale shifting agriculture can still be found in some tropical rain forest areas where there is only a sparse cultivating population. Unfortunately, however, increase in population and the advent of large-scale plantation-type agriculture has been responsible for much larger areas being cleared for much longer periods. Soil which had developed through countless millennia in a dim, moist microclimate has been opened up, at one fell swoop, to the ferocious glare of the midday tropical sun.

This change in microclimate, along with the loss of the products of leaf-fall, has had a devastating effect on some types of tropical red soil. In many places the soil's structure has been changed completely. Within a few years of forest clearance, unless copious amounts of organic material are supplied to offset the cessation of leaf-fall, a tropical soil can soon become completely inorganic. No more mineral nutrients are released by humus decay—the nutrient cycle is broken! Along with this, the fierce heat of the sun and the drying effect of the wind ensures an upward movement of solutions in the soil after only a day or so without rain. In certain types of soil, sesquioxides and colloidal silica are deposited from these solutions when they reach the upper layers and the water solvent evaporates. A hard, impervious crust thus develops as lateritic substances cement the soil particles. This crust does not absorb rain-water quickly as did the mulch of leaves on the original forest floor; with any degree of slope there is, in consequence, a much increased run-off with consequent erosion. Many furrowed, barren hillsides in the West Indies, West Africa and eastern Brazil bear witness to the effectiveness of this type of exploitation.

Even when this over-exploited land is abandoned by cultivators it is not rapidly re-invaded by forest as were the small clearings of the primitive cultivator. In extreme cases, where a thick, lateritic crust has formed, even the trees of the regeneration subsere seem unable to obtain a foothold and much more xerophytic shrubs and herbs dominate the surface for a long period. In most cases, however, regeneration species such as

Cecropia spp. and *Musanga cecropioides* invade quickly and form communities which are commonly referred to as 'secondary forest'. Vast areas of former tropical rain forest are now covered with this secondary forest even where human populations are not exceptionally heavy. Over great stretches of lowland and foothills in Malaya, for instance, it is now very difficult to find true primary rain forest. If left undisturbed there is some evidence that most of these areas of secondary forest would revert to the climatic climax state but it seems likely that this would take a very long time. The changes in soil structure and soil chemistry which have occurred appear to prevent the successful development of the seedlings of the true climax species. Fortunately there are some types of tropical red soils which are not so profoundly transformed by agricultural exploitation but since the nature of the changes just outlined is not understood, the risk of 'opening up' any new area of tropical rain forest for large-scale agriculture is always great unless there is very careful supervision.

The factors limiting tropical rain forest

Apart from those places where it is bounded by the sea and where human interference has altered the vegetation pattern, tropical rain forest gives way everywhere, under natural conditions, to one of three distinct formation-types. In the lowlands of the continental interiors and on the west side of continents, generally speaking, it gives place to some kind of seasonal forest; on the eastern sides of continents, where wet climates extend continuously outwards to cooler latitudes, it gives way to simpler types of evergreen rain forest; on the lower slopes of mountains it gives way, through a transition zone of sub-montane forest, to a 'tropical montane' or 'mossy forest' formation (Chapter XX). Tropical rain forest extends as far as areas where, as a formation, it cannot exist because of climatic conditions, but the precluding factors are different in each of the three cases.

The typical evergreen species of all the synusiae of tropical rain forest require copious and reliable rainfall throughout most of the twelve months of the year. Wherever, even for a short season, rainfall becomes either light or unreliable, this formation-type rapidly gives way to one which, as a whole, is better adapted to periodic water deficiency. The latter is dominated by trees of which a greater or lesser proportion are deciduous (Chapter XVI). In any particular instance, however, the point at which the change begins to take place and the exact nature of the factors causing the change, are most difficult to determine. An abrupt change from one formation to the other is rarely seen; usually a broad zone of ecotone intervenes. Even in large parts of eastern Amazonia and in the most rainy areas in Trinidad, the forests are transitional or ecotone forests[1] and not true tropical rain forest. In these parts of Trinidad there are only three months

[1] These are often referred to as 'evergreen seasonal forests'.

with a mean precipitation of less than four inches and all months have more than 2·5 inches; at Manaos, on the middle Amazon, again there are only three months with less than four inches and only one with less than 1·5 inches. Apart from the fact that up to 3% of the individuals composing the A-layer are deciduous, these ecotone forests are very similar to tropical rain forest in all respects. The ecotone forests merge gradually into quite distinct seasonal formations, however, and in the Americas this usually occurs where the drier season becomes longer than three months, each with an average of less than four inches rainfall. It must be emphasised, however, that this is no more than the statement of a general correlation; almost typical tropical rain forest can be found in places where the drier season is much more pronounced. At Akilla in southern Nigeria, in areas where tropical rain forest of almost typical structure appears to flourish, there is a continuous season of five months, each with less than four inches precipitation; three months have less than two inches. Here a persistently high atmospheric humidity, even in the drier season, offsets the rainfall shortage. Also, along many river sides within areas of quite pronounced dry season, there are ribbons of 'gallery forest' almost identical to tropical rain forest. Here the reliable ground-water supply throughout the whole year makes the vegetation independent of actual rainfall. On the other hand, outcrops of very pervious rock may support only seasonal forest in areas where, otherwise, the rainfall is adequate for tropical rain forest. In Trinidad, for instance, seasonal forest is found on limestone in areas where all other types of outcrop carry a transitional forest very similar to tropical rain forest. It is clear that the boundary between tropical rain forest and seasonal forest is extremely complex and diffuse and that, even in one place, it will never be definable in simple, purely climatic terms.

In eastern Asia, eastern Australia, southern Brazil and the Greater Antilles–Gulf States area, tropical rain forests give way gradually to the broad-leaved evergreen forests and mixed evergreen forests of middle latitudes (Chapters V and VI). Although luxuriant, these have a much simpler structure. They are dominated by species which are capable of withstanding generally lower temperature conditions but exactly which aspects of temperature operate to preclude most of the typical dominants of tropical rain forest has not been decided. It is possible that the generally lower temperatures of the winter season so debilitate many tropical species that they can neither grow healthily nor produce seeds. It seems more likely, however, that the occasional winter frosts, or the occasional very low temperature that may occur only once in every decade or so, may be more important in preventing the spread of tropical plants. Two facts have been ascertained beyond doubt; firstly it is known that many tropical species cannot withstand the slightest frost (though others appear capable of doing so); secondly, tropical rain forest, with its characteristic three-tiered structure, is only found in areas which are completely frost-free. The transitional evergreen forests of north-eastern New South Wales (30° S.) and southern Florida (25° N.) are found to contain an admixture of tropical and mid-latitude species and the three-tiered structure is usually absent. Today it is very difficult to arrive at firm conclusions about the original nature of the vegetation along these well-watered eastern margins however; human interference has been profound and, in the case of southern China, of very long duration.

The dominant trees of tropical rain forest require protracted sunshine as well as high temperatures and a plentiful water supply. It is doubtful if most of them could maintain their rapid rate of growth and development in persistently dull, cloudy conditions. This is almost certainly one of the reasons why typical tropical rain forest does not extend very far up the flanks of mountains. Decrease in temperature and increase in atmospheric humidity and windiness should not be dismissed however. The height at which typical tropical rain forest begins to give way to other types of community varies very greatly from one locality to another. Generally speaking the transition begins at progressively lower levels the nearer one approaches the coast; it also tends to be lower on small, isolated mountains, and higher on the flanks of great mountain masses (Chapter XX). The average height at which tropical rain forest begins to give way to transitional 'sub-montane' forest is about 3,500 feet and the average height of the base of the true 'montane' or 'mossy forest' formation-type is about 5,000 feet. In countries like Malaya and Trinidad these limits are much lower, while on the eastern flanks of the Andes and the western flanks of East African mountains they are higher.

CHAPTER XVI

Tropical Seasonal Forests

THE general characteristics of tropical rain forest are so distinct that it has been possible to recognise it as a distinct plant formation-type and to map its potential distribution relatively accurately. The same cannot be said for those forest communities to which tropical rain forest gives place where seasonal drought is experienced. Nevertheless there are obvious similarities in structure between those tropical seasonal forests found in America, Asia, Australasia and Africa so that a tentative classification on very broad lines can be attempted. The separate formation-types recognised here, because their distributions are known only imperfectly, can only be shown as a single category on small-scale maps however (Appendix I).

I. THE SEMI-EVERGREEN SEASONAL FOREST

It has already been shown (Chapter XV) that tropical rain forest undergoes a slight change as one passes to areas with only very short drier seasons but that this transitional or ecotone forest is very similar in structure and appearance to the typical tropical rain forest. As the drier season becomes more pronounced, however, the transitional forest merges into another, quite distinct formation-type. Because its dominant trees are partly evergreen and partly deciduous, this has been referred to in recent literature as 'semi-evergreen seasonal forest'[59]. Forests of this kind have been identified on all the continental areas which extend into the tropics.

The American formation

The American semi-evergreen seasonal forest has been described in its undisturbed state in Trinidad [4]. Here it attains a height of about seventy feet and, in contrast with tropical rain forest, has a relatively simple structure of only two tree storeys. Although the lower of the two is still composed predominantly of evergreen trees, some 20% to 30% of the individuals in the upper one are regularly deciduous. Even more important is the fact that most of the evergreen species in the upper layer have the facility of losing their leaves in an abnormally dry season; they are thus said to be 'facultatively deciduous'. This formation also differs from tropical rain forest in its minor characteristics. Very few of the larger trees in the seasonal formation are buttressed and, in spite of their evergreen habit, a large number of lower-storey trees are small-leaved or 'microphyllous' species. Furthermore, although the semi-evergreen seasonal formation is rich in lianes, epiphytes are noticeably scarcer than in tropical rain forest. The increase in microphyllous species and the recession of epiphytes is an obvious reflection of the fairly prolonged drought which is experienced from time to time. Epiphytes in particular, might be expected to become scarce with the slightest dry season; the aerial collections of humic material upon which they are almost completely dependent for water, will obviously dry out much more quickly than the soil.

The nutrient cycle in the semi-evergreen forest is very different from that in tropical rain forest. The seasonal formation has an almost wintry aspect in an abnormally dry season; the ground is covered by a deep litter of freshly-fallen leaves which does not begin to decay to any degree until the following wet season; the light also is able to penetrate to the ground through the partly leafless canopy almost as it does during the winter in the summer deciduous forest of higher latitudes. Some writers have emphasised an entirely different aspect of the winter appearance of this forest however. This is the season when many species, particularly in the lower storey and the undergrowth, produce their flowers; in a number of associations, one can encounter a great blaze of colour, often on a completely leafless plant. This is in marked contrast to the generally deep shade and sombre colours of the lower layers of tropical rain forest.

Semi-evergreen forests such as those in Trinidad are widespread throughout the West Indies and South America. They have been described in Venezuela, the western interior of Guyana and the north-eastern interior of Colombia; they extend in a wide belt along the southern rim of the Amazon Basin and, in a second belt, parallel to the coast of south-eastern Brazil from Recife to the southern end of the Brazilian Plateau; they also occupied large areas in southern Panama, east-central Cuba and other smaller islands from which they have now been almost completely removed.

The American semi-evergreen seasonal forest is mainly found where the mean annual rainfall is between thirty and fifty inches but, in view of the fact that a similar formation is found in Asia where the rainfall is much heavier, this is probably of little significance. It is more important to note that the American formation is generally found where there is a drier season of five months, each with a mean rainfall of less than four inches but more than one inch. Drier seasons of this nature appear to be equally effective in favouring this formation at the expense of tropical rain forest, regardless of the amount of rain which falls in the wet season.

The Indo-Malaysian formations

The semi-evergreen forests of Asia and northern Australia are, in fact, the wetter varieties of the vegetation type which has classically been referred to as 'monsoon forest'. In spite of the name, however, these must be regarded merely as formations of the much more widespread formation-type. The Indo-Malaysian tropical rain forest, which extends from the Western Ghats to the south Pacific, is flanked to the north and south, in many places, by this semi-evergreen formation. It is widespread in north-eastern India, Burma, Thailand and Indo-China on the one hand, and stretches from eastern Java to northern Australia on the other. Because of this wide separation, the Australian forests are very different in species-content from those of Burma and India. It is mainly in Burma and India that this forest has been studied intensively and many generalisations about 'monsoon forest' are based on observations made in these two countries. Even here a great deal of doubt remains regarding the original, undisturbed distribution of 'monsoon forest'. This is well illustrated by the experiences of foresters in India in the nineteenth century. In an effort to conserve the more valuable deciduous timber trees in the seasonal forests, reserves were created. Grazing and cultivation were excluded and every effort was made to prevent fires which, up to that time, had swept through the forests at frequent intervals. The aim was to permit the trees to re-generate naturally and thus to ensure a constant supply of valuable timber. Regeneration was thus much improved but not always in the way that had been intended. It was discovered in several places that the saplings which appeared in the protected areas belonged to predominantly evergreen species of tropical rain forest. This demonstrated quite clearly that the existing forest had not been true climatic climax 'monsoon' forest at all; it had been a plagioclimax community whose existence had been permitted by a degree of burning and human interference. Apparently the thicker bark of the teak (*Tectona grandis*) and other deciduous trees makes them resistant to this kind of treatment whereas the thinner-barked species of tropical rain forest cannot survive it. Because of this discovery the exact position of the boundary between climatic climax tropical rain forest and semi-evergreen seasonal forest cannot be mapped with certainty, particularly in heavily populated areas like north-eastern India and Java. There are some parts of Burma, however, where climatic climax semi-evergreen forest has been studied in almost completely undisturbed conditions [71].

This Burmese part of the formation appears to contain two widespread but quite distinct associations. The first of these is dominated by the pyinkado (*Xylia xylocarpa*), a deciduous leguminous tree which can attain a height of 120 feet when growing in soils on pervious rocks. No detailed descriptions of the structure of this type of forest are available but it seems that a large proportion of the trees associated with the pyinkado are evergreen and that the species composing the lower layers of the forest are almost entirely so. The association is found predominantly in western and southern Burma in those places where the mean annual rainfall is between eighty and ninety-five inches but where there is a regular dry season from December to April and where three or four of these dry months have less than one inch.

The second association is usually referred to as 'moist teak forest' because of its importance as a source of this valuable timber tree. The teak (*Tectona grandis*) rarely occupies more that 10% of the stand, however, and is usually associated with an even larger proportion of pyinkado. Moist teak forest is rarely found where the mean annual rainfall exceeds eighty inches; with wetter conditions than this the pyinkado tends to replace it completely. On the other hand the moist teak forest is replaced by more xerophytic communities where the rainfall falls below about sixty inches. This association is thus widespread on mountain slopes in Tenasserim and around the 'dry belt' in interior Burma.

Although teak and its associates can withstand up to about five or six months' drought, it should be noted that their growth régime is not governed directly by rainfall. Not only does the teak retain its leaves during the first month or two of the dry season, it normally sprouts new ones a month or more *before* the beginning of the wet season. This demonstrates quite clearly that its régime is governed by an innate, rhythmical impulse; it does not respond directly to the stimulus of an increase in wetness. This species, like many others in the same communities, is inherently adapted to a growing season of a particular length and is not capable of adapting its régime to growing seasons which are either very much shorter or very much longer. This may be one of the reasons why teak does not profit by being grown in plantations in areas of more rainfall and longer growing season; here it merely becomes hollow and abnormal in development.

In structure, the wet teak forest is very similar to the semi-evergreen seasonal forest of America. Normally the dominant teak and pyinkado grow to a height of between eighty and ninety feet but may achieve 120 feet on particularly favourable soils such as those on very pervious sandstones on the Pegu Yoma. Even where the forest achieves this greater height, however, it usually retains its simple, two-layered structure. The shrub layer beneath the trees is discontinuous though bamboo thickets do occur at frequent intervals. Finally, although lianes are an obvious element in the forest structure, they are by no means as plentiful as in the contiguous tropical rain forest; epiphytes also are inconspicuous.

Of the semi-evergreen seasonal forests of Indonesia and northern

Australia, those of eastern Java appear to have been described in greatest detail [67]. Again the teak is one of the main dominants but there is considerable controversy regarding its presence here. Because of the distance separating the Javanese seasonal forests from those in Burma and Thailand, some authorities have doubted whether this tree can be indigenous to Java; it is possible that it was introduced centuries ago because of its value as a timber tree. This view is supported by the fact that some of the east Javan forests (*vide infra*) contain almost pure stands of teak. Since there are alternative explanations for this, however, the question must be regarded as being far from settled; many ecologists would certainly hold that the teak is far too well established in Java to be a mere introduction.

The African 'dry evergreen forest' formation

There appear to be no large areas of forest in Africa which are closely analogous to the semi-evergreen formations of the other continents. In all probability this is due to the fact that extensive savannas and savanna grasslands occupy most of the areas both to north and south of the tropical rain forest belt (Chapter XVIII). The nearest approach to semi-evergreen seasonal forest is afforded by the so-called 'dry evergreen forest' of Nigeria and other parts of West Africa. This extends in a discontinuous belt along the northern side of the West African tropical rain forest into which it merges imperceptibly. Although this seasonal forest has some features in common with the semi-evergreen seasonal forest of Trinidad, it is different in several ways [59]. The most important of these are that it is usually three-layered and that its deciduous trees do not lose their leaves at one and the same time so as to give the community a very distinct seasonal aspect. These points must not be over-emphasised, however; the A-layer of this forest is very discontinuous and composed predominantly of deciduous trees; furthermore, during the drier months from November to March, the lower layers of the forest are better illuminated than at other times of the year because of a general thinning of the canopy. The two lower tree layers and the discontinuous shrub layer are almost entirely evergreen but it has been noted that, in contrast with the contiguous tropical rain forest, quite a large percentage of the herbaceous plants on the forest floor die down completely during the dry season. Another reflection of the drier conditions which have to be withstood is that, although lianes are important in the structure of this type of forest, epiphytes are quite rare.

Most of the 'dry evergreen forest' which remains today has been severely modified by shifting cultivation, so that a good deal of inference is necessary in order to arrive at a picture of its former appearance. Furthermore, it seems likely that, in the next few decades, the fragments and vestiges which remain will be completely swept away. In view of this, it seems probable that other areas with similar climate, particularly to the north and south of the Congo Basin, have also carried this kind of forest but have had it transformed or completely removed. Since no careful survey of forest reserves over these vast areas has yet been made, much uncertainty persists.

2. THE DECIDUOUS SEASONAL FOREST

As one passes from semi-evergreen seasonal forest into areas of even more protracted dry season, so one encounters plant communities which are even better adapted to withstand drought. In many places, however, these communities can still be described as 'forests' since they form a complete tree cover over the ground. Once again, communities of some-what similar structure are found on all the main continental areas and, on the basis of this, they have been classed as a single formation-type called the 'deciduous seasonal forest'.

The American formation

Typical deciduous seasonal forest has been described, in what appears to be an almost completely undisturbed state, on the island of Tobago off the north coast of South America [59]. It is composed of two distinct storeys, the upper one not exceeding sixty feet in height and the lower one varying between ten and thirty feet. The upper storey is discontinuous and is composed predominantly of deciduous trees while the lower storey is more continuous and almost entirely evergreen. Apart from being of smaller stature, these trees also tend to be of quite a different form from those composing the semi-evergreen formation. Whereas the latter are mainly very straight with well-formed trunks, the taller trees in the more xerophytic deciduous formation are often more gnarled and crooked and throw out branches quite near to the ground. Furthermore, while the dominant trees in the semi-evergreen formation usually have compact, conical or rounded crowns, those of the deciduous forest have crowns which are normally umbrella-shaped or divided into distinct horizontal 'layers' (Plate XXVIa). The reasons for this characteristic shape have been much debated. A reflection of the drought-resistant nature of this community is the fact that large numbers of the trees composing both the strata are microphyllous. The forest is also very poor in lianes and almost devoid of epiphytes. The fact that herbaceous plants, particularly grasses, are only scantily represented on the floor of this forest is probably very significant (Chapter XVIII).

This type of forest is found locally in Venezuela and in the western parts of Central America from western Panama to Guatemala, as well as in many rain-shadow areas on islands in the West Indies. Generally speaking, it is found where there is a season of five consecutive dry months, each with less than four inches of rain of which two have less than one inch. Nevertheless, it is frequently found intermixed with semi-evergreen forests where very pervious outcrops provide a habitat more subject to drought.

The Indo-Malaysian formations

In the Asiatic-Australasian zone, decidous seasonal forest very similar to that of South and Central America once covered very large areas. Much of it has now been removed or greatly modified in India and southern Burma but extensive tracts still remain in central Burma, Thailand and Indo-China. It is apparent that a forest type, closely related in structure and life-form but phylogenetically very different, is also found in nothern Australia. The latter has usually been classed with the semi-evergreen forests, however, and referred to as 'monsoon forest'.

Several forest associations in Burma appear to be similar in form to the deciduous seasonal forest of Tobago; one of the most widespread is the 'dry teak forest'. Here the teak (*Tectona grandis*) is a main dominant just as it is in the wet teak forest but it is associated with an almost entirely different set of species. In the dry teak forest the great majority of associated tree species are deciduous. Two of the most prominent of these associates are the 'eng' or 'in' tree (*Dipterocarpus tuberculatus*) and a species of sal (*Shorea obtusa*). These species, along with many more, form a complete canopy about seventy or eighty feet high beneath which a discontinuous storey of lower trees is usually present. While the A-layer trees are predominantly deciduous, the B-layer ones are nearly all evergreen. Shrubs, particularly the bamboo (*Dendrocalamus strictus*), often form quite dense thickets beneath the trees but, once again, in undisturbed forest, grasses and other herbaceous plants are not important. Dry teak forest is not found where the mean annual rainfall is below about forty inches but it can be found right up to the seventy-five-inch isohyet. In fact, both its distribution and its luxuriance are very much influenced by soil and the underlying rocks; on good, loamy soils it gives way to wet teak forest with a mean annual rainfall as low as fifty inches, whereas on poor sandy soils the more xerophytic community intrudes into much wetter areas [59].

An association very similar in structure to the dry teak forest of Burma is found in eastern Java on impervious sub-soils which dry out very quickly at the beginning of the dry season. This is the 'djati-forest' in which the teak[1] is even more prominent than in the analogous association in Burma but where a variety of other tree species are still important. A detailed and vivid description [67] has been given of the structure and appearance of this type of forest. The canopy is usually sixty or seventy feet high, being composed of the crowns of deciduous trees like the teak, the wattle (*Acacia leucophloea*) and *Albizzia procera* (*Mimosaceae*) along with those of a small minority of evergreen trees like *Butea frondosa* (*Leguminosae*), *Albizzia*

[1] 'Djati' is the Javanese name for the teak (*Tectona grandis*).

stipulata and *Schleichera trijuga* (*Sapindaceae*). Smaller trees, many of which are evergreen, form a distinct under-storey and bamboos and other shrubs are abundant throughout. Many of these shrubs belong to the pea family (*Leguminosae*) and carry brilliant flowers, particularly when the penetrating light is at its most intense in the dry season. A variety of attractive herbaceous plants are also scattered over the floor of the forest, but except for small areas where for sporadic reasons the soil is exceptionally thin, sandy and dry, grasses are poorly represented. In these dry places there are patches of alang-alang grass (*Imperata cylindrica*) and glagah (*Saccharum spontaneum*). As in the dry teak forest of Burma, epiphytes are rare; the only ones occurring with any degree of frequency are shrubby species of fig (*Urostigma spp.*). On the other hand, semi-parasites of the mistletoe family (*Loranthaceae*) are exceptionally common on the taller trees.

Apart from the typical deciduous forests of Burma and Java, there are other, more xerophytic, associations in south-east Asia which should probably be included in the same formation. These are well represented in Burma and, generally speaking, are found in areas where climate and soil conspire to create more unfavourable conditions than those in which the dry teak forest is found. Indaing ('in' or 'eng' forest) is nearly always found on soils where there is some serious impediment to root development. In sandy soils this is caused by an iron-rich 'pan' but very shallow, droughty soils on limestone and soils on top of lateritic 'cuirasses' (Chapter XIX) on other types of rock also support this community. Because it is so closely associated with these soil types, the indaing has a wide climatic range with rainfall of between twenty-five and eighty inches. These shallow soils can retain little water for the dry season regardless of the amount of rain which falls in the wet season. The indaing is like the dry teak forest except that the more demanding species, like the teak itself, are missing. The dominant tree above laterite horizons is the 'in' (*Dipterocarpus tuberculatus*) but this may become sub-dominant to a great variety of other species elsewhere. The height of the canopy on laterite is only thirty to forty feet and is formed by trees whose form is mainly very gnarled and spreading; only the 'in' itself maintains a fairly upright growth. In several ways this type of forest differs from the typical deciduous seasonal forest. Firstly, it usually has no recognisable under-storey of smaller trees; a tangled mass of shrubs is usually the only kind of woody undergrowth. Secondly, epiphytes, particularly orchids, are normally abundant on the branches of the dominant trees especially where the rainfall is heaviest. Thirdly, a grassy, herbaceous layer is frequently, though by no means always, present. The extent to which the indaing communities are climatic climax has been much debated; large parts of it may very well be secondary vegetation on areas which have suffered human interference.

There are two further communities of low deciduous forest which often grow alongside the indaing around the 'dry belt' in Burma; they have the same general height and structure as the latter but are dominated by different species. The 'te' forest is a community in which the 'te' (*Diospyros burmanica*) is the dominant tree, though it is associated with several other species. This forest type is found on soils which are particularly light and sandy. The 'than-dahat' forest, on the other hand, is found on very heavy clays where drainage impedence is marked in the wet season but where drought is profound throughout the dry season. It is dominated particularly by the dahat (*Tectona hamiltoniana*) and the than (*Terminalia oliveri*). Both 'te' and 'than-dahat' communities reach an average height of no more than thirty feet and form canopies which are almost entirely deciduous. They are found mainly where the mean annual precipitation is between thirty-five and forty inches and thus have to withstand a long and intense dry season. Both also are characterised by an almost continuous grassy cover over the forest floor; they can therefore be referred to as 'grassy forest' or 'savanna woodland' (Chapter XVIII).

The African formation

Although stands of tall deciduous trees occur over vast areas of tropical Africa, nowhere can it be said that there is continuous, true deciduous seasonal forest such as that already described in America, Asia and Australia. Woodland, variously described as 'dry forest', 'miombo forest' or 'tree steppe', is the prevalent vegetation in a great belt extending from Angola across Zambia and Malawi into western and central Tanzania and including most of the Katanga Province of the Congo and the northern strip of Rhodesia. Very similar woodland is also found more locally in the southern Sudan, central and northern Nigeria and many other parts of West Africa. Indeed, as will be seen later (Chapter XVIII) there is no way of making a clear distinction between this type of vegetation and 'savanna woodland'. Almost everywhere in the 'dry forest' one can see through the trees to a distance of between 100 yards and half a mile [69]. Quite clearly this cannot be regarded as closed forest. Apparently woody plants are never so dense that a grassy undergrowth is completely precluded (Plate XXVI*a*).

No clear stratification is apparent in these communities and although shrubs are scattered beneath the trees in many places, they are by no means universal. The dominant trees are usually no more than sixty feet in height and are mainly flat-topped and spreading. It is noteworthy, however, that quite a number of these species also occur in the 'dry evergreen' closed forest where they adopt a much more compact and upright form. The grasses dominating the undergrowth are mainly tall and belong to a large number of genera.

Throughout its range, this vegetation type experiences a prolonged seasonal drought of between four and seven months—a drought which,

PLATE XXVII

A. THE CAATINGAS. A nineteenth century drawing showing the generally low, open cover of thorn trees and cacti, the sparse herbaceous vegetation and the occasional occurrence of taller trees like the 'barrigudos' (*Cavanillesia arborea*), and the palm (*Cocos coronata*). State of Bahia, Brazil.

(After Martius in Schimper, 1903.)

B. AUSTRALIAN MICROPHYLLOUS WOODLAND. Scrub-woodland dominated by mulga (*Acacia aneura*).

(From a photograph by Sir T. W. E. David in *Géographie Universelle*, Vol. X, 1930.)

A. AMERICAN CACTUS SCRUB. Dominant cactus (*Cereus ingens*) and *Agave sp.* in Mexican semi-desert. Non-succulent herbaceous plants are sparse and recessive. (From a photograph by Stahl in Schimper, 1903.)

B. AFRICAN SEMI-DESERT SCRUB. A drawing of a small area of semi-desert near Kihuiro near the base of Kilimanjaro. Succulent and thorny plants are dominant. (From a figure by Volkens in Schimper, 1903.)

PLATE XXVIII

in many places, is almost absolute for four or five months. On the average between thirty and forty inches of rain falls during the wet season. During the prolonged drought the vast majority of trees and shrubs lose their leaves and the grasses die down completely so that a great mass of combustible material forms a continuous cover over the ground. Because of this, fire sweeps through the forest at very frequent intervals; in many places this may be an almost annual event. Few saplings of the trees and shrubs survive this holocaust and it is noteworthy that the mature trees which dominate these communities all have a thick, corky bark which effectively insulates them. These circumstances lead one to suspect that fire has a most important influence and may be the main factor determining the density of the tree cover. As will be seen later (Chapter XVIII), human interference and fire are probably the main reasons why typical, closed-cover seasonal forests are only local on the African continent. Nevertheless, one cannot rule out the possibility that a certain amount of grasses and herbaceous plants may have been present in the undergrowth under quite natural conditions. Since trees can grow continuously across these areas, however, it is perhaps more realistic to think of the communities as 'grassy forests' rather than 'treed grasslands'; they have been mapped here as seasonal forests (Map 9).

CHAPTER XVII

Microphyllous Forest, Semi-desert and Desert

No abrupt change in vegetation type is found where the deciduous seasonal forests of America and Asia are flanked by even drier regions. Quite often, as the dry season becomes longer and the mean annual rainfall becomes less, the dominant deciduous trees decrease in stature and adopt a more gnarled and spreading habit. Also more and more species disappear from the stands so that, although the communities still compose a closed-canopy forest, this is very low-growing and floristically depleted. Ultimately, through an ecotone of greater or lesser width, the deciduous forest gives way to a different formation-type. Though this is lower-growing it is still dominated by woody plants. This type of community is dominated by low, bushy trees and tall shrubs[1] many of which have small leaves; consequently this type of vegetation has been referred to as 'microphyllous forest'. The term 'thorn forest' has also been applied to this kind of community in many places, and this is certainly appropriate in Africa and the Americas where most of its dominant plants are normally armed with a fearsome array of thorns and spines. It is inappropriate in Australia however; consequently the former term is now thought to be the most suitable one for the formation-type as a whole.

I. MICROPHYLLOUS FOREST

The American formation

The microphyllous forest formation of the Americas is variously referred to as 'thorn woodland' and 'thorn scrub'. If the general cover of thorny shrubs and bushy trees is overtopped here and there by an isolated, taller tree, the former term is used. Since woodland merges into scrub at such frequent intervals, however, and since the same species of thorny, woody plants compose the overall cover in both types of community, it seems

[1] A woody perennial with only one stem springing from the root is spoken of as a 'tree' and one with several stems as a 'shrub'.

reasonable to regard them as belonging to the same formation. The formation is widespread on the northern coastlands and ranges of Venezuela and Colombia as well as in the interior lowlands of the latter and around the Gulf of Maracaibo. It covers a vast area in north-eastern Brazil where it is called the 'caatingas' ('light-forests'). These occupy most of the State of Ceara, the western part of Rio Grande do Norte, Pernambuco and Bahia, much of Piaui and the eastern strip of Goias. Thorn forest is also found further south on isolated limestone outcrops in Minas Gerais. The same formation covers many areas in the West Indies, in particular on outcrops of limestone. The most extensive of these are in interior Hispaniola, western Jamaica, central and western Cuba and most of the Bahamas. Finally, it occupied the northern strip of the Yucatan Peninsula, the driest parts of the eastern coastlands of Mexico from the Isthmus of Tehuantepec northwards to the tropic, and the western coastlands from Guatemala to the fringes of the Sonora Desert.

The thorn forests of Venezuela have been described in some detail [67]. Generally speaking they vary in height between ten and thirty feet and are dominated by a mixture of deciduous and evergreen species, most of which are thorny. It is important to note that the evergreen species have leaves which are entirely different from those of the trees in tropical rain forest. In this very dry environment the evergreen leaves, apart from being small, are hard and tough and possess a variety of extreme xeromorphic adaptations. It seems as if the evergreen habit may be a positive advantage in these areas where the wet season may be extremely short and unreliable; during the short, moist growing periods no time has to be wasted in growing new leaves before photosynthesis can begin. In a very different climatic context, these xerophytic, evergreen trees may be the ecological counterpart of the evergreen conifers of high latitudes (Chapter IV). Species of the pea family (*Leguminosae*), both evergreen and deciduous, are particularly well represented in these thorn forests. Evergreen mimosas (*Mimosoideae*) and deciduous species of *Cassia* have been mentioned [67], the latter having very showy, yellow flowers during the season when they are leafless. Scattered amongst the other thorny plants there are also several species of cacti (*Cereus spp.*) which may grow to a height of thirty feet.

Other synusiae, though not plentiful in this formation, are nevertheless represented. Some slender twining plants are present, though they are by no means so conspicuous as the great woody lianes of the wetter forests. Also, contrary to what might be expected, a number of species of epiphyte are quite common. Of necessity these have remarkably efficient adaptations to reduce transpiration and one of them, a species of Spanish moss (*Tillandsia recurvata*), has small, hard leaves very densely covered with hairs. A few species of remarkably xerophilous epiphytic orchid are also present, one of which (*Jonopsis sp.*) has very showy flowers. Of greater significance, however, is the fact that, in many areas, the herbaceous ground layer is only sparsely represented; grasses in particular may be almost absent [59]. Quite often there is only a sparse cover of succulent plants along with some species of the family *Bromeliaceae*.

The caatingas of north-eastern Brazil, though very similar to the Venezuelan thorn forests, differ in some particulars. The general, though discontinuous, cover of thorny shrubs is here overtopped by isolated taller plants which are often of peculiar form. Tall columnar cacti of various species rise like great candelabra above the general level, as do certain xerophytic palms (*Cocos coronata*). Most striking of all are the 'barrigudos' or Brazilian bottle-trees (*Cavanillesia arborea*), whose great trunks, often swollen to a diameter of as much as fifteen feet, act as water reservoirs (Plate XXVII*a*). Beneath these the thorny shrubs and small trees are predominantly deciduous so that during the long dry season the landscape appears almost entirely lifeless; only the succulent stems of thorny cacti (*Cereus spp.* and *Opuntia spp.*) provide a touch of dull green [67]. Amongst the dominant shrubs, species of mimosa and acacia (*Leguminosae*) preponderate. Epiphytes are rare in this community though slender lianes are fairly frequent. Again, however, over large areas, grasses are rare and the rosettes of prickly, sharp-edged *Bromeliaceae* grow almost unchallenged. This is particularly significant in view of the fact that the caatingas form a much more open community than do some other thorn forests. Attention should be drawn to the fact that much of the typical caatinga is on very pervious, siliceous soils developed on granites, gneisses and similar rocks. Thus, even in the wet season, when the vegetation is lush, green and well-watered, water penetrates downwards very rapidly; only a few hours after rain has ceased, the upper layers of the soil are again perfectly dry. The possible significance of this point will become apparent later (Chapter XVIII).

This apparent success of thorn forest on pervious outcrops is re-emphasised further south in Minas Gerais where, in areas generally dominated by semi-evergreen seasonal forest, limestone outcrops carry a vegetation very similar to the caatingas except that taller trees are more infrequent. This 'thorn bush' usually consists of a dense cover of low, thorny bushes and trees nearly all of which are deciduous and through which an abundance of slender cacti and other succulents push upwards to the light. A very similar type of thorny vegetation was also found on the Tertiary limestones of the West Indies and northern Yucatan though much of this has now been much modified by fire and cultivation. Also, in the lowlands of both eastern and western Mexico, a thorn forest very similar to the caatingas, though often more dense, covers much of the land.

Although the American thorn forest is dominated by the same life-forms throughout and, in consequence, must be regarded as belonging to the same formation-type, it varies greatly in height and luxuriance. This is to be expected in view of the considerable differences in physical circumstances experienced throughout its range. It is found, at its most typical, where the mean annual rainfall is twenty to twenty-five inches or a little less and where six or seven months of the year have negligible rain. It has penetrated into areas where both the mean annual rainfall and the length of the wet season are much greater, however, for two main reasons. Firstly, very pervious rocks can cause really intense drought even in areas like eastern Minas Gerais where the dry season is neither long nor pronounced. Secondly, a great variation in rainfall amounts and length of the wet season can preclude the growth of more demanding trees even where the mean monthly figures suggest that tall trees should be successful; a particularly intense and prolonged dry season occurring only once every ten years is quite sufficient to prevent the invasion of semi-evergreen or even tall deciduous seasonal forest. It is for this reason that thorn forest is found in parts of northern Ceara where the mean annual rainfall is more than sixty inches.

The Asiatic formation

Thorn forests very similar to those of the American continent have been described in India, Burma and Thailand. Because of human interference over a long period of time, the original nature and distribution of the formation in India is difficult to assess but it appears to have been widespread in the dry belt extending south-south-eastwards from the Satpura Range to the coast of Madras as well as along the valleys and depressions of the Punjab and lower Indus Valley. In the dry belt of Burma and the lowlands of south-western Thailand, however, sufficient of the original vegetation remains to give some indication of the original nature of the climax.

Much of the dry belt of Burma was originally covered by two related types of community. The first of these is referred to as the 'sha-dahat thorn forest' because its commonest dominants are the sha (*Acacia catechu*) and the dahat (*Tectona hamiltoniana*). Apart from the latter, nearly all the dominant trees and shrubs are thorny, the tallest of them attaining a height of about thirty feet. Woody climbers infest many of the trees and it appears that, even when undisturbed, this community possessed a considerable grassy undergrowth. The second type of community, referred to as 'sha thorn scrub', is found in even more arid parts of the dry belt. Here the sha becomes quite predominant but grows to a height of no more than six feet. Even the dahat, which is still present as a sub-dominant, only develops into a bush a mere three feet in height. Grasses and some rosette plants again form an undergrowth but there is a great deal of completely bare ground beneath the shrubs.

The thorn communities of Burma are found on both light and heavy soils but the former permit a much greater luxuriance of growth quite regardless of climate. Where sands and clays are in close contiguity, thorn trees are found on the former and bushes on the latter; where the rainfall is sufficient to support thorn trees on the heavier soils, dry teak forest is present on the pervious ones [70]. A patchwork of communities related to geology has thus arisen and the boundaries between communities are frequently very abrupt. As a consequence of these relationships, sha thorn scrub usually occurs on sands where the mean annual precipitation is less than thirty inches, whereas it probably never occurs with less than thirty inches on clays.

The African and Australasian formations

Although plant communities dominated by microphyllous trees and shrubs are of very frequent occurrence in both Africa and Australasia, they do not seem to cover large areas so continuously as in the continents already discussed. Though various species of acacia, along with other trees of similar form, are almost universal in a belt south of the Sahara from Guinea to Kenya, as well as in vast areas down the east of the continent from Somalia to the Union, they are predominantly in open communities in which grasses predominate. Dense thorn scrub or thorn forest, occurring extensively, has only been described in Tanzania, interior Natal and south-western Madagascar. Elsewhere, in generally semi-desert areas, it often occurs in fringing zones along river banks and in 'wadis'; it also occurs sporadically within general areas of savanna (Chapter XVIII). The 'bush woodland' of Tanzania, as described by Engler [67], appears to be very similar to the American thorn forest. It is dominated by spiny trees, some of which are deciduous but many of which are evergreen with heavily cutinised leaves; it is also quite rich in slender climbing plants and even possesses a few species of xeromorphic epiphytes. Significantly also, this forest has been described as being quite dense with only sparse undergrowth.

In Australia again, in spite of the great range of certain species of very xeromorphic, microphyllous shrubs and trees, the communities in which these occur appear to be mainly open and grassy. North of the area covered by sclerophyllous mallee (Chapter X), however, in most of the western interior of the continent, isolated patches or thickets of trees, usually lacking thorns and up to thirty or forty feet in height, can be found (Plate XXVII*b*).

In many places these are dominated by the mulga (*Acacia aneura*); elsewhere, particularly in the interior of Queensland, by the brigalow (*Acacia harpophylla* and *A. glaucescens*). Numerous species of wattle (*Acacia spp.*) along with the weeping myall (*Acacia pendula*), the pepper tree (*Drimys aromatica*) and others are all found in these communities. Locally also, particularly in Queensland, these communities are overtopped by isolated taller trees. One of these gives the landscape an appearance which is particularly reminiscent of the caatingas of Brazil; this is the Australian bottle-tree (*Sterculia spp.*) whose shape and functions are very similar to those of its South American equivalent.

2. SEMI-DESERT SCRUB

On its dry side, tropical microphyllous forest gives way to semi-desert scrub. In this formation-type, succulent plants are often more important, the individual plants are more widely spaced and, normally, the average height of the dominant plants is considerably lower than in microphyllous forest. In many places the latter gives way imperceptibly to the semi-desert scrub through a wide zone of ecotone, but this should not be permitted to obscure the fact that many of the dominants of the two formation-types are quite distinct both phylogenetically and physiologically. The dominants of microphyllous forest are characteristically deep-rooted; they compete most successfully on permeable soils into which the water from occasional showers can penetrate quickly and deeply. The typical dominants of semi-desert formations are quite different in that, regardless of soil type, they use the majority of their roots to exploit only the uppermost few inches of the soil. In many of these areas the sub-soil and underlying rock are rarely if ever moistened, so that the plants which can most successfully utilise the water in the uppermost, moistened layer, before it is lost by evaporation, are most likely to survive. The large areas of apparently untenanted soil between the individual dominants are quite misleading; in reality the whole of the area is exploited at very shallow depth by a comprehensive network of roots. It is only where impeded drainage maintains a water-table at shallow depth that deep-rooted species can predominate.

The American formations

The semi-desert scrub of south-western U.S.A. which has already been described (Chapter X) is a sub-tropical extension of formations which are widespread in tropical America. Because of the importance of cacti (*Cactaceae*), these formations have often been referred to as 'cactus scrub' [59] but it must be remembered that numerous other plants, particularly low-growing, thorny shrubs, are almost universally co-dominant. In South America it is extensively represented along the fringes of the Atacama Desert, from southern Ecuador to central Chile, and by an outlier on the opposite side of the Andes in north-western Argentina. This formation is also found locally in Venezuela where it covers much of the small island of Patos [59]. The North American formation covers most of the lowlands of Lower California, the north-eastern part of the Mexican Plateau and the Sonora Desert from Mazatlan to the lower Colorado River [35].

These American formations occur in all stages of luxuriance, from an almost complete cover of flourishing succulents and shrubs to sparse communities which leave most of the ground bare. It is noteworthy, however, that communities of all stages of luxuriance can be found in which grasses are rare or absent. In South America cacti, predominantly of the genera *Opuntia* and *Cereus*, are dominant along with dwarf, small-leaved shrubs amongst which species of the family *Leguminosae* are particularly frequent. In North America many genera of cacti, along with species of *Agave*, dominate the communities (Plate XXVIIIa). In both formations, however, the lowest layer of the vegetation often consists of xerophytic shrubs and succulents rather than grasses.

The semi-deserts of Mexico are particularly interesting in that they occur in two distinct types of environment. On the one hand there are the frostless lowlands of the Sonora Desert and, on the other, the much cooler slopes and basins of the mountains and the Plateau. In spite of differences in species necessitated by this contrast, however, these two sets of communities are very similar. The *Cactus-Agave* vegetation between Oaxaca and Tehuacan, at a height of above 5,000 feet at the southern end of the Plateau, was described in some detail by Karsten in the late nineteenth century [67]. Here, although there are very hot days, the nights are cold even in summer; frosts are frequent in the winter half of the year and even snow falls on occasion. In spite of this the vegetation is very rich in species; many species of cactus belonging to the genera *Echinocactus, Mamillaria, Cereus, Opuntia* and *Echinocereus* are very common along with other succulents such as species of *Agave, Sedum* and *Echeveria*. Most of these are now propagated in many parts of the world as garden and pot plants. Numerous thorny and hairy shrubs are usually co-dominant with these succulents, representatives of the *Leguminosae* (*Mimosa spp.* and *Cassia sp.*) and *Compositae* being particularly frequent. In spite of this great variety of plants, however, grasses are scarce and, to a greater or lesser degree, bare ground intervenes. It is only after infrequent heavy rain that these communities appear at all grassy. This is because of the presence of a number of ephemeral species which complete their life cycles in a very short period and whose seeds are then capable of lying dormant for a long time.

The south-east Asian semi-desert

Semi-desert scrub is represented in south-eastern Asia by a 'formation' of only limited extent. It is only in the driest parts of the dry belt of Burma that it is found and, even then, only on alkaline soils. It seems to be localised mainly where a gypseous pan has developed in a clay sub-soil. These communities are usually referred to as '*Euphorbia* semi-desert' because the dominant plants are mainly thorny, succulent species of the family *Euphorbiaceae*. These low bushes are often far apart with a good deal of intervening bare soil, but grasses are much more frequent and of much ranker growth than in any part of the American formations. Because of its localisation on heavy soils with impeded drainage, however, one must have some reservations regarding the ecological status of *Euphorbia* semi-desert; there can be little doubt that it would be replaced by deeper-rooting sha thorn scrub if the soil were freely drained. Since the American formations are found over a large range of soil types, these local Asian plant communities must be regarded as of questionable status.

The Australian formation

Much of the dry interior of Australia, where the mean annual precipitation is below fifteen inches, is characterised by a discontinuous vegetation of semi-desert scrub. Indeed, much of the area which is often referred to as 'desert' has a permanent vegetation of perennial plants. Many of these have remarkable powers of drought-resistance, being capable of surviving over periods of two or three years of almost absolute drought. Although relatively little has been written about the exact distribution and nature of these semi-desert communities, it is clear that they differ considerably in life-form from their equivalents on other continents. None of the common dominants here has adopted the compact, succulent life-form; almost entirely, survival has been achieved by the development of thick, corky bark, the reduction of leaf-size and the reduction of the rate of transpiration by various other means. Species of mallee (*Eucalyptus spp.*) and wattle (*Acacia spp.*) are very common throughout, but these are associated with large numbers of other species. Also of very frequent occurrence is the porcupine grass or spinifex (*Triodia hirsuta* and *T. irritans*). This plant has become infamous for the almost ferocious way in which it protects itself from disturbance by animals; its sharp-edged, finely-pointed leaves (Plate XXIX*a*) can inflict deep cuts not only in human flesh but also in that of cattle and horses. It is mainly because of the drought-resistant spinifex that large areas of central Australia, with a mean annual rainfall of less than ten inches, have a substantial vegetation cover. It can invade areas of drifting sand thus giving them some degree of stability and, since it may be the only native plant that can withstand such sandy conditions, it should perhaps be regarded as the true climatic climax dominant on this type of surface. Distinct areas of spinifex are intercalated with areas of semi-desert shrub in many places and, since the spinifex is neither a shrub nor a succulent plant, it will be understood why semi-desert vegetation in Australia presents an apparent anomaly.

The Indo-Saharan formation

Between Mauritania and the Thar Desert there lies an unbroken dry tract where tropical semi-desert plant communities occupy a far greater area than on the whole of the remainder of the earth. These communities are often very sparse and impoverished but, wherever they are well developed, they show remarkable similarities in structure and in the species of which they are composed. Furthermore, their dominant life-forms are very similar to those composing the cactus scrub of the Americas; thorny and succulent plants predominate throughout. There appears to be almost every reason for regarding this as a formation of the same formation-type. Late in the nineteenth century the German botanist Volkens gave a vivid description of this scrub as it occurs in eastern Africa between northern Tanzania and the southern half of the Red Sea [67]. He presents a picture of an open community with its dominant plants no more than four to six feet in height in which many of the species are veritable caricatures (Plate XXVIII*b*). There are 'trees' such as *Euphorbia tirucalli* which have trunks as thick as one's body but which bear crowns of branches, pendant and quite leafless, which are no thicker than the finger. These are intermixed with thorny bushes belonging to the genera *Acacia, Salvadora* and *Gymnosporia (Celastraceae)* over which sprawl vines and creepers like *Cissus quadrangularis* and *C. rotundifolia.* Thorny, cactus-like plants such as *Euphorbia heterocproma* are almost universal along with other succulents of a great variety of forms. Probably the strangest form of all is represented by *Pyrenacantha malvifolia (Olacaceae)* and *Adenia globosa (Passifloraceae).* Each of these plants consists of a huge tuber which is often about three feet in diameter and projects above the ground to a height of about three feet; its surface, according to species, is brown and leathery or dark green and granular. From the crests of these structures, shoots are thrown out at the beginning of the moister season; these grow rapidly and twine themselves up and over the surrounding bushes. A great deal of the ground between these bizarre dominants is bare but a thorny herb, *Blepharis togodelia (Acanthaceae),* almost covers the ground in many places.

This formation is best developed where the rainfall, though slight (in many places less than ten inches), nevertheless occurs regularly in a distinct moister season. In the northern Sahara this tends to be the winter and early spring while in the southern Sahara it is the summer. Most of the plants have a definite régime; they are either deciduous or, even if evergreen, they throw out flowers and produce seeds during the most favourable part of the year.

The south-west African formation

A distinct formation of semi-desert vegetation also borders the Namib Desert of south-western Africa on its eastern flank. It extends from the coast in central Angola as far as the northern Karroo. Though the dominant species here are not closely related to those in the Indo-Saharan formation, nevertheless they belong to the same life-forms. Succulents belonging to the *Euphorbiaceae, Crassulaceae* and *Compositae* are frequent, the melkbosch (*Euphorbia sp.*) being a particularly conspicuous and drought-resistant one which flourishes up to the very edge of the driest desert. These are intermixed, to a greater or lesser extent, with xerophytic shrubs belonging predominantly to the *Leguminosae* and *Compositae*. On the Karroo plateau, succulents are less prominent and shrubby plants predominate; the Karroo bush (*Pentzia spp.*) in particular dominates over wide areas and provides valuable natural forage for sheep and goats. In most communities of this formation the ground is almost bare of grasses and other herbaceous plants.

3.   TROPICAL DESERTS

Complete 'deserts' of any considerable extent are very rare in tropical regions; here and there one can find areas of drifting sand or bare rock which, to all intents and purposes, are devoid of plant and animal life but usually, within any few square miles, there is the odd hollow or crevice where life has found a foothold. Some of the great sandy deserts or 'ergs' of the Sahara, the Rub' al Khali of southern Arabia, Gibson Desert in Western Australia and parts of the interior valley of the Atacama Desert in northern Chile are almost lifeless over large areas however. The term 'desert' has thus never been applied rigidly and exclusively to areas which are literally lifeless. There are vast 'desert' tracts which, for much of the time, appear to be completely bare but which are revealed to contain the dormant organs of many plants by a closer examination. These plants only become obvious during periods of a few weeks at a time following infrequent showers of rain.

A distinction must be made between two quite different types of desert plant. Firstly there are those, to which reference has already been made, which are nourished directly by rain and, in consequence, must spend most of their lives in a dormant state. Secondly there are those which occupy depressions in the surface from which it is possible for roots to penetrate to the water-table. These two types must be adapted, both morphologically and physiologically, to quite different environmental conditions.

Of the rain-dependent desert plants some are annuals and some are perennials. These are similar in the depth of their rooting systems, both types rooting quite superficially in the layer which is wetted by very slight showers. Because of the great rapidity with which they are able to complete their life cycles, some of the annual plants are quite remarkable for their lack of obvious xeromorphism; their slender stems and roots, thin leaves and delicate flowers are not in the least what one might expect of small desert plants. They owe their survival entirely to the ways in which their seeds are protected against desiccation. In many cases this is effected by means of an almost completely impervious fruit- or seed-coat; in others the entire dead remains of the plant form a protective mechanism around the seeds. A striking example of the latter adaptation is found in the 'Rose of Jericho' (*Anastatica hierochuntica*) which is found in the Middle East and the eastern Sahara. After the fruits have ripened, the whole of this plant rolls up into a tight ball and remains in this attitude so long as there is no rain; as soon as thorough moistening occurs, however, the structure uncoils, the fruits open and the seeds are released on to the moist soil surface (Fig. 26).

Several other plants have been described which protect their seeds in an exactly similar way and, coupled with the fact that the whole structure frequently breaks free from its root and trundles over the desert before the wind, this makes it one of the most effective methods of seed protection

Fig. 26. Dead plant of the 'Rose of Jericho' (*Anastatica hierochuntica*) in a moist condition (left) and a dry condition (right). (From Marilaun and Oliver, 1904.)

and dispersal that has been observed. Other annual desert plants survive by storing water in succulent leaves and stems during the short moist period and using it up in the production of flowers and seeds during the beginning of the ensuing drought. Saharan plants like *Mesembryanthemum crystallinum* and several species of the *Resedaceae* and *Cruciferae* have adopted this mode of life. Most of the perennial rain-dependent plants are also herbaceous. Indeed, a very large proportion of those in the Egyptian Sahara appear to be bulbous *Monocotyledones*. These are able to survive the long, rainless periods by storing water and food in an underground bulb, heavily protected with scale leaves, and, on the basis of this, to send up shoots and flowers with great rapidity if stimulated by a moistening of the soil. In other deserts these perennial plants seem to have developed other kinds of underground storage organs. In the Namib, for instance, underground tubers and root-stocks are more frequently encountered.

The desert plants which are dependent upon ground-water cannot, in a certain sense, be regarded as true climatic climax species. They are too bulky and demanding to survive on the actual rainfall that falls on any area that could possibly be exploited by means of a single rooting system. It is only where relief causes run-off and seepage water to concentrate underground that they are found. These plants are thus very localised in depressions and wadis, at spring and seepage lines and on low-lying plains and alluvial fans. Many of the larger areas formerly occupied by these plants, such as the Saharan oases and the 'little Egypts' of Peru, have now been so completely transformed by cultivation that the exact nature of their original vegetation is obscure. Nevertheless many smaller and less-favoured areas are sufficiently undisturbed to provide valuable information. During and immediately after short, heavy rain storms, water rushes into the Saharan wadis in enormous quantities but very little, if any, reaches the sea or even a major lake; almost the whole of it sinks into the floor of the watercourse which rapidly becomes 'dry' again. For a considerable subsequent period, however, a large amount of subterranean water is available there, and this can be exploited by any plant which has sufficiently long and efficient roots. In many places, however, all the available water is likely either to have been used or to have seeped away before the next storm occurs. Consequently, although water is available to these ground-water plants for a much greater length of time than is the case with the rain-dependent ones, the former must also be able to survive complete drought; in most cases they are highly xeromorphic and adapted to long periods of dormancy.

The great majority of these plants are perennials and many are spiny shrubs. Quite a number of them appear, from their above-ground organs, to be small and squat; it is only when their roots are examined that their great length is discovered. In many species the length of the rooting system is at least twenty times as great as that of the stems. Typically the leaves of these plants are either very small or replaced completely by thorns; all the green organs are heavily protected by thick cuticles, waxy coverings or hairs. There are several exceptions to this general rule, however, of which the Saharan colocynth (*Citrullus colocynthis*) is particularly noteworthy. This plant of the gourd or cucumber family (*Cucurbitaceae*) has long, soft shoots bearing large, broad leaves which it retains throughout the whole of the summer. During this growing season it also produces several fruits the size of large melons. On encountering this plant in the midst of its xeromorphic neighbours one would almost be prepared to accept the postulate that it is protected from excessive transpiration in some unknown, almost magical way; it does not seem feasible that such a plant could remain firm and unwilted throughout the savage heat of the Saharan day. In fact it is able to do so only because of its unbelievably efficient and extensive rooting system which taps the water-table permanently and supplies water to the stems and leaves sufficiently rapidly to offset the enormous transpiration rate. Though this case is unusual, it serves to illustrate how a plant with an efficient rooting system can afford to lose water at the maximum rate while the majority of its neighbours have overcome the main problem of the environment in exactly the opposite way—by modifying their above-ground organs to restrict transpiration.

It is in this desert environment, where water is available in depth, that several of the most grotesque and weird members of the plant kingdom have evolved. Probably the strangest of all is *Welwitschia mirabilis* (*Gnetaceae*) which grows on flat, stony plains in the northern part of the Namib Desert (Plate XXIX*b*). Its aerial organs consist of a woody, hollow structure covered with reddish bark which attains a height of no more than six inches but which may have a circumference of ten or twelve feet. From the rim of this huge, woody stem there grow two large, ribbon-like leaves which maintain continuous growth from their bases throughout life. The flowers are produced inside the great, saucer-like depression on top of the stem and a tap root projects downwards from it to a considerable depth. This remarkable plant may attain an age of 100 years or more and because of its obvious adaptation to environment and the vast differences between it and its nearest relatives, it is thought to have been evolving in a desert environment for millions of years. A second freak of the plant kingdom is the naras (*Acanthosicyos horrida*) which is often closely associated with *Welwitschia* in the Namib. The naras is a thorny shrub of the family *Cucurbitaceae* and it bears fruit superficially resembling large oranges. Frequently it appears to have taken root on the crest of a mobile sand-dune but observation has shown that this is very far from the truth. Normally it roots on the flat floor of the stony desert into which it sends down long roots to tap the water-table. When once this is achieved the naras is capable of very rapid growth; frequently it is swamped by drifting sand but soon grows to the surface and re-forms a bush at the crest. In this way a dune gradually develops around the obstruction afforded, in the first place, by the naras bush; ultimately this is quite important in obstructing the drift of sand into the interior. Although a naras bush may appear to have a height of a mere four or five feet, the total length of the plant may thus be something like thirty or forty feet, more than three-quarters of which is root.

In dealing with the plant-life of deserts one is concerned with individual plants as distinct from plant communities. In this environment a plant can be regarded as being in direct relationship with its physical environment whereas, elsewhere, plants grow so close together as to modify the environment for each other. Even in semi-desert scrub, where much bare ground is normally apparent, wind speed, humidity and even temperature near the ground are different from what they would be were there no plant cover. Furthermore there is certainly vigorous root competition beneath the soil in semi-deserts; this is usually unimportant beneath desert. Indeed, if a physiological distinction between climatic climax desert and semi-desert is to be made, this last criterion might well be regarded as the most satisfactory.

CHAPTER XVIII

Tropical Savanna

Nature and distribution

In the foregoing chapters only passing reference has been made to tropical savanna; indeed much emphasis has been laid upon the fact that grasses are rare or absent in many types of community ranging from the tropical rain forest to the desert. This may appear very strange in view of the fact that vast areas of savanna do exist in the tropics. The term 'savanna' probably had its origin in Central America as a Carib word for marshes, scrublands or any area not covered with forest [59]. It is now used throughout the tropics but its sense has been narrowed to apply to plant communities in which grasses (*Graminae*) and sedges (*Cyperaceae*) are important. Even so it has been used to refer to a great range of communities from unbroken, treeless grassland to woodland in which trees and shrubs form an almost continuous cover but where there is a continuous grassy undergrowth. It is probably not without significance that there are all gradations from completely herbaceous communities to those in which a forest cover is interrupted by no more than the occasional opening of the canopy.

It has been found convenient by some authorities to divide savannas into three categories on the basis of the grass cover that they support [69]. It must always be remembered, however, as already mentioned, that the amount of tree-cover varies enormously within each category. Furthermore the three types are distinct neither in structure nor in species; they grade imperceptibly into each other. The most luxuriant type of savanna has been referred to as 'high grass–low tree savanna' [69]. It occurs extensively only in Africa. The name 'elephant grass savanna' has also been used since the elephant grasses (*Pennisetum spp.*) are common dominants; they are nearly always mixed with other tussock grasses, however, belonging primarily to the genera *Andropogon*, *Imperata* and *Hyparrhenia*. These grasses are never less than five or six feet high at the end of the growing season, indeed they commonly form a dense jungle ten or twelve feet in height. The trees in the elephant grass savanna are predominantly deciduous and no more than thirty or forty feet in height but where they occur in larger groves, beneath which the growth of grasses is somewhat suppressed, greater heights may be achieved and evergreen species, typical of the African 'semi-evergreen' or 'dry' forests, often appear. The high grass–low tree savanna occupies two main belts; the first extends eastwards from Portuguese Guinea and Sierra Leone across northern Ghana, central Nigeria, the northern Congo and southern Sudan to Uganda and the second one across the south-western Congo and northern Angola. Both zones thus flank the tropical rain forest belt with which they interdigitate considerably.

The second type has been referred to in Africa as 'acacia–tall grass savanna' but analogous communities in other parts of the tropics have various local names. The dominant plants again are tussock grasses which form an almost continuous cover over the ground even beneath the trees. At the height of the growing season the grasses attain a height of two to five feet. They comprise a large number of species, each tropical continental area having its characteristic communities. The trees may be either deciduous or evergreen. Deciduous species of *Acacia* and *Combretum* are commonest in Africa (Plate XXX*a*), evergreen *Eucalyptus* in Australia and species from a great range of families in South America. In the last case many of the species are evergreen and large-leaved and capable of growing to a height of fifty or sixty feet [14]. In Africa the acacia–tall grass savanna occupies a continuous belt from Gambia and Senegal eastwards across northern Nigeria to central and south-eastern Sudan and is also widespread in southern Kenya, eastern Tanzania, Mozambique, Rhodesia, and in Botswana up to the fringes of the Kalahari It also occupies much of western Madagascar. In South America the extensive 'campos' of the Mato Grosso and much of Goias (Plate XXX*b*) are similar in some ways and the Llanos of the Orinoco Basin (Plate XXVI*b*), though more deficient in trees, have a similar grassy vegetation. In Australia a great belt of similar savanna extends southwards from east-central Queensland far beyond the tropic into New South Wales and northern Victoria. Finally, smaller enclaves of savanna, with grasses of medium height, are found within generally forested areas or between intensively cultivated lands in southern and south-east Asia; they are of common occurrence on the north-east of the Deccan in India and are found even within areas of tropical rain forest in Borneo and New Guinea.

The third and final category of savanna is characterised by a discontinuous cover of highly xerophilous desert grasses along with a scattering of small thorny trees or bushes. Much of the soil is completely bare. In Africa this type of community has been referred to as 'acacia–desert grass savanna' or 'orchard steppe'. Here deciduous species of *Acacia* predominate over other woody plants and species of *Aristida* are by far the commonest grasses. An almost continuous belt runs along the southern fringes of the Sahara from northern Senegal to Eritrea and

A. PORCUPINE-GRASS. Typical tussocks in central Australia.
(From a photograph by Sir T. W. E. David in *Géographie Universelle*, Vol. X, 1930.)

B. SOUTH-WEST AFRICAN DESERT. *Welwitschia mirabilis* on a sandy plain in
South-west Africa.
(From a photograph by Schenck in Schimper, 1903.)

PLATE XXIX

A. ACACIA–TALL GRASS SAVANNA. A typical landscape near Meru in Kenya.
(From Shantz and Marbut, 1923.)

B. SOUTH AMERICAN SAVANNA. Typical campo cerrado on plateau surface
near Lagoa Santa, Minas Gerais.
(From Monica M. Cole, 1961.)

PLATE XXX

extensive areas are also found in south-eastern Ethiopia and northern Kenya. In southern Africa much of the Kalahari in Botswana and South-West Africa is covered by very similar vegetation. This most arid type of savanna is also to be found on all the other continents which extend into the tropics. In Australia it is extensive in interior Queensland, Northern Territory and in the western and north-western parts of Western Australia. In many places here the grasses which accompany the shrubs and low trees make relatively good natural forage; indeed, in places, woody plants almost disappear and unbroken sweeps of pure tropical grassland are to be found. Parts of the Barkly Tableland and Kimberley Plateau are of this nature. In India there are large areas of desert grass savanna on the driest parts of the western and south-eastern Deccan and, on the Mexican Plateau of North America, the true semi-desert scrub gives way in many places to a vegetation with a strong admixture of desert grasses.

Climatic correlations

Because of its great extent, many nineteenth-century plant geographers were of the opinion that savanna must come into existence under the influence of a particular type of climate. Schimper [67] expressed himself quite categorically on this point; he said that savannas generally are found where there are clearly-defined wet and dry seasons, the latter being characterised by a very dry soil and a very dry atmosphere and the former being reliable and little broken by dry intervals. He thus explained the predominantly woody vegetation of the caatingas of the Sertão in terms of the unreliability of the wet season in that part of north-eastern Brazil; here the soil is apt to dry out completely at the height of what should be the growing season so that shallow-rooted grasses cannot compete with deeper-rooted woody perennials.

Subsequent surveys and research have shown quite conclusively that no such 'savanna climate' exists. It has already been demonstrated that communities dominated by woody perennials and almost devoid of grasses are to be found in all types of tropical climate from the perennially wet to the semi-desert. This being the case, there can be no question of a particular sort of climate automatically causing the development of savanna.

The possible effects of fire

During the present century, an increasing number of authorities have asserted that frequent fires over a long period of time would be quite sufficient to reduce forest to some kind of savanna. It may seem inconceivable that vast savannas such as those in Africa can owe their existence to human activity but, so far as large sections of them are concerned, there is much evidence in favour of this. It is perhaps significant that savanna is more widespread in Africa than in any other continent since

Rvs

archaeological evidence indicates that man, at a relatively advanced stage
of culture, has been present here for at least 10,000 years. There is even
an eyewitness account of great fires covering the landscape as early as the
fifth century B.C.[1] Deciduous and semi-evergreen forests with dense
undergrowth burn easily and fiercely during a prolonged dry season and
it is doubtful if more than a very few species in climatic climax forest
could survive such treatment. It seems likely that most of them would
perish and their place be taken by herbaceous plants and by the few
indigenous woody plants whose bark was sufficiently thick or fire-resistant
to permit them to survive. Indeed, in such a situation, it is possible that
particularly fire-resistant species of tree and shrub which were not present
in the original community, would be permitted to invade. When once this
change had been effected and a dense cover of grasses, shrubs and low
trees had taken the place of the original vegetation, the fires would tend to
be even more intense than before. It is an established fact that the high
and tall grass savannas of Africa today are burned off almost universally
during the dry season. In many places this is done purposely to clear away
the old, dead material in order to ensure a better growth of young grass
in the following season. There can be no doubt regarding the fire-resistant
nature of the majority of savanna trees. Many writers have noted the great
frequency with which palms occur in American savannas (Plate XXVI*b*)
and how 'grass trees' (*Xanthorrhoea sp.*) often stand isolated in Australian
grasslands. All these trees are particularly fire-resistant because of the
peculiar structure of their trunks. The baobab (*Adansonia digitata*), with
its thick bark and sponge-like, water-saturated wood, usually stands
isolated in the tall grass African savannas and nearly all the other common,
woody species of the high and tall grass savannas have very thick, corky
bark. In a situation where fire sweeps through the vegetation at frequent
intervals it is obviously very difficult for young trees and bushes to become
established; any woody plants which are no higher than the surrounding
grasses when a fire occurs are likely to be destroyed. It is only when a
young tree has managed to grow above this level that its chances of
survival are anything more than negligible; a remarkable series of
accidents must be necessary for a sapling to survive wherever the grass
grows tall and dense. In certain situations it is clear that a certain degree
of grazing by native cattle must be favourable to woody growth; the
animals reduce the grass in the wet season so that there is less to burn in
the dry season. If the cattle are too numerous, however, they themselves
will kill young woody plants by persistently removing their foliage. It is
probably because of this that thorny, repellant trees and shrubs such as
acacias have become so numerous even in areas where the wet season
appears to be sufficiently pronounced and prolonged to support much
less xerophytic species of tree. Because of the complex interaction of fire,

[1] The voyage of Hanno of Carthage, *circa* 480 B.C.

grazing and numerous other factors, one can see why the nature of the savanna may have become so different from place to place even within the same region. In one locality circumstances have combined to leave the landscape completely treeless while, not so far away, groves and trees survive in profusion.

A further observation which has caused some authorities to abandon the idea of 'climatic savannas' is the extreme sharpness of the present boundary between savanna and forest. Where, within a few yards, one can pass from open grasslands to almost closed forest, there is obviously no question of climate being responsible. This has been noted in a number of places in Africa where 'high grass–low tree savanna' frequently gives way abruptly to 'dry evergreen forest' or even to tropical rain forest. It is conjectured that the forest canopy was opened by persistent native cultivation [38]; this permitted grasses and other herbs to invade on to the abandoned fallows, fires became more frequent and, in consequence, tree regeneration became very difficult. Most of the thin-barked species of tropical rain forest were incapable of surviving in these conditions and thicker-barked ones intruded. Examples of the latter are the African mahogany (*Khaya senegalensis*) and *Lophira alata*.

In an effort to clinch the 'pyrogenous theory' a number of foresters and ecologists in Africa have carried out ecological experiments and made careful observations of ecological successions during the past half century. In the Zamfara forest reserve south of Sokoto in north-western Nigeria it was noted that a very considerable increase in woody plants took place spontaneously in enclosures of grassland which were protected from fire [43]. In the same area it has been observed that steep and rocky slopes, over which fire cannot run and where cultivation has never taken place, are often dominated by woody plants when the more level areas near by are predominantly grasslands. Further observations of this kind are required over a much wider area before inferences can be drawn with any degree of confidence. Even if the results of such observations should be mainly negative, however, this still would not prove, of necessity, that the savanna is true climatic climax. Many areas have been so denuded of forest trees that there are now few, if any, left to provide seed for the re-invasion of the savanna. Furthermore, profound alterations in the composition and structure of the soil could have taken place because of the removal of a former forest cover so that conditions are now quite unsuitable for the germination of tree seeds or the establishment of seedlings. If this is the case it might take centuries or even millennia for weathering processes and ecological development to prepare the surface for the re-establishment of forest. Indeed, in some cases, the whole of the present land surface might have to be dissected and removed by natural erosion before conditions could again be suitable for the climatic climax vegetation (Chapter XIX). It seems likely, however, that over wide areas no such drastic transformation of the soil has taken place. Observations made in the Bahr el Ghazal and Equatoria Provinces of southern Sudan have shown that only a short period of protection from fire permits tree seedlings to establish themselves and to grow freely in many parts of what are now savanna grasslands [54]. The Dinka tribesmen here burn the dead grass with great thoroughness and regularity during each dry season and nearly all trees and shrubs succumb before reaching maturity. Wherever a small patch of ground, by chance, escapes the fire for one or two seasons, however, it has been noted that tree and shrub seedlings, including those of the evergreen 'mahogany' (*Khaya sp.*), grow quite freely.

On a purely floristic basis, one can divide the savannas of West Africa into three main zones running from east to west parallel to the coast. From south to north, these are referred to as 'the Guinea zone', 'the Sudan zone' and 'the Sahel zone'. Throughout the Guinea zone the larger groves of trees contain many evergreen species but the trees scattered as individuals throughout the savanna are mainly very much more xerophilous. In the Sudan zone, evergreen species become unimportant and broad-leaved deciduous ones such as the baobab (*Adansonia digitata*) are found on the more favourable sites. Nevertheless, fine-leaved, thorny shrubs and small trees, particularly species of *Acacia*, are very common. Finally, in the Sahel zone, fringing the southern Sahara, low, thorny trees become absolutely predominant amongst the woody plants and, because of the thinner vegetation cover, fire cannot spread easily. In this zone, in consequence, grazing animals are much more important ecological agents than fire [72].

There is quite a close (though by no means perfect) correspondence between the extent of these three zones and that of the three structural categories of savanna already outlined. The Guinea zone is predominantly high grass–low tree savanna, the Sudan zone acacia–tall grass savanna and the Sahel zone acacia–desert grass savanna. If one accepts the thesis that savannas are mainly plagioclimax communities, it is very tempting to assume that the Sahel zone is an area of climatic climax thorn forest, the Sudan zone one of deciduous seasonal forest and the Guinea zone one of 'dry evergreen' and tropical rain forest. Until much more is known about ecological successions in these areas, however, such ideas as these can be no more than sheer speculation. On the distribution maps presented here (Appendix I), a purely structural classification of savannas has been used. The existing vegetation has been shown, the savannas being sub-divided basically according to the salient characteristics of the trees which they contain.

Edaphic savannas

Despite the growing evidence in favour of the plagioclimax status of much savanna, there are clearly many tropical grasslands where fire and grazing are of little or no significance. In all cases, however, they appear to be due to local soil conditions and not to peculiarities of climate. Along the sides of many rivers within generally forested areas in the Congo Basin and in South and Central America, and even on larger areas such as the Orinoco Delta, there are savannas whose existence is obviously due to the condition of the soil. Some of them have rarely if ever been known to burn; waterlogging of the soil for most if not all of the year prohibits the invasion of forest trees. A further significant point is that the grasses and woody plants composing these 'water savannas' are mainly different from those in the drier savannas in the same regions; sedges, mosses and other marsh plants are plentiful and species like the American buriti palms (*Mauritia spp.*) are of frequent occurrence. It is quite clear that all these savannas are no more than stages or subclimaxes in hydroseres. The existence of further large areas of savanna has also been attributed to some degree of waterlogging. Among these the Llanos of the Orinoco Basin are notable. Both Schimper and Carl Sachs [67] have given detailed descriptions of the Llanos; the area is predominantly grassland but is very frequently interspersed with groves and has a liberal sprinkling of isolated trees. The sub-soil of much of the Llanos is heavy and not very permeable so that, on this level surface, very wet conditions are experienced during the pronounced rainy season. A number of authorities have suggested that this wetness, contrasted with the extreme dryness of the soil during the dry season, may prevent the germination and development of most species of tree. It is not without significance, however, that the Llanos are burned off at very frequent intervals by the transhumant graziers in order to stimulate the growth of new grass during the following wet season. Furthermore, one of the commonest species of tree scattered over the Llanos is the very fire-resistant palm, *Capernicia tectorum* (Plate XXVI*b*). It is thus possible that both fire and grazing may be contributory factors in maintaining the Llanos as savanna.

A similar problem is encountered regarding some upland savannas in Trinidad and British Honduras and others on some low plateau areas in Surinam. In all of these the rainfall appears to be sufficient to support semi-evergreen forest if not tropical rain forest but the level land and very pervious soils carry only grassland with a scattering of bushes or, in British Honduras, pine trees (*Pinus caribaea*) [59]. Similar kinds of savanna have been described in several places in south-east Asia where the surface is level and pervious. In all cases these appear to be old erosion surfaces which have been exposed to weathering and leaching for a very long time (Chapter XIX). In consequence it is possible that the soil has become too poor in mineral nutrients to support the rich forest vegetation which the climate would otherwise allow. A poorer, more open forest or 'heath forest' (Chapter XV) may therefore have come into existence. Because of the well-drained soil and the greater degree of penetration of the sun's rays, such a type of vegetation would dry out much more frequently and thoroughly than normal rain forest. Fires would thus be fiercer and more frequent and the forest would be reduced to savanna. An anomalous type of forest, owing its existence to soil conditions, would thus be transformed into savanna through the agency of fire.

Reconciliation of conflicting theories

Although the concept of 'climatic climax savanna' finds few proponents at the present day, conflicting opinions are still held by those who seek to explain the nature and distribution of savannas. Some have emphasised the role of fire while others have insisted that a combination of climatic, hydrological and soil conditions may be quite sufficient without any firing whatsoever. In the case of formerly cultivated savannas within generally forested areas in Africa, there can be little doubt that evidence is on the side of the former; so far as the 'water savannas' are concerned, the latter are obviously right. As regards the large majority of savanna areas, however, it seems as though the truth will only be reached by following up some of the views expressed by both schools of thought. Although heavier soils are innately more suitable for grasses than lighter ones, particularly in regions experiencing a seasonal rainfall régime, it seems, nevertheless, that trees are able to occupy the ground in most cases unless some disturbing factor like fire tilts the scales against them.

The campos of central Brazil provide an interesting example. It has recently been asserted [14] that these savannas are an 'edaphic climax' vegetation, developed on the large areas of old, undissected peneplain. This argument is very persuasive if one concentrates on distributions; savannas of very similar structure and containing some of the same dominant species are found on this type of surface ranging as far north as quite separate and distant areas north of the Amazon in north-eastern Brazil. Interesting questions arise when one examines the structure and composition of the campos savannas however. These are by no means homogeneous; they grade from almost closed-canopy, semi-evergreen woodland (cerradão), through a treed savanna (campo cerrado) to a very grassy savanna in which trees and shrubs are few and scattered (campo sujo). Added to this there is the fact that many species of tree are common to all these three grades of savanna; indeed some trees of the closed forest (mata), which covers much of the sloping, dissected land, are found in savanna.

Some of the more extreme proponents of the concept of the 'climatic climax formation' have maintained that, given a struggle for dominance between two completely different life-forms under undisturbed conditions, one will ultimately become completely dominant over the other [81]. Many authorities would probably agree, on the other hand, that this cannot be accepted as a universal ecological law; nevertheless there is sufficient sound, ecological reasoning and experience behind it to cause one to look with some suspicion on any landscape, part of which carries trees (scattered or in groves) but with the intervening areas completely treeless. Not only is such vegetation suspect as climatic climax; it is also difficult to see how its nature can be explained by using such a term as 'edaphic climax'. Savanna such as the Brazilian 'campo cerrado' (Plate XXXb), occupying a smooth, featureless surface, poses awkward ecological problems. At a particular place one observes a well-grown, quite healthy tree but, on all sides, the ground is occupied almost completely by grasses. It is not obvious why a tree should have been able to grow to maturity at one point while much of the rest of the ground has not been invaded by trees. Clearly the trees involved here do not *require* to grow at some considerable distance from each other because of competition at root level, since the very same species are found growing quite close together in the cerradão. Until basic problems regarding the physiological requirements of the trees concerned, the nature of their rooting systems and the frequency and nature of fire scars on them have been investigated, no satisfactory conclusions can be reached.

Even the observation that the savannas of central Brazil are limited to the ancient erosion surfaces while forest occurs on all dissected land can be interpreted in different ways. On the one hand it might be maintained that the impoverished soils and the underlying lateritic pan of the plateaux can support only savanna while the slopes, better drained in the wet season and moister in the dry season, are capable of supporting luxuriant seasonal forest. On the other hand, while conceding that a poorer vegetation might be expected on the plateaux, a very strong case can be made for the efficacy of fire in the thinner plateaux vegetation and for the failure of fire to penetrate deeply into the moister, closed-canopy forests. Again, little advance can be expected until a great deal more is known about the nature of both vegetation and soil. In view of the type of vegetation found in a rather similar type of environment in southern Angola and Zambia however (Chapter XVI), one cannot help suspecting that the true climatic climax on these Brazilian plateau surfaces is a kind of 'cerradão'. According to descriptions of the 'cerradão', it is dominated by broad-leaved, predominantly evergreen trees of similar life-form to those in the 'miombo forests' except that the latter are predominantly deciduous. In both these extensive communities, trees cover much of the ground but, because the canopy is thin, a continuous undergrowth of heliophilous grasses is permitted. It is because of this last point that it is almost a matter of taste as to whether they should be called 'savannas' or merely 'grassy forests'.

The Australian savannas

The difficulty of differentiating between forest and savanna is also well illustrated by the belt of vegetation which extends from east-central

Queensland southwards across New South Wales to north-central Victoria. Here the first white settlers encountered a vegetation which was different from the rain forest which they had found on much of the seaward side of the Great Dividing Range. This interior vegetation was dominated by tall eucalyptus trees whose branches formed a canopy over the ground in many places. This canopy was so thin, however, that a great deal of direct sunlight reached the ground even when the sun was quite low in the sky. This high light intensity permitted the growth of light-demanding grasses over the whole of the forest floor (Plate VI*b*). The eucalypts are ecologically peculiar in that, in spite of their great size, they are markedly xerophilous. Their leaves are small and their rooting systems exploit the lower horizons of the soil with very great efficiency. It is perhaps because of this last point that an under-storey of shrubs may not have been a common feature in these communities; the grasses, exploiting the uppermost soil horizons, thus had no further competitors for space and light.

Though most of these open eucalyptus communities have now been destroyed by grazing and cultivation, sufficient evidence is available upon which to assess their ecological status. There seem to be good reasons for regarding them collectively as a true climatic climax formation. It is possible, however, that a certain amount of firing, either natural or man-induced, may have caused the undergrowth to be less shrubby and more grassy than it would otherwise have been. Whether the formation should be referred to as 'forest' rather than 'savanna' is, again, a matter for quite subjective decision. At the present time ecological nomenclature is quite non-committal; a certain amount of grass in the undergrowth does not preclude the use of the term 'forest' nor does the presence of an almost closed cover of trees above grassland automatically preclude the use of the term 'savanna'. If the very tentative views expressed in this chapter ever become universally acceptable, however, it might be most useful to reserve the term 'savanna' for purely seral or plagioclimax communities of inter-mingled grassland and trees, and the term 'forest' for all true climatic climax communities in which trees form a complete canopy, quite regardless of the nature of the undergrowth.

Some wider evolutionary implications

It has been suggested that very large areas of savanna may have been brought into existence by the activities of man and that thousands of square miles of continuous savanna now exist in Africa where formerly grasses were subsidiary or almost non-existent. If this is the case, it must have entailed a vast expansion in numbers and range of numerous herbaceous species; many which, formerly, were merely prisere plants or plants of the forest ground-flora, became dominants. Such a transformation would have wide implications regarding the fauna. When Africa was first explored by Europeans it was found to have a fauna which was extremely rich in large, herbivorous mammals; great herds of many species roamed over vast areas. Many of these animals were hooved, fast-running species, adapted to an open environment; they were therefore concentrated in the savannas and open woodlands. All these species must have taken millions of years to evolve. If the extent of savanna and the openness of much of the forest has been increased by man during the past 10,000 years, there are only two alternative possibilities: either these animals were formerly very much fewer and of more limited range or else they have adapted their habits to the new, man-made type of environment. There is evidence that both types of change have taken place. Animals like zebras (*Equus spp.*) and giraffes (*Giraffa camelopardalis*) seem to be adapted, in their protective coloration, to open woodland and the forest edge rather than to open grassland; indeed some of them still inhabit the more woody types of environment. On the other hand, some of the gazelles, for instance, may formerly have been animals of the semi-desert who were able to extend their range when woody vegetation was reduced. Unfortunately it is very difficult to test these theories since, during the past 20,000 years, great changes in climate, with consequent effects on vegetation distribution, are known to have taken place. Even if archaeological and palaeontological evidence can show that a particular animal had a range very different from its present one some 10,000 or 20,000 years ago, this may be due entirely to the fact that climatic change has caused a wholesale shift of vegetation belts; on the other hand it may be due partly to human agency. Much of the Sahara Desert appears to have been well treed during parts of the Pleistocene Period. Climatic change has thus complicated investigations into vegetation history in the tropics, as it has in higher latitudes.

CHAPTER XIX

Tropical Soils subject to Seasonal Drought

CONTRASTS in soil type in tropical regions are just as great as contrasts in vegetation. In between the red soils of tropical rain forest and the bare rock pavements and drifting sands of the tropical deserts, there is both complexity and diversity. Although it is most desirable to study vegetation and its underlying soil as an entity, unfortunately, in much of these areas, this is not yet possible. In the first place far too little is known about the processes of soil formation in tropical regions; the inter-actions between soil and vegetation are little understood. Secondly, on the evidence available there appears to be little correlation between the distribution of soil types on the one hand and of vegetation types on the other. Partly because of the vast extent of various grades of savanna and grassy woodland, the clear distinction between grassland soils and forest soils cannot be made as it can in higher latitudes. At the present state of knowledge one obtains the impression that parent-material is more important as a soil determinant than in the extra-tropical regions of North America and Eurasia. Vast areas of a single zonal type such as the chernozems of Russia and the Great Plains in the U.S.A. do not seem to occur, whereas a geological outcrop of a particular lithological type frequently seems to be co-extensive with a patch of a particular soil type. As knowledge grows and more realistic classifications are made, however, this impression may very well be dispelled.

Generalised soil maps

The very generalised world soil maps that have appeared have made little contribution to realistic thinking about tropical soils. Not only have they shown continuous belts of soil extending across tropical continents, they have applied names to these which imply close relationship to, if not identity with, zonal soils of higher latitudes. Thus belts of tropical 'chern-ozem' and 'chestnut-brown soil' have been shown extending from east to west across Africa just south of the Sahara [69]. In some cases such close relationships may exist; in others, appearances are almost certainly illusory. In all cases, any attempt to state wider relationships is premature. Although the so-called tropical 'chernozems' often come into existence beneath a grassy vegetation in areas subject to prolonged seasonal drought, nevertheless the climate they experience and the types of grass present are very different from those in the chernozem zones of North America and Eurasia. Indeed, a closer examination of the soils within the 'tropical chernozem belt' reveals that many of them are neither black nor calcareous. In fairness to the authors concerned, it must be stated that they have admitted that much of the soil distribution shown on their maps is inferential; they have merely noted the climate and the geology and assumed, on the basis of this, that certain types of soil should be present. This has been done where no soil survey has ever taken place. Thus, speaking of the 'chernozem' in Africa, Shantz and Marbut [69] state: 'On the basis of this climate-topography relationship this group of soils has been extended on the map over wide areas of which we have no definite information whatever as to the nature of the soil, reliance being placed entirely on the rainfall and the apparent character of the surface relief.'[1] Inference and extrapolation on such a vast scale are never safe procedures; they are hazardous in the extreme when the phenomena involved are so imperfectly understood. Such maps may be meaningful to those who produce them; they can be very misleading to those who have never visited the areas to which they refer.

Tropical pedocals and related soils

Many tropical soils, even in areas with a long dry season, are acidic pedalfers. The capillary rise of soil solutions and the deposition of soluble salts during the dry season is completely offset by the torrential rain and soil saturation of even a relatively short wet season. Nevertheless, large areas of freely-drained pedocals do occur in areas where parent-material and climate collaborate to permit the accumulation of lime somewhere in the soil profile. As might be expected, these soils are concentrated in the semi-deserts and in the drier parts of microphyllous forests and tall grass savannas. Some of the 'brown soils' of the Sahelian zone in Mauritania and the Sudan appear to be in this category as are some of the 'plains soils' in Kenya and Tanzania. Some profiles from the latter have been described [53]. They occur beneath both microphyllous forest and grass and may be anything from whitish grey to light red or red-brown in colour. Many contain a continuous horizon of calcareous concretions at a depth of no more than twelve inches. There is also some evidence that soils very like the North American chernozem, with a clearly marked horizon of lime-accumulation, occur in free-drained areas around Bulawayo in Rhodesia and in Queensland, Australia. These tropical black soils are the

[1] Op. cit., p. 162.

centre of much controversy, however, and will receive separate treatment (*vide infra*).

Complexity arises from the fact that many freely-drained soils which occur in just the same climatic regions as the tropical pedocals are non-calcareous. Thus, in Kenya and Tanzania, the calcareous 'plains soils' are matched by comparable areas of non-calcareous 'plains soils'. This is remarkable in that both categories have formed in similar conditions of precipitation and ground hydrology. As far as can be seen, precipitation is so infrequent that it is nearly all lost either by run-off or evapotranspiration; little percolates deeply and the sub-soil is almost permanently dry. There are two possible explanations why some of the soils can remain non-calcareous in such conditions. Either the parent-material is so poor in calcium that there is none to accumulate, or the soil has been developing for such a long time, through various periods of different climate, that all the weathered horizons have been leached of all bases, including lime, during former, wetter epochs. Probably the latter is the likely explanation in the great majority of cases. On the basis of these East African examples therefore, it is clear that one should not expect to find lime-accumulating soils in the tropics merely because the climate alone appears to be favourable.

Tropical black earths

Many of the most valuable soils in the tropics are black, dark grey or very dark brown in colour and can be referred to collectively by the purely descriptive term of 'tropical black earths'. It is very probable, however, that they have evolved in quite a number of different ways. Several of these black soils have been referred to as 'tropical chernozems' but there is still considerable doubt as to whether this term should be applied to any of them. The true chernozems of higher latitudes are all freely-drained, pervious soils which have developed beneath grasses on a variety of types of parent-material; they increase in calcium content downwards, culminating in a horizon rich in calcium carbonate concretions near the base of the profile (Chapter IX). When examined in detail, nearly all the tropical black earths are found to be different from the true chernozems in one or more particulars. Of those which have been described, the tropical soils which correspond most closely to the true chernozems are found in Rhodesia and eastern Australia. The former, as described near Bulawayo, have a profile eight feet or more in depth, the horizons below about four feet being impregnated with calcium carbonate and of a much lighter colour. Some of these soils, occurring on top of sandstone mesas, appear to be freely-drained but the majority are found in depressions or on flat areas with impeded drainage. The latter are referred to as 'vlei soils' and are obviously very different from chernozems (*vide infra*). The Australian black soils occur in many localities in interior Queensland and more continuously, in a wide belt from south-central Queensland to north-central New South Wales. Though many of them are derived from calcium-rich, basic igneous rocks, some occur on Carboniferous sedimentaries with various types of lithology and, in one area, on acid granodiorite [53]. They all appear to possess the horizon of calcium carbonate accumulation, characteristic of true chernozems, and to occur beneath predominantly grassy vegetation. Furthermore they all appear to be freely-drained. Because of these characteristics, the Australian black soils appear to be more like the chernozems of Eurasia and North America than any of the other tropical soils and it is perhaps not without significance that they occur in that part of sub-humid Australia where the rainfall is most evenly distributed throughout the year.

The great majority of tropical black earths are obviously very different from the above. Characteristically they are heavy soils with some drainage impedence and quite often they occur in low-lying, flat areas or in actual hollows where the regional water-table is high. Though a very dark brown in colour, they do not have the characteristic hue of the true chernozems. Furthermore, although they are fairly homogeneous and of considerable depth, this does not appear to be due to the deep penetration of fine rooting systems as in the chernozems; it is due primarily to the development of deep cracks during the dry season. Particles from the uppermost horizons fall into these crevices so that, over a relatively short period of time, the whole profile is thoroughly mixed. A zone of calcium carbonate accumulation is by no means always present but, when it occurs, it is significantly at a depth which corresponds to the maximum depth of penetration of the seasonal cracks. Rapid leaching to this depth takes place at the beginning of the wet season before the heavy soil swells and the cracks close.

All these heavy, black soils appear to be very alkaline throughout the whole of their profiles. Even in their uppermost horizons, the majority seem to have a pH of at least 8 which is in excess of that found in most chernozems. This strong alkalinity appears to be due either to the highly calcareous nature of the parent-material or to a ground-water rich in lime; it appears to have no close relationship to climatic conditions. Although nearly all these black soils are found where there is a marked alternation of wet and dry seasons, this does not cause the regional development of this soil type. Nor is a particular vegetation type a prerequisite. Although the majority seem to have formed beneath pure grassland or savanna, a number of quite typical black earths appear to have developed beneath forest.

In northern Java, black soils of various depths are found on basic lavas and volcanic tuffs and on limestones. Though many are now cultivated, they were originally beneath forest in some places and beneath grassland in others [53]. In most of them a thick zone of calcium carbonate accumulation is found at the base of the profile (Plate XXXI*b*). Although they occur in regions of fairly heavy mean annual precipitation, these Javanese soils nearly all experience a dry season in which three or four months have less than two inches of rain. The regur soils or black cotton soils of the Deccan are similar in many ways although, in the main, they experience a lower mean annual precipitation and a much more pronounced dry season. They occur typically on the Deccan trap rock or basalt but extend on to contiguous sandstones and gneisses. Predominantly they are rich, heavy soils but they pose many problems in that they become very much more sandy and better drained in many places and yet retain their dark coloration. It was mainly because of this that some authorities in the past were tempted to regard the regur soils as close relatives of the true chernozems. Recent work has gone far to refute this however. It has been pointed out that the dark colour of these soils is not just due to the presence of humus as in the case of the true chernozems. Many samples of regur have been found to contain less than 1% humus. Furthermore, the soils which are richest in humus are not necessarily those with the darkest colour. In fact the colour is probably due to a particular type of base-saturated soil colloid which consists of clay and humus in particularly intimate molecular association. A great deal remains to be discovered about the origin and structure of these soils however.

Soils very similar to those in Java and on the Deccan are found in Africa. Some of the 'plains soils' of north-eastern Uganda are blackish, heavy and highly calcareous. They crack deeply during the dry season and have a zone of calcium carbonate nodules at a depth which varies between six inches and two feet. They have formed in the deep alluvium of a drained lake basin beneath a cover of acacia–tall grass savanna. Black soils of a similar nature have been found in a number of other localities in East Africa. It seems likely that the 'black turf soils' of the Transvaal are also closely related. They are very heavy and always contain a horizon of calcium carbonate nodules somewhere in the profile. It has recently been emphasised that, wherever these soils occur *in situ*, they are on basic igneous rocks; in particular they are found on outcrops of the ultra-basic rock called norite. The connection between this type of rock and this particular type of soil is maintained throughout quite a range of climatic conditions in South Africa.

Because so many similar soils of this type are found over wide areas in the tropics, some authorities in the past have felt the desirability of allotting them to one of the great zonal soil classes. There now seems to be a growing opinion that, although many of the tropical black earths are related to each other, they are quite distinct from other groups, particularly those of higher latitudes. Consequently, a special term has been used to refer to them. Because they are predominantly heavy and highly calcareous they are now referred to as 'margalitic soils' (Latin *marga*=marl) [53]. At the present time it is useful to have a special term such as this which is quite non-committal regarding wider relationships.

Lateritic soils

It is in those tropical regions with pronounced wet seasons alternating with definite seasons of drought that horizons of laterite are found extensively. The processes by which these have been formed are not fully understood (Chapter XV) but it seems likely that, in many places, periodic drying out of the uppermost soil horizons has been a prerequisite. On some old erosion surfaces, even where the climate is favourable for evergreen forest, however, deep weathering and leaching over a long period of time may have been sufficient to impoverish the whole of the rooting medium. Because of this it is possible that the vegetation deteriorated, more efficient drying of the surface occurred, and a thick crust or 'cuirass' of laterite developed in the strongly laterised horizons. In other areas, enormous in extent, gradual climatic deterioration may have caused the rain forest to retreat from the maximum position that it may have achieved at the height of the last pluvial period.[1] Some of these former rain forest soils, under less luxuriant vegetation and with a seasonal rainfall régime, may have developed a laterite crust. In parts of the tropics it is even possible that such crusts are much older than the Pleistocene [86]. Finally, it is quite certain that very rapid development of laterite crusts has taken place because of human activity. Certain types of friable red soil change their structure remarkably quickly if the shade of the canopy and the products of leaf-fall are denied them. The change appears to be most rapid if a dry season of several months is experienced but recent observations suggest that this is by no means a necessity.

In these three ways, and possibly in a number of others, massive laterite crusts have developed. They may be many feet in thickness (Plate XXXI*a*) and, because they are often not so well jointed as most rocks, they tend to cause poor drainage conditions in a wet season. It will be realised from the foregoing, however, that because of climatic change, laterite can now be found in areas which are far too dry to permit laterite formation at the present time. These crusts are thus entirely fossil, and completely different kinds of soil may now be forming due to the weathering of their upper surfaces. Thick laterites are even to be found in some of the driest parts of the tropical deserts. Some tentative and generalised maps of laterites and soils derived from laterites in Africa, India and Australia, indicate how extensive these fossil structures must be (Fig. 27).

There are many lateritic soils in these same areas which do not possess

[1] At several periods during the Pleistocene, the climate of some tropical areas which are now dry, appears to have been much wetter. These have been called 'pluvial periods'.

an actual rocky, sesquioxidic crust however. Friable red soils, with or without disseminated nodules of laterite, are found in many places. For one reason or another, laterisation has not caused complete desilicification of the clay minerals in these soils. They are developing on level or gently sloping surfaces of a variety of ages; they are found beneath a variety of types of vegetation; they vary greatly in colour, some being red, some brown and some yellow. It is assumed that, in most of them, the laterisation process is in operation at the present day.

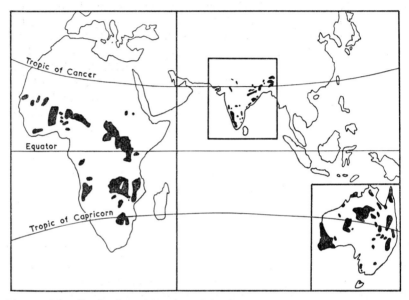

Fig. 27. The distribution of laterites, and soils derived from laterites, in Africa, India and Australia. (Adapted from Mohr and Van Baren, 1954, after Prescott and Pendleton, 1952.)

A different type of lateritic soil is found on moderate or fairly steep slopes in areas where a plateau, capped with a laterite crust, is in process of dissection. These soils are normally red in colour and contain fragments of laterite. The latter have not been formed *in situ* however; they are derived from the crust which is being eroded at the top of the slope (Fig. 28). Since such crusts are rock-like and resistant, it is quite common for them to act as scarp-formers around the eroding fringes of ancient surfaces [86]. The eroded laterite thus moves down the slope along with all the other material derived directly from the rock itself.

The mapping of tropical soils

The majority of tropical soils seem to belong to the lateritic, margalitic and freely-drained pedocal groups. There are numerous other types

PLATE XXXI

A. A LATERITE 'CUIRASSE'. A section through a crust several feet in depth.

B. A MARGALITIC SOIL. A deep, black soil underlain by a thick horizon of lime accumulation under conditions of impeded drainage in Java. (Both from Mohr and Van Baren, 1954.)

A. TROPICAL MONTANE FOREST. Arborescent ericaceous plants mainly belonging to the genus *Philippia*, dominating in the upper part of the mossy forest on Kilimanjaro. Although the mossiness here is not so marked as it is at lower levels, it is still quite a feature both on the ground and on the vegetation. (Photograph by A. F. Braithwaite.)

B. ALPINE SHRUBLANDS. Arborescent *Lobelia volkensii* (centre foreground) and *Senecio johnstonii* (left and right) in the upper Mobuku Valley at about 12,500 feet on Mount Ruwenzori. (Photograph by R. F. Peel.)

PLATE XXXII

however. In many areas with impeded drainage, solontshak and solonetz soils are found which are similar to those in higher latitudes (Chapter X); podzolic soils also have been identified on many freely-drained sites (Chapter XV). Furthermore, many soils have been described whose affinities are far from obvious. In some areas there are unbroken expanses of a single soil group but it is much more usual to find a patchwork of different types. Experience has shown that the pattern of these soil mosaics is determined largely by two factors; the lithology of the parent-material is very important and the relief of the surface perhaps even more so. In areas which are in process of dissection, two or three types of soil are

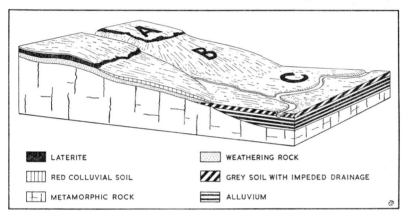

LATERITE   WEATHERING ROCK

RED COLLUVIAL SOIL   GREY SOIL WITH IMPEDED DRAINAGE

METAMORPHIC ROCK   ALLUVIUM

Fig. 28. Diagrammatic representation of a common type of catena encountered in the southern Sudan with a lateritic cuirasse capping the undissected plateau (A), red colluvial soils on the dissected slopes (B) and dark-grey soils on the flat valley floors where drainage is impeded (C).

usually found which are distinct both genetically and from the point of view of innate fertility. They occur in a mosaic which is intimately related to the pattern of land-forms [52]; usually one type of soil is found on the undissected surface, another on the dissected slopes and a third on the gently-sloping or flat floors of the excavated valleys. One type of mosaic, with three distinct facets of this type, has been found in a number of places in Africa and the Americas. An example from the southern Sudan has been described particularly well [54]. Here a thick crust of laterite caps the undissected plateau while the valley sides carry lateritic soils, often containing derived fragments of weathered laterite but with no continuous crust. Much of the valley floors, on the other hand, have dark grey or blackish soils which appear, from their description, to belong to the margalitic group.

Not only are the soils here closely related to relief; all the similar relief facets also carry the same type of vegetation. Such an inter-related suite of land-forms, soil types and plant communities is now usually referred to as a 'catena'. The concept of 'the catena' has enabled some advance to be made in the mapping of both soil and vegetation in areas where, as yet, the types involved are only imperfectly understood. Without committing oneself to any statement of wider relationships, it is possible merely to describe the profiles of the individual soil types in the catena and to show the conditions of relief on which they occur. It must be understood, therefore, that soil mapping on a catena basis does not, of necessity, make any progress towards a zonal soil map on a continental scale. Whether this will ever be produced at a valid and realistic level for tropical areas remains to be seen.

CHAPTER XX

The Vegetation of Tropical Mountains

It is appropriate that a world study of vegetation and soils should culminate in some account of their distributions on mountains in the tropics. Environmental extremes, from the very hot to the very cold, are here brought into close proximity; tropical rain forest may envelop the lower slopes while the peak is perpetually snow-clad. One might almost expect to find many of the main vegetation zones of the earth repeated here in miniature with tropical forests giving way upwards to sub-tropical forest, mid-latitude forest, boreal forest and, finally, to tundra. Although some suggestion of this kind of sequence is frequently encountered, nature cannot be said to have repeated herself with such absolute faithfulness on any tropical mountain. In spite of the gradual decrease in air temperature with increase in height, one should not expect such analogies. Even on the coldest peaks of tropical mountains, the sun reaches a high elevation at midday throughout the whole of the year. The pronounced division of the year into a relatively warm season and a much cooler one is not experienced and yet it is to such a régime that the plants of higher latitudes have adapted themselves. Furthermore, on many tropical mountains, increase in height entails an increase in precipitation; because of the uplift given to impinging air masses, the slopes at certain heights are inordinately wet. These two factors, along with several others, produce climatic conditions which are peculiar to tropical mountains. These conditions reappear time and time again in characteristic patterns. Plant formations have evolved which are adapted to these climates and adaptation seems to have followed remarkably similar lines in all parts of the tropics. Because of this convergent evolution, some general statements can be made about the peculiarities and patterns of tropical mountain vegetation.

Soils on tropical mountains

It would, on the other hand, be unwise to attempt similar general statements about the distribution of soils on tropical mountains. There is a general tendency for lateritic soils to be produced on the lower slopes and podzolic ones on the cooler, upper slopes, but beyond this, generalisations are misleading. The complexities due to contrasts in slope and parent-material are very great. Not only are many tropical mountains of volcanic origin, each one being composed of its own peculiar suite of igneous materials, but also the predominance of steep slopes entails the rapid removal of top soil by soil creep so that permanent immaturity is ensured. Unweathered minerals are kept near to the surface and the effect of parent-material thus remains strong. Quite apart from this, however, the lower slopes of different tropical mountains experience the complete range of tropical lowland climates from the perennially wet to the permanently dry. Any comprehensive statement about the soils on these slopes would thus involve repetition of almost everything that has already been said about tropical soils. Indeed, when the upper slopes also are taken into consideration, nearly all the major soil types of the world are found to be represented here.

Vegetation zones

A similar succession of formations is to be found in nearly all the places where a tropical mountain rises from a lowland clothed in tropical rain forest. Although few detailed ecological studies have yet been carried out, the general descriptions available are sufficient to indicate the nature of the communities that are encountered. It is apparent that only slight differences in relief within the foothill zone have been sufficient to cause differences in forest composition. As long as they retain their height and three-tiered structure, these communities are regarded merely as different associations of tropical rain forest. Wherever the mountain is sufficiently high, however, tropical rain forest ultimately gives way to a lower type of forest, often only sixty or seventy feet in height, with only a two-layered structure. Very often these communities contain species which are identical with, or closely related to, those in near-by extra-tropical regions. Thus, in Malaya and Java, representatives of the Japanese and Chinese floras make their appearance and, in the mountains of Mexico, broad-leaved and coniferous trees similar to those in the forests of the southern U.S.A. become increasingly common with increase in height. These communities are often referred to as 'sub-montane' or 'transitional' forests. Some ecologists regard them as separate formations because of the difference between them and the contiguous tropical rain forest; others look upon them as mere zones of ecotone between tropical rain forest and true montane forest. The sub-montane forest belt usually corresponds with the climatic zone in which mean cloudiness and relative humidity increase rapidly upwards.

The true montane forest has a distinct structure and composition wherever it occurs (Plate XXXIIa). Its height is often no more than thirty or forty feet and only a single stratum of trees is usually present. The trunks and boughs of these dominants are also much contorted. An even more striking characteristic of this forest is its extreme richness in epiphytic mosses and lichens; boughs and twigs are usually festooned with these plants and the actual bark of individual trees may be completely obscured. The term 'mossy forest', so often applied, is not inappropriate. The density and wetness of these forests is emphasised in a most graphic description of their appearance on Mount Kilimanjaro in Tanzania by Volkens [67]:

'The characteristic feature of the forest . . . is that it consists from the ground upwards of a compact mass of leaves which not only prevent our view through it horizontally and vertically, but also almost completely cover all axial parts, branches and stems. We see leaves below us, around us, above us, and wherever we look, whereas in all forests at home, besides the green colour of the foliage, the brown, grey and black tints of the trunks and branches play their part. All the space occupied by the forest is filled by a mass of interlacing branches which, year in and year out, are uniformly foliaged. In addition to this is a further peculiarity, to which all travellers agree in giving prominence, and which may thus count as the most striking feature. I refer to the dense envelopment of nearly all woody plants by cryptogamous epiphytes. We see them hanging down in draperies metres in length, or resting at the extremities of the branches as spherical bird's-nest-like balls as large as the human head, or clothing younger erect branches like distended stockings, covering thicker, horizontal branches with a cushion-like mass, which, to borrow a simile from Holst, threatens to topple over on both sides. They are mostly lichens, mosses and *Hymenophyllaceae*, the first chiefly high up, where sunlight reaches them, the others in the shade of tree-crowns.'

These forest formations, at once stunted and luxuriant, have developed in this way because of peculiar climatic conditions. For nearly all the year they are enveloped in cloud so that the sun is completely obscured. Extreme leafiness is thus encouraged. Windiness as well as dearth of solar energy also restricts growth but the extreme development of epiphytes is due to the fact that the vegetation is very rarely dry.

The floras of the montane or mossy forests have characteristics peculiar to themselves. A few of the species are identical with those in tropical rain forest but large numbers are different. In particular there are many species which are closely related to those in the moist sub-tropical forests. Thus, in the montane forests of Malaya and Indonesia, evergreen oaks (*Quercus spp.*) are often dominant just as they are in the lowland forests of southern Japan. Conifers also are of frequent occurrence but they are usually very different from those which dominate the montane forests of higher latitudes. It is only in Mexico that the familiar spruces (*Picea spp.*) and firs (*Abies spp.*) are common and it is noteworthy that here one is approaching the fringes of the tropics where a pronounced division of the year into a warm and a cool season is apparent. In New Guinea the montane conifers are very close relatives of those which dominate in the forests of northern New Zealand and sub-tropical South America. Among the commonest are *Agathis alba*, a close relative of New Zealand's kauri pine (*Agathis australis*), and *Araucaria cunninghamii*, belonging to the same genus as the 'monkey puzzle' or araucaria pine of Chile. Species of *Podocarpus*, a genus of conifers so widely distributed in the sub-tropical parts of New Zealand, South Africa and South America, are also very common. Although these conifers may occur in almost pure stands in tropical montane forests, it is much more usual to find them intermixed with broad-leaved angiosperms.

Maximum precipitation and cloudiness on tropical mountains is nearly always achieved somewhere between 6,000 and 10,000 feet. Above this level rainfall is less heavy and the climate is more sunny. Where the forests persist to a sufficiently high altitude they may, in consequence, show an increase in height and a decrease in 'mossiness'. This has been noted in the montane forests of New Guinea, particularly in sheltered valleys at an altitude of about 11,000 feet [6]. Ultimately, however, the montane forests give way upwards to an alpine zone of shrubs or grasses. This is often reached through an ecotone zone of dwarfed trees often referred to as 'elfin woodland'. Some of the dominant species here are identical with those in lower parts of the montane forest even though, superficially, they appear very different. Quite often the community is no more than two or three feet in height. It appears that the prevailing high wind speeds prevent any vertically disposed shoots from surviving; they wither and die back because of an excessive rate of transpiration. Only those twigs which grow out horizontally near to the ground, where wind speed is much reduced, are able to survive. On the other hand, other trees in these communities adopt the low-growing, creeping habit quite regardless of the direct effect of wind. All the trees thus creep along near to the ground which may be quite obscured by the dense growth. It is sometimes possible to walk over the top of this 'forest', so dense is the structure produced by the interwoven branches.

Many types of community are found in the alpine zone on high tropical mountains. Quite often the elfin woodland gives way to a community of shrubs but open grassland also is frequently encountered [60]. There is some evidence that, under undisturbed conditions, two quite distinct formations would be present—a lower one dominated by shrubs and an upper one of grasses. Even at these heights, human interference over a long period may have had a profound effect. In eastern New Guinea, for instance, the alpine grasslands are burned regularly by the natives during their wallaby hunts [59]. Whereas the uppermost forests now give way abruptly to grassland, in many cases this may formerly not have been the case.

Both the alpine shrublands and the alpine grasslands are of absorbing interest floristically. These frost-tolerant communities within surrounding areas of tropical vegetation have been evolving in isolation for millions of years. They therefore contain two particularly interesting categories of plant. First, there are those species which have been able to survive here because of the lack of competition from taller plants. Secondly, there are those which have evolved on isolated mountain masses and which are found nowhere else on the earth. Thus, in the shrublands near the snow caps on the tops of East African mountains like Ruwenzori and Kiliman-jaro, there are tall groundsels (*Senecio johnstonii*) and lobelias (*Lobelia volkensii*) which are limited to these small areas (Plate XXXII*b*). Analogous to these there are those arborescent species of *Espeletia* and *Culcitium*, belonging to the daisy family (*Compositae*), which are of common occurrence in the alpine zone of the tropical Andes. Today they often stand isolated in low scrub or alpine grassland and are referred to by the natives as 'frailejon' [67]. On the other hand, quite a number of the alpine grasses, particularly those which grow very near the edge of the perennial snow caps, belong to the same species as some of the heath and moorland grasses in middle and high latitudes in the northern hemisphere. It is rather surprising to find our own tufted hair-grass (*Deschampsia caespitosa*) listed as a common species on Kilimanjaro and its near relative, the wavy-hair grass (*D. flexuosa*), as a common plant amongst the rocks near the crest of Mount Kinabalu (13,698 feet), the highest mountain in Borneo. It is clear that some of these uppermost plant communities on tropical mountains are very similar in appearance to the tussocky grass moors of some British uplands.

The altitude of vegetation zones

Because of the scarcity of comprehensive ecological studies and the diverse ways in which different observers have defined the different plant communities on tropical mountains, it is difficult to compare the heights of the main vegetation zones in different places. On the basis of available information it is possible, nevertheless, to detect certain similarities. If three large mountain masses in the tropics are compared (Fig. 29), the limits of the main zones appear to occur at similar heights. The eastern mountains of New Guinea, the eastern slopes of the Peruvian Andes and the western slopes of the Ruwenzori massif all rise from hot, moist lowlands covered with tropical rain forest. On Ruwenzori and in New Guinea this tall, three-storeyed forest gives way to a lower, two-storeyed one at an altitude of between 5,000 and 6,000 feet while a similar change takes place on the Andes at between 4,000 and 5,000 feet. There is no obvious reason why the transition should be lower in Peru. It is possible that the few observations that have been made, provide too small a sample upon which to base a valid generalisation and that future observations will show a closer conformity. This is by no means a foregone conclusion however; it may very well be that the tolerances of many of the dominants in western Amazonia, for some unknown reason, are different or that physical conditions on the eastern Andean slope are, in some way, peculiar.

In all three areas the 'transitional' or 'sub-montane' forest gives way to 'montane' or 'mossy' forest at between 6,000 and 7,000 feet. The boundary appears to be much more variable in New Guinea than in the other two areas but this may merely be a reflection of the greater degree of precision with which the observations have been presented. It must be remembered that different degrees of exposure and different types of soil-parent-material must cause great irregularities in the zonal boundaries on all mountains. Consideration of the more pronounced of these aberrations is usually omitted from general statements.

In all these three cases the montane forest extends upwards to an average altitude of about 12,000 feet. Though the formations differ very much in detail, they are similar in basic structure. They are usually no more than thirty or forty feet in height and are extremely wet and mossy, particularly at their lower levels. Bamboos are common in all three formations but only in East Africa has a zone in which bamboo (*Arundinaria alpina*) is dominant been described [38]. Considerable contrasts are also encountered in the upper sections of the three formations. On Ruwenzori and Kilimanjaro [65] arborescent species of the heather family (*Ericaceae*) are prominent (Plate XXXIIa); in New Guinea anomalously tall forests dominated by conifers (*Podocarpus spp.*) occur, particularly in sheltered valleys, while on the Andes no one type or family of trees has been singled out for particular mention.

Some of the more outstanding peculiarities of the alpine vegetation have already been noted. These communities, generally speaking, become lower-growing and more impoverished as one proceeds upwards. Perennial snow is not encountered in eastern New Guinea but it is reached at about 15,000 feet on Ruwenzori and, on the average, at about 16,500 feet in central Peru.

Although analogous vegetation zones seem to occur at very similar altitudes on the larger mountain masses in the tropics, an examination of mountains of different sizes reveals great contrasts. It appears that the smaller and more isolated the mountain, the lower its vegetation zones tend to be. A small, isolated mountain, particularly if it is near to the sea, is exposed to higher mean wind speeds and to moister air masses; cloud and wind thus reduce vegetation luxuriance at relatively low levels. Great mountain masses, on the other hand, provide sufficient obstruction to atmospheric flow to modify profoundly their own climates. This phenomenon was first noted in Europe and has been referred to frequently as the 'Massenerhebung Effect'. The magnitude of this effect is very apparent if five mountains of the Indo-Malaysian region are viewed as a

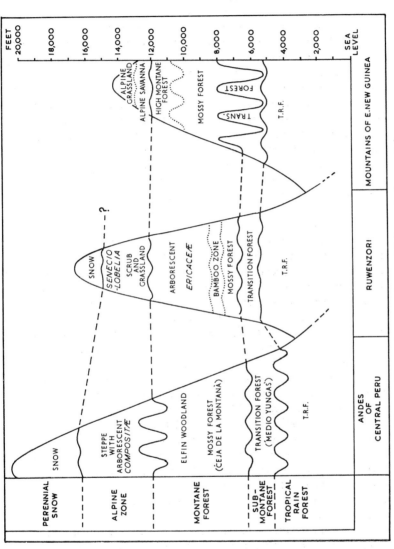

Fig. 29. Vegetation zonation on high tropical mountains.

series (Fig. 30). Each of these has been studied in some detail and the altitudes of their vegetation zones determined with some degree of accuracy. A similar relationship is seen between American mountains; although true montane forest is not encountered below 6,000 feet on the western Andes of Peru, a structurally similar forest is reached at little more than 2,000 feet on the windward side of the mountains of Trinidad.

It should be noted however that some writers with a wide experience of tropical montane vegetation have cast doubt on the existence of true montane forest at such low levels [61]. They maintain, very rightly, that any plant formation must be defined according to floristics and physiognomy as well as structure. They suspect, therefore, that much of the low level 'montane' forest that has been described on mountains near the coast (particularly as low as 3,000 feet) is merely mist-stunted sub-montane forest whose true nature will be revealed by future research into species-content and morphological characteristics. In view of these reservations it must be emphasized that the picture presented here (Fig. 30) is based in great part on the gross structure of the communities involved.

Dry tropical mountains

The series of hygrophytic or moisture-loving formations already described does not appear on all tropical mountains. Xerophilous communities often intervene in some way to complicate the 'ideal' pattern. Many mountains rise from arid or semi-arid plains and plateaux so that all their lower slopes carry only desert vegetation or scrub. It is not until a height of several thousand feet is reached that increased rainfall permits forest growth. Many of the East African mountains rise from plains covered with thorn forest or acacia savanna; mossy forest may ultimately be attained at a height of around 8,000 feet but this is through a transition zone of xerophilous seasonal forest and not through hygrophytic sub-montane forest. The most extreme examples are the mountains which rise from tropical deserts. The highest mountains of the Ahaggar Plateau (Tahat, 9,840 feet) and the Tibesti Plateau (Emi Koussi, 11,201 feet) remain sufficiently dry throughout their entire heights to be incapable of supporting hygrophilous forest. Above about 5,000 feet the vegetation they carry is very similar to the evergreen sclerophyllous forest and scrub of the Mediterranean coastlands. Most of the air streams which pass over these highlands are so dry as to yield little cloud and precipitation even when forced to rise.

Even within generally moist areas, dry belts are sometimes found at intermediate heights on mountain slopes. On the eastern edge of the Congo Basin, although the lowlands are covered with tropical rain forest and the higher slopes with montane forest, a narrow zone of xerophilous savanna often intrudes between the two. Although climatic information for these mountain areas is very scarce, it is thought, from their position, that

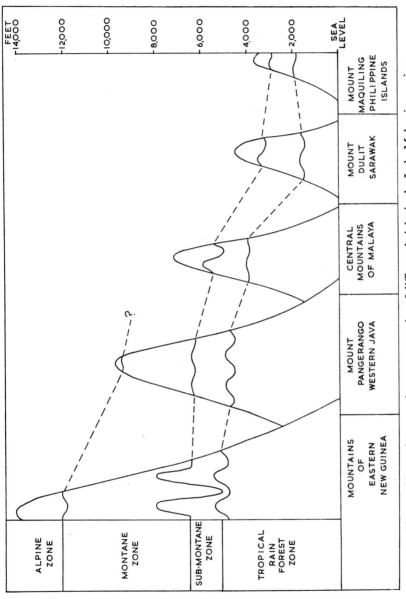

Fig. 30. Vegetation zonation on mountains of different height in the Indo-Malaysian region.

these xerophilous communities owe their existence to the föhn effect. The slopes along the side of a mountain mass must experience descending air whenever winds blow from the direction of the high land. If the prevailing wind is from this direction, descending air which has been adiabatically warmed and dried, must be of frequent occurrence. As soon as this air reaches gentler slopes or the foot of the mountain, however, the adiabatic drying effect ceases, the air begins to pick up moisture by evaporation from the surface, and thus ceases to have such a desiccating effect on the vegetation.

It is only on the western side of the Andes, from northern Peru to north-central Chile, that a tropical mountain slope is seen which has desert or semi-desert vegetation throughout its entire height. The semi-desert also extends eastwards on to the high Andean plateau in Bolivia. Where not almost completely plantless, these dry slopes carry very open communities dominated by dwarf shrubs and desert grass. The former belong predominantly to the *Compositae* and, amongst the latter, an extremely xerophilous feather grass (*Stipa jehu*) is predominant. These desert and semi-desert areas are referred to as the 'puna'. It occupies the whole of the west Andean slope in these latitudes because of the perman- ence of extremely stable air. Throughout the whole year the thermal structure of the atmosphere is such as to prevent air from being pushed uphill from the Pacific Ocean. On the other hand, moist air which pushes westwards from the Amazon lowland loses almost all its moisture during the ascent up the eastern slopes and is then warmed and dried consider- ably with every thousand feet of descent down the western slope. The puna is unique in tropical regions; even the higher slopes of mountains in the Sahara, in Arabia and in central Australia have a substantial shrub vegetation, if they are not actually forested.

Tropical mountains and ecological problems

On most tropical mountains one encounters a greater variety of plant communities than in any other areas of comparable size on the earth. The plant formations merge through zones of ecotone just as they do elsewhere but these are generally much narrower and the changes are more abrupt than in lowland areas. In consequence many fundamental problems concerning the distribution of communities and the relationship between vegetation and climate can be studied most advantageously on tropical mountains. The complication of differences in insolation between north- facing and south-facing slopes, which is of such significance in higher latitudes, only attains any degree of importance as one approaches the fringes of tropical regions.

This innate attraction of tropical mountains as areas for research is offset to a certain extent by difficulties of terrain. Because of their wetness and the prevailing cloudiness, many tropical mountains have been unattractive from the agricultural point of view; population has remained sparse in consequence and communications are undeveloped. Consequently these areas are amongst the most difficult in the world in which to set up climatic stations and research centres. Wherever this can be overcome, however, the lack of development and the sparseness of population is a great advantage. Vegetation has often been disturbed only slightly; indeed there are places where almost unmodified formations can be found. As most of the foregoing chapters have shown, this is an advantage which is rare indeed on the surface of the earth.

CONCLUSION

CHAPTER XXI

The Outlook for Wild Nature

'*These wild creatures amid the wild places they inhabit are not only important as a source of wonder and inspiration, but are an integral part of our national resources and of our future livelihood and wellbeing.*

In accepting the trusteeship of our wild life, we solemnly declare that we will do everything in our power to make sure that our children's grandchildren will be able to enjoy this rich and precious inheritance.'

JULIUS NYERERE
CHIEF FUNDIKIRA
TEWA SAIDI
Arusha; September, 1961.

THE fundamental concept upon which this book is based is that of the climatic climax plant community. No apology is made for this in spite of the fact that human communities have now modified the plant cover almost beyond recognition over wide areas; indeed, the very fact that this transformation has taken place very recently (in terms of the evolutionary time scale) means that it would be unrealistic to take cognisance of the present situation alone. The fauna of the earth evolved within a pattern of purely natural climax and seral communities; if these communities are completely removed or transformed, although some species of animal may be sufficiently flexible to adapt themselves, there can be little doubt that large numbers of more specialised ones will perish.

Man has now reached a stage of technological advancement where he can reasonably visualise a not-too-distant future in which it would be *physically* possible for all men to have much more leisure to appreciate their surroundings. The days have gone when man, as a relatively puny, superstitious being, was bound to live in fear of his wild environment and its fierce denizens. The rapid growth, in richer countries, of a great nature-loving public with a desire for the wild, wide-open spaces, suggests that this might grow to attain vast proportions in a theoretical, emancipated future. The future wellbeing of life on the earth may well be in

Tvs

danger, however, because a large majority of those responsible for keeping society informed seem to have accepted a philosophy in which economic and social advancement are bound inevitably to a mechanical kind of technology: forests are regarded as mere 'undeveloped' areas and the trees within them can be viewed in only two possible ways: either they contain useful material, in which case the mechanical saw is the appropriate impliment; or they occupy space which should be producing crops, in which case the bulldozer is called for. Within the framework of such a philosophy, the best kind of scientific education for the young is one in which physics and chemistry alone are pre-eminent; not because physics and chemistry are a sound basis upon which to build biological understanding, but because engineers must have them before they can design bigger and better bulldozers. Ecological biology is merely a pleasant adjunct to a liberal education—almost an 'arts' subject!

Before this unfortunate attitude can be discarded, it will probably be necessary to convince the legislature that ecologists are not unrealistic people who wish to return the whole earth to a pristine state of natural glory. No-one would be so perverse as to deny our *partial* dependence, present and future, upon bulldozers and other mechanical contrivances or to ignore the elementary fact that vast areas must be devoted to crops. Indeed the ecologist would be the first to point out that human interference can be positively beneficial from the point of view of conservation. The conservationist's first aim is to promote the secure existence of the maximum number of species—he aims at diversity. Up to a certain point, man's removal of relatively monotonous forest and its replacement with a varied patchwork of crops and grass, stimulates greater diversity of animals as well as plants. One only need consider the profusion of wild mammals, birds and insects adapted to life on British farmlands to realise the truth of this; quite obviously a large range of species, which previously may have been relatively rare, were able to expand and flourish when man opened up the forest. It is only when the original plant community is reduced to negligible proportions or completely swept away that impoverishment occurs. Usually the larger mammals and birds suffer first, either because they are predators or merely because they are conspicuous or good to eat. The final and most insidious threat arises when the invading human communities begin to destroy the soil they have occupied and to poison the ground, the water and the atmosphere with effluent or with chemical insecticides and weed-killers.

There are thus two main grounds for misgiving when one reviews present ecological trends over the earth as a whole. First, there is the purely practical, economic consideration of wasting soil resources; secondly, there is the more aesthetic and intellectual viewpoint regarding the loss of species and the elimination of wild nature. Though there are, doubtless, many people who would regard only the former as being worthy of serious consideration, many others would take the wider view. In any case, the conservation of the soil and the conservation of wild species are inextricably interrelated.

The purely practical fears that have been expressed are very well grounded. When man removes a forest, either by felling or by depasturing his domesticated animals, he inevitably ensures soil deterioration unless he immediately replaces the wild vegetation with a cultivated one (or with a cropping system) which will be equally capable of maintaining the richness of the nutrient cycle. In parts of Britain the nutrient cycle has certainly been maintained, or even improved, by replacing the forest with permanent grass, but trends such as this have been exceptional. Over vast areas forests have been replaced by much less demanding vegetation. Although climatic change may have been responsible for some of this deterioration, there is conclusive evidence that many dusty, unproductive semi-deserts have been brought into existence by human agency.

If human activity has produced such effects in the past, is it not possible that future trends may be even more disastrous? The remaining virgin lands are much less extensive and the encroaching human populations are expanding at an almost explosive rate. Indeed many authorities seem to be resigned to the fact that most large mammals and many large birds, particularly in the tropics, are likely to disappear during the next few decades apart from the few which may be amenable to permanent preservation in zoological gardens [34]. There can be little doubt that many species of plant will suffer the same fate.

In his long struggle against the elements, man has come to equate 'progress' with the 'conquest' of wild nature. In recent times the advent of tractors, insecticides and rifles has ensured that this process of 'conquest' can be a very rapid one. It is doubtful if these methods can provide anything more than a very temporary easing of the population problem however. Unless the areas to be cleared are selected with extreme care and the methods of subsequent cultivation are skilfully supervised and rigorously controlled, productivity will fall. Under certain conditions almost complete sterility can be achieved in a catastrophically short time. Since, in most countries, there is a great dearth of people who are suitably qualified to administer and supervise such development schemes, the enormous dangers are apparent. Even from the purely practical viewpoint therefore, it might be much more desirable to concentrate effort and capital into raising the productivity and fertility of areas already under cultivation. The initial capital outlay and the social problems encountered in such a course of action would certainly be great; nevertheless it is probably the only one which can ensure ultimate success.

The great open spaces which remain on the earth are the safety valves for *Homo sapiens*. With the present rate of population increase, there might be little room left for expansion in another hundred years. There would be some excuse for continued expansion if there were evidence to suggest that population increase would have ceased by the time the safety-valves had been closed. There is no such evidence; indeed demographic trends over the past 150 years indicate that the reverse is more likely to occur. Like Mr. Micawber, however, it appears that the administrations of the world are proceeding in the hope that something is bound to turn up!

Even if the unlikely happens and population equilibrium is established without causing widespread sterility of land, thoughtful people of the twentieth century should ask themselves if they are reconciled to the inevitable elimination of wild nature. Apart from the aesthetic stimulation to be gained from the contemplation of a wilderness where the blind forces of nature can interact without human interference, it is both salutary and useful to have such wild areas available for study. It is not so very long since man emerged from the primeval forest in which he had evolved. The reason why our doctors, educationalists and administrators fail us on occasion is perhaps due to the fact that they have not recognised or remembered this. They often seem to regard our present environment as 'normal' whereas, in fact, it is very new and abnormal. A few thousand years is a very short time for a forest organism to become completely adapted, organically and psychologically, to the noises, smells and foods of a twentieth-century town. Furthermore, opportunity to observe examples of control by competition in a wild environment can be a positive advantage to agriculturalists and biochemists. Many major problems in agriculture arise from the fact that crop plants and domesticated animals are collected together in large numbers where disease of one kind or another can spread so easily. Also many crops and animals are grown in environments which are very different from those in which their wild ancestors evolved in the not very distant past. The ills which arise from these circumstances will probably never be cured by the wholesale and uncritical application of insecticides and fungicides; indeed the problems may ultimately be aggravated by such practices.

Apart from the large areas of potentially cultivable land found in Amazonia, the Congo Basin, New Guinea and Borneo, smaller patches of land not profoundly modified by man are found here and there on all the continents. They are most numerous, however, in tropical regions within the frontiers of countries where population increase and lack of wealth are perniciously associated. Even the more responsible politicians and administrators in such countries could probably not be expected to lend their support to any suggestion of enforced 'neutralisation' of considerable parts of their territories. At best they would probably regard such suggestions as irresponsible romanticism; they might even be excused if they suspected that the educated and privileged people in the economically-emancipated countries were aiming to create playground reserves on land which was needed for food production. They would certainly be justified in pointing out that the inhabitants of Great Britain rid themselves of bears and wolves, and the forests that had sheltered them, several centuries ago.

Nevertheless, the force of the argument already put forward is not diminished by such internationa suspicions and it is most encouraging to find no less a person than the President of Tanzania (*vide supra*) pleading for a more enlightened approach to large-scale conservation. Such attitudes might very well grow and spread if a more realistic sense of values could be made to permeate through international politics and if international schemes of aid to under-developed countries could be placed on a more liberal, less mechanical basis. The present population of the world, along with a substantial increase, could very well be fed, clothed and housed from the land which it now occupies, if human effort were dedicated to the intensification of agriculture and the increase of manufacturing industry. Instead of this, however, nations are much more concerned in 'effectively' occupying the territories that they possess so that empty spaces will not be a cause of envy and, ultimately, of aggression by more numerous and crowded neighbours.

In the light of the present state of international politics, the views expressed here cannot be regarded as anything but idealistic. Obviously the nations of the world will not suddenly adopt a universal policy in which conservation has absolute priority. Nevertheless, if man is to survive as a numerous species on the earth, he will, in the near future, have to conserve what he has. Those people of all nations who have the aptitude to perceive the great dangers which will shortly beset our species, might be expected to use their influence to ensure that the soil we use for food-production is not wasted and that, whenever specific plants, plant communities and animals are threatened with extinction, reserves of adequate size are created to preserve them. In the near future, it is also to be hoped that governments will evolve long term policies for preserving larger tracts. If mankind does survive to enjoy a more leisured and emancipated future, the descendants of the present impoverished peasantry of Asia, Africa and Latin America will hold the present generation responsible, if their legacy is a world in which there are no rain forests, swamps and scrublands, with their associated fauna, to contemplate and to marvel at.

Vegetation Maps of the Continents

ALL the main vegetated land areas of the earth are shown on the ten maps presented here. Thirty-four major vegetation types have been recognised, each being represented by a characteristic symbol. The theoretical climatic climax vegetation has been indicated in all areas where its nature can be inferred on the basis of sound evidence. In places about which there is still some difference of opinion, the existing wild vegetation or that which is known to have existed in the past, has been shown. A general key has been prepared in which climatic climax formations and ecotones are shown as such, and where the doubtful status of other types of vegetation is clearly indicated. Each map has also been given a separate key on which communities are given their local names. The symbols used in the general key are retained throughout, however, in order to demonstrate inter-continental relationships.

(C C F = Climatic Climax Formation-Type; C C E ≗ Climatic

EXTRA-TROPICAL VEGETATION

Type No.	Designation	Ecological Status	Symbol
1.	Tundra and alpine vegetation	C C F	⸎
2.	Boreal forest dominated by larch (*L. dahurica*)	C C F	↑
3.	Boreal, sub-alpine and montane coniferous forest	C C F	⩑
4.	Coast and lake forest	C C F	⩑
5.	Mixed boreal and deciduous forest	C C E	⩖
6.	Mixed boreal and lake forest	C C E	⩖
7.	Mixed lake, boreal and deciduous forest	C C E	⩖♀
8.	Mixed lake and deciduous forest	C C E	⩖
9.	Deciduous summer forest	C C F	♀
10.	Blanket bog alternating with deciduous forest	C C E	-♀-
11.	Blanket bog alternating with mixed forest	C C E	-⩖-
12.	Mixed southern pine and deciduous forest	(?)	⩖
13.	Southern pine forest	(?)	⩕
14.	Broad-leaved evergreen forest	C C F	♀
15.	Evergreen mixed forest	C C F	⩖
16.	'Forest-steppe'	(?)	₁♀₁
17.	Steppe	(?)	ıʲı
18.	Semi-desert scrub and woodland	C C F	Ψ
19.	Sclerophyllous scrub	C C F	▲
20.	Sclerophyllous scrub with desert grass	(?)	▲v
21.	Australian sclerophyllous forest	C C F	ϒ
22.	Australian sclerophyllous savanna	(?)	ϒᵧ

KEY

Climax Ecotone; (?) = ecological status doubtful)

TROPICAL VEGETATION

Type No.	Designation	Ecological Status	Symbol
23.	Tropical rain forest	C C F	Y
24.	Tropical seasonal forest	C C F	⬆
25.	Microphyllous forest and woodland	C C F	⊂
26.	Semi-desert scrub	C C F	⌒
27.	Desert	C C F	⠴⠄
28.	Broad-leaved tree savanna	(?)	⬆ɣ
29.	Microphyllous tree–tall grass savanna	(?)	⊂ɣ
30.	Microphyllous tree–desert grass savanna	(?)	⊂v
31.	Semi-desert scrub with desert grass	(?)	⌒v
32.	Desert alternating with porcupine grass semi-desert	C C F	⠴⌒v⠄
33.	Tropical montane forest	C C F	⬆
34.	Tropical montane forest with conifers	C C F	⬆

Perennial ice

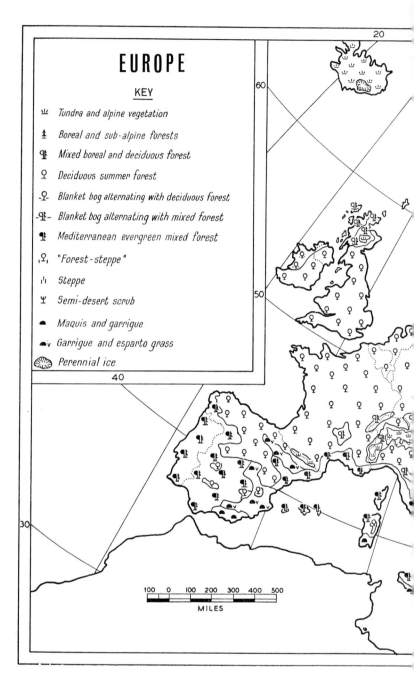

EUROPE

KEY

⊻   *Tundra and alpine vegetation*

⊥   *Boreal and sub-alpine forests*

ⵕ   *Mixed boreal and deciduous forest*

♀   *Deciduous summer forest*

-♀-   *Blanket bog alternating with deciduous forest*

-ⵕ-   *Blanket bog alternating with mixed forest*

ⵕ   *Mediterranean evergreen mixed forest*

,♀,   *"Forest-steppe"*

ı'ı   *Steppe*

Ψ   *Semi-desert scrub*

●   *Maquis and garrigue*

●ᵥ   *Garrigue and esparto grass*

◉   *Perennial ice*

100   0   100   200   300   400   500

MILES

M

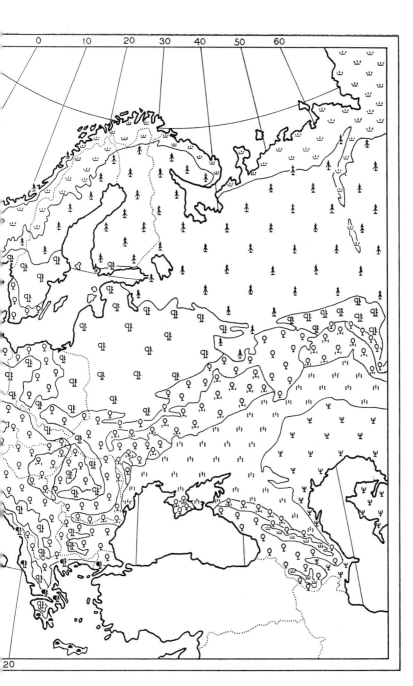

E

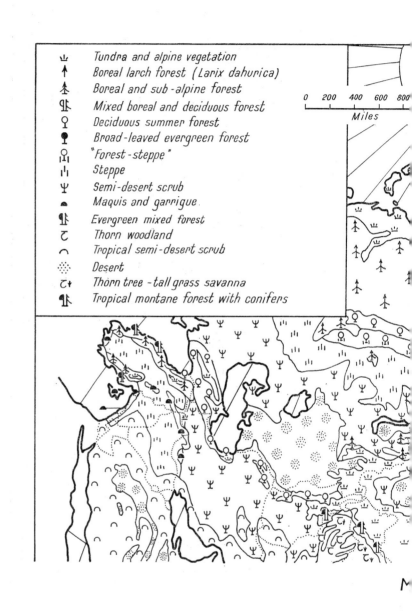

⊻	Tundra and alpine vegetation
↑	Boreal larch forest (Larix dahurica)
⚘	Boreal and sub-alpine forest
⚘↑	Mixed boreal and deciduous forest
♀	Deciduous summer forest
♀	Broad-leaved evergreen forest
♀ı̣ı	"Forest-steppe"
ı̣ı	Steppe
Ψ	Semi-desert scrub
●	Maquis and garrigue
⚘♀	Evergreen mixed forest
౽	Thorn woodland
⌒	Tropical semi-desert scrub
∴	Desert
౽↓	Thorn tree-tall grass savanna
⚘↓	Tropical montane forest with conifers

0   200   400   600   800

Miles

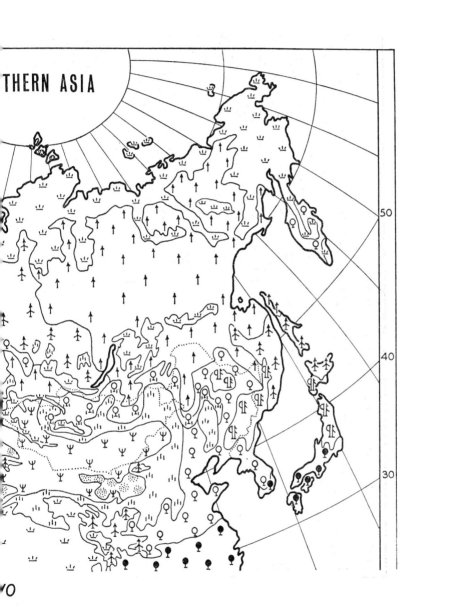

THERN ASIA

50

40

30

O

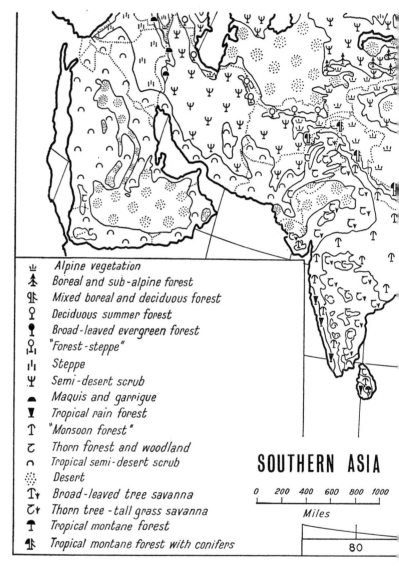

⸌	Alpine vegetation
⍒	Boreal and sub-alpine forest
ⸯⵏ	Mixed boreal and deciduous forest
♀	Deciduous summer forest
♀	Broad-leaved evergreen forest
ⸯⵏ	"Forest-steppe"
ᵢᴵᵢ	Steppe
Ψ	Semi-desert scrub
▬	Maquis and garrigue
ⵏ	Tropical rain forest
↑	"Monsoon forest"
ꜿ	Thorn forest and woodland
⌒	Tropical semi-desert scrub
⸬	Desert
↑ᵧ	Broad-leaved tree savanna
ꜿᵧ	Thorn tree - tall grass savanna
♠	Tropical montane forest
ⸯⵏ	Tropical montane forest with conifers

SOUTHERN ASIA

```
0   200  400  600  800  1000
└─┴─┴─┴─┴─┴─┴─┴─┴─┴─┘
         Miles
```

80

M

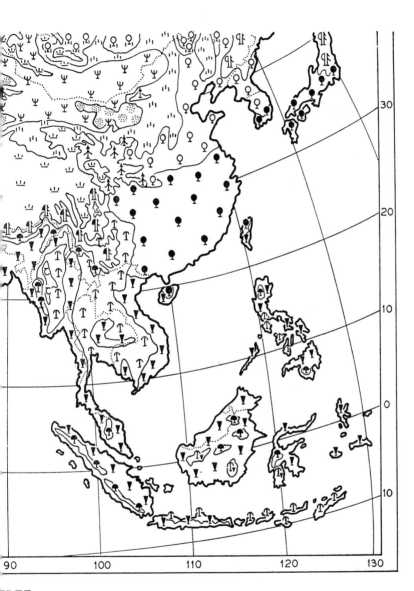

30

20

10

0

10

90   100   110   120   130

REE

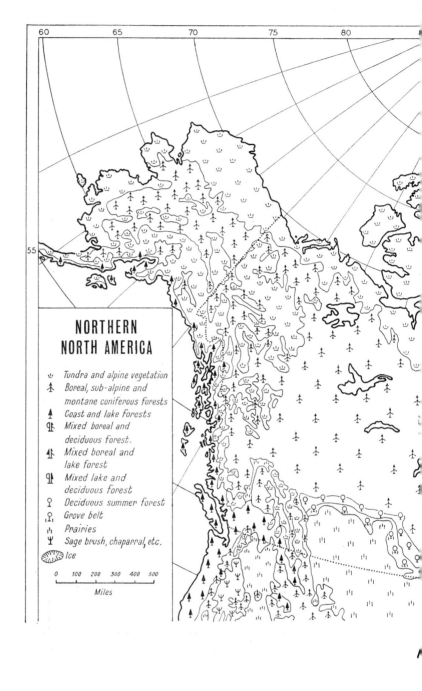

**NORTHERN
NORTH AMERICA**

ᴗ Tundra and alpine vegetation
↟ Boreal, sub-alpine and
    montane coniferous forests
♠ Coast and lake forests
ٵ Mixed boreal and
    deciduous forest.
↟ٵ Mixed boreal and
    lake forest
ٵ Mixed lake and
    deciduous forest
♀ Deciduous summer forest
ٵ Grove belt
ᵢₗᵢ Prairies
Ψ Sage brush, chaparral, etc.
⬯ Ice

0   100   200   300   400   500

Miles

85   80   75   70   65

60

55

50

45

UR

s

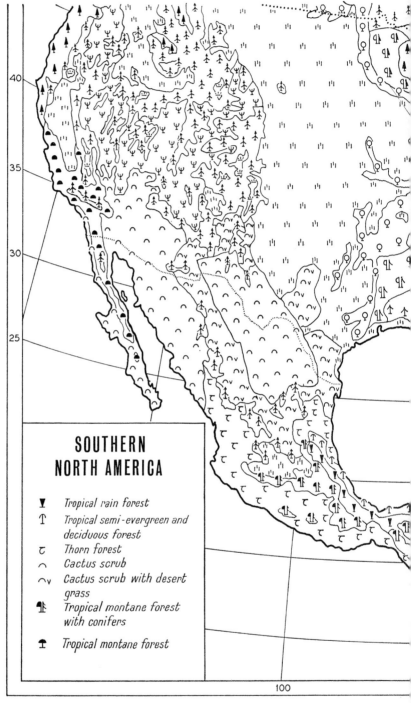

SOUTHERN
NORTH AMERICA

Ψ	Tropical rain forest
⇑	Tropical semi-evergreen and deciduous forest
ᴄ	Thorn forest
∩	Cactus scrub
∩ᵥ	Cactus scrub with desert grass
⬆	Tropical montane forest with conifers
⬆	Tropical montane forest

100

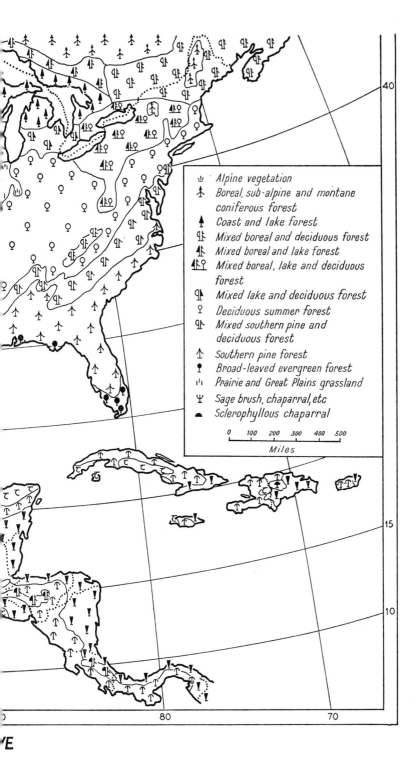

⊻	Alpine vegetation
⋏	Boreal, sub-alpine and montane coniferous forest
⋏	Coast and lake forest
⊊	Mixed boreal and deciduous forest
⊊	Mixed boreal and lake forest
⊊⚲	Mixed boreal, lake and deciduous forest
⊊	Mixed lake and deciduous forest
⚲	Deciduous summer forest
⊊	Mixed southern pine and deciduous forest
⋏	Southern pine forest
●	Broad-leaved evergreen forest
�ιⅼ	Prairie and Great Plains grassland
Ψ	Sage brush, chaparral, etc
●	Sclerophyllous chaparral

0   100   200   300   400   500

Miles

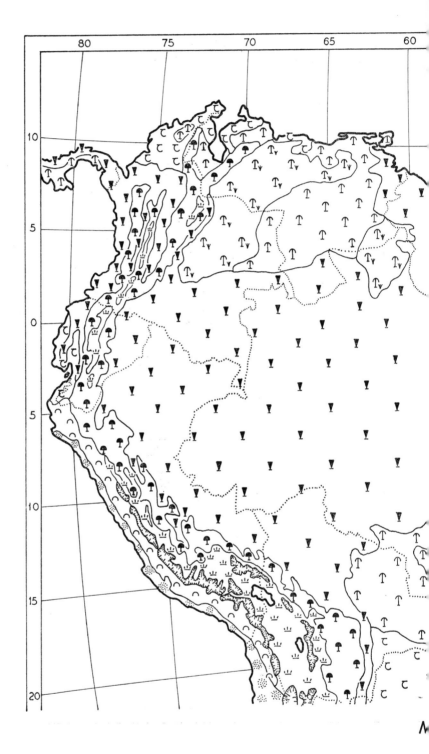

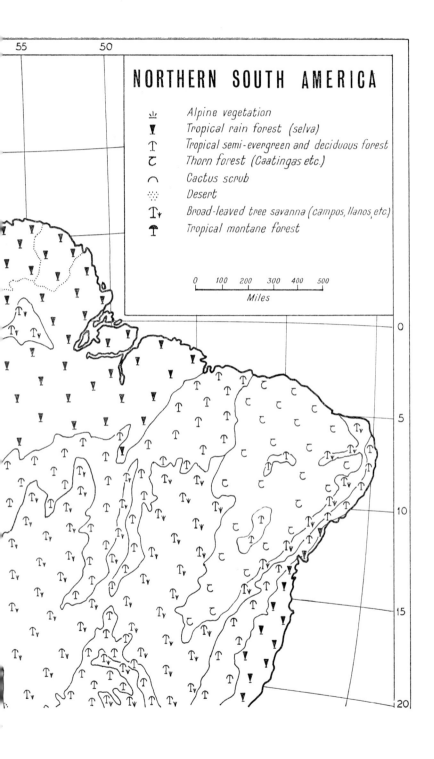

NORTHERN SOUTH AMERICA

⅄	Alpine vegetation
⅄	Tropical rain forest (selva)
⅄	Tropical semi-evergreen and deciduous forest
⅄	Thorn forest (Caatingas etc.)
⌒	Cactus scrub
⁖	Desert
⅄	Broad-leaved tree savanna (campos, llanos, etc.)
⅄	Tropical montane forest

0   100   200   300   400   500
Miles

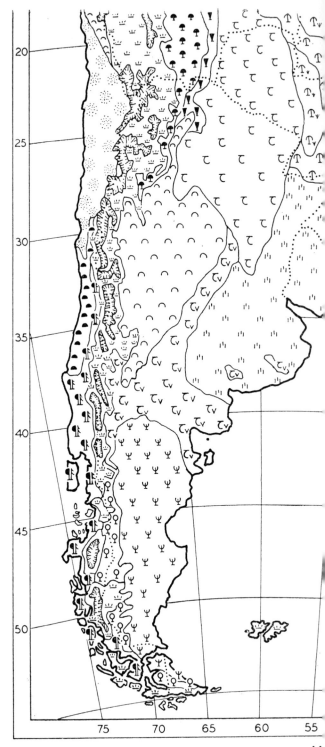

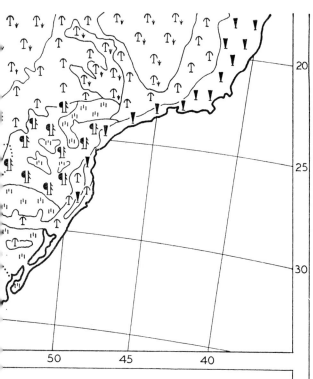

20

25

30

50   45   40

SOUTHERN SOUTH AMERICA

⊻	Alpine vegetation
♀	Deciduous "beech" forest
⫯⨝	Evergreen mixed forest
�	Pampas and other grasslands
Ψ	Patagonian semi-desert scrub
◖	Chilean sclerophyllous scrub
❘	Tropical rain forest
T	Tropical semi-evergreen forest
ζ	Thorn forest
∩	Cactus scrub
⋰	Desert
T⋎	Broad-leaved tree savanna (campos)
ζv	Thorn tree - desert grass savanna
⊤	Tropical montane forest

0   100   200   300   400   500
Miles

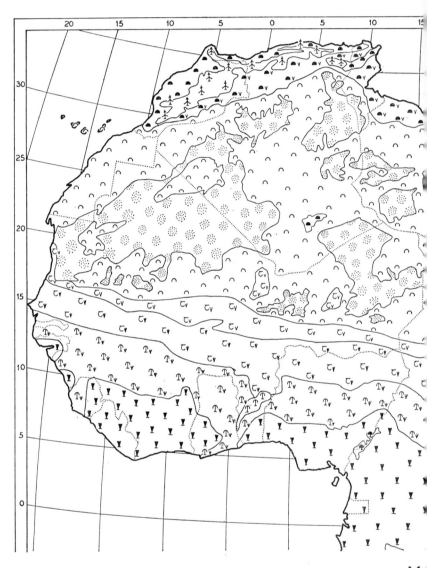

MA

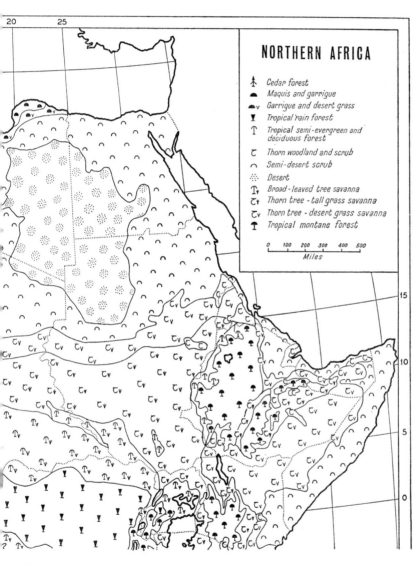

NORTHERN AFRICA

⋏	Cedar forest
♠	Maquis and garrigue
♠ᵥ	Garrigue and desert grass
ⵟ	Tropical rain forest
⋔	Tropical semi-evergreen and deciduous forest
ᴛ	Thorn woodland and scrub
∩	Semi-desert scrub
⋰	Desert
ⵟ₊	Broad-leaved tree savanna
ᴛ₊	Thorn tree - tall grass savanna
ᴛᵥ	Thorn tree - desert grass savanna
ⵟ	Tropical montane forest

0   100   200   300   400   500
Miles

20   25

15

10

5

0

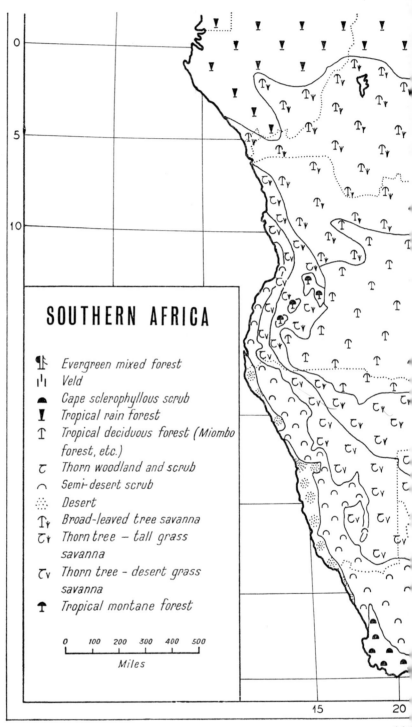

SOUTHERN AFRICA

- 🍃 Evergreen mixed forest
- ⵏ Veld
- ⬤ Cape sclerophyllous scrub
- ⵏ Tropical rain forest
- ⵏ Tropical deciduous forest (Miombo forest, etc.)
- ⵀ Thorn woodland and scrub
- ⌒ Semi-desert scrub
- ⋯ Desert
- ⵏ᷇ Broad-leaved tree savanna
- ⵀ᷇ Thorn tree – tall grass savanna
- ⵀᵥ Thorn tree – desert grass savanna
- ⵏ Tropical montane forest

```
0    100   200   300   400   500
|    |     |     |     |     |
            Miles
```

0

5

10

15   20

MA

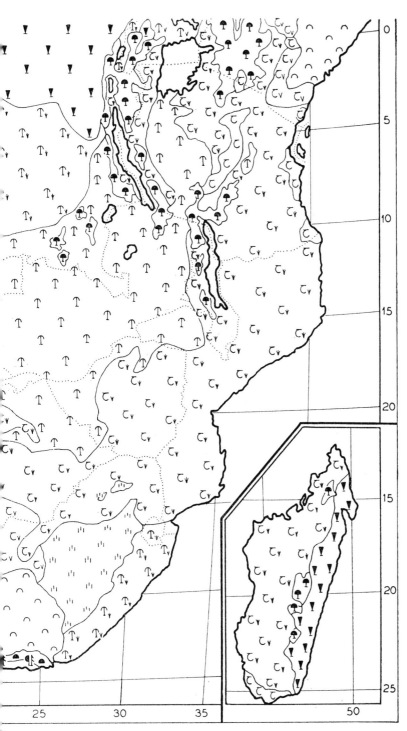

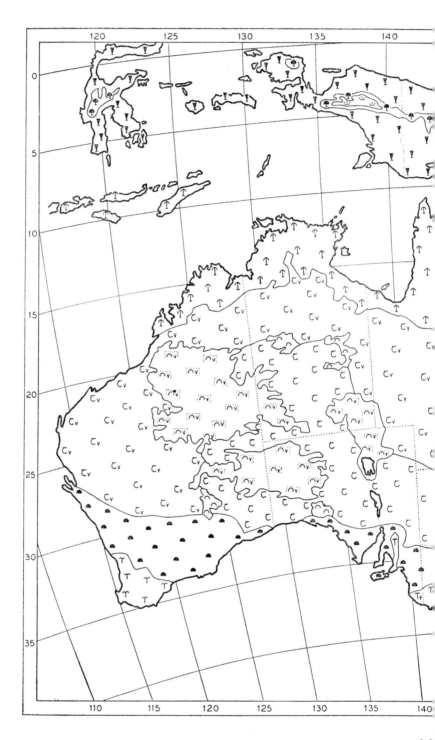

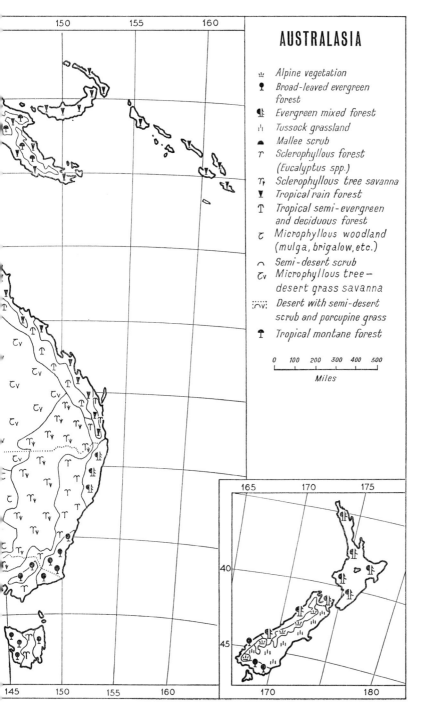

Alpine vegetation
Broad-leaved evergreen forest
Evergreen mixed forest
Tussock grassland
Mallee scrub
Sclerophyllous forest (Eucalyptus spp.)
Sclerophyllous tree savanna
Tropical rain forest
Tropical semi-evergreen and deciduous forest
Microphyllous woodland (mulga, brigalow, etc.)
Semi-desert scrub
Microphyllous tree–desert grass savanna
Desert with semi-desert scrub and porcupine grass
Tropical montane forest

0   100   200   300   400   500

Miles

Climatic Correlations with Vegetation

A. CLIMATIC RÉGIMES IN EXTRA-TROPICAL REGIONS: PRECIPITATION (1), MEAN DAILY MAXIMA (2) AND MEAN DAILY MINIMA (3)

(Temperatures in °F. and precipitation in inches)

		Jan.	Feb.	Mar.	Apr.	May	Jun.	Jul.	Aug.	Sep.	Oct.	Nov.	Dec.	Ann.
TUNDRA														
Baker Lake	(1)	0·2	0·2	0·3	0·3	0·2	0·7	0·9	1·1	0·8	0·5	0·3	0·3	5·8
(64° N. 96° W.)	(2)	−24	−22	−7	10	29	43	60	59	43	24	5	−11	
	(3)	−38	−31	−20	−6	16	30	39	41	30	9	−12	−24	
Thule	(1)	0·1	0·1	0·1	0·1	0·1	0·2	0·5	0·5	0·4	0·1	0·1	0·2	2·5
(77° N. 69° W.)	(2)	1	−4	−2	9	28	41	46	43	33	23	13	−1	
	(3)	−16	−21	−19	−10	16	30	36	33	21	8	−3	−17	
Deception Island	(1)	2·3	2·1	2·7	2·0	0·2	0·3	0·6	1·0	0·9	4·3	3·8	2·0	22·2
(63° S. 61° W.)	(2)	37	37	33	31	27	23	21	22	26	31	32	36	
	(3)	31	31	29	24	19	14	11	12	17	23	25	30	
BOREAL FOREST														
Nitchequon	(1)	1·3	1·4	1·7	1·5	2·7	3·7	4·1	4·1	3·1	3·1	2·4	1·7	30·8
(53° N. 71° W.)	(2)	0	5	18	32	44	58	65	62	52	40	23	7	
	(3)	−21	−18	−6	10	27	40	48	47	39	28	11	−9	
Surgut	(1)	0·5	0·4	0·4	0·4	1·3	2·5	2·5	2·9	1·9	1·3	0·8	0·7	15·6
(61° N. 73° E.)	(2)	−9	1	15	29	43	58	67	63	51	31	8	−4	
	(3)	−19	−11	−2	13	31	46	54	51	40	23	−1	−15	
Verkhoyansk	(1)	0·2	0·2	0·1	0·2	0·3	0·9	1·1	1·0	0·5	0·3	0·3	0·2	5·3
(68° N. 134° E.)	(2)	−54	−41	−13	19	42	60	66	58	43	12	−31	−52	
	(3)	−63	−56	−39	−10	23	48	47	40	27	−3	−40	−56	
DECIDUOUS SUMMER FOREST														
Oxford	(1)	2·4	1·8	1·5	1·8	2·1	1·7	2·3	2·2	2·2	2·5	2·7	2·2	25·4
(52° N. 1° W.)	(2)	45	46	51	56	62	68	71	71	66	58	50	45	
	(3)	35	34	36	40	45	50	54	53	50	44	38	35	
Penza	(1)	1·0	0·9	0·8	1·1	1·8	2·3	2·1	1·8	1·9	1·8	1·6	1·2	18·3
(53° N. 45° E.)	(2)	11	16	26	45	64	71	76	73	59	46	27	16	
	(3)	3	6	14	31	46	54	57	54	44	33	21	12	
Nashville	(1)	4·6	4·1	5·1	4·3	3·8	4·1	4·0	3·6	3·4	2·6	3·5	4·0	47·1
(36° N. 88° W.)	(2)	47	50	59	69	78	86	89	88	82	72	58	49	
	(2)	31	33	40	49	58	67	70	68	62	50	40	33	

		Jan.	Feb.	Mar.	Apr.	May	Jun.	Jul.	Aug.	Sep.	Oct.	Nov.	Dec.	Ann.
COAST AND LAKE FOREST														
Sitka	(1)	7·8	6·7	6·1	5·5	4·2	3·3	4·3	7·2	10·4	12·8	10·2	9·1	87·4
(57° N. 135° W.)	(2)	38	40	43	48	54	59	61	62	59	52	45	40	
	(3)	27	28	30	34	39	44	49	49	45	40	33	30	
Eureka	(1)	7·1	6·5	5·2	3·3	1·8	0·7	0·1	0·2	1·0	2·3	5·2	6·3	39·7
(41° N. 124° W.)	(2)	53	53	54	56	57	59	60	60	61	60	57	54	
	(3)	41	41	43	44	47	50	51	52	51	48	45	42	
Escanaba	(1)	1·4	1·4	1·9	2·2	3·0	3·3	3·3	3·3	3·3	2·7	2·1	1·6	29·5
(46° N. 87° W.)	(2)	23	24	33	46	59	70	76	73	65	53	40	28	
	(3)	7	6	16	30	42	52	58	55	49	39	27	16	
SOUTHERN HEMISPHERE EVERGREEN MIXED FOREST														
Port Elizabeth	(1)	1·2	1·3	1·9	1·8	2·4	1·8	1·9	2·0	2·3	2·2	2·2	1·7	22·7
(34° S. 26° E.)	(2)	78	78	76	73	71	68	67	68	68	70	72	75	
	(3)	61	62	60	55	50	45	45	47	50	54	57	59	
Valdivia	(1)	2·6	2·9	5·2	9·2	14·2	17·7	15·5	12·9	8·2	5·0	4·9	4·1	102·4
(40° S. 73° W.)	(2)	73	73	69	62	56	52	52	54	58	63	65	69	
	(3)	52	51	49	46	43	42	41	40	41	44	46	50	
Auckland	(1)	3·1	3·7	3·2	3·8	5·0	5·4	5·7	4·6	4·0	4·0	3·5	3·1	49·1
(37° S. 175° E.)	(2)	73	73	71	67	62	58	56	58	60	63	66	70	
	(3)	60	60	59	56	51	48	46	46	49	52	54	57	
BROAD-LEAVED EVERGREEN RAIN FOREST														
Hokitika	(1)	10·3	7·5	9·4	9·3	9·6	9·1	8·6	9·4	8·9	11·5	10·5	10·3	114·4
(43° S. 171° E.)	(2)	66	67	65	61	57	53	53	54	56	59	61	64	
	(3)	53	53	51	47	42	38	37	38	42	46	48	51	
Gabo Island	(1)	2·8	2·7	2·9	3·3	3·9	4·2	3·4	2·8	3·0	3·1	2·6	2·4	37·1
(38° S. 145° E.)	(2)	70	71	70	66	62	59	57	60	61	63	65	68	
	(3)	60	61	60	56	52	48	47	47	49	52	55	58	
Pensacola	(1)	4·0	4·4	4·8	3·9	3·3	4·6	6·6	7·6	5·7	4·1	3·9	4·5	57·4
(30° N. 87° W.)	(2)	60	62	67	73	80	85	87	87	85	77	68	61	
	(3)	46	48	54	61	67	74	75	75	72	62	53	47	
SCLEROPHYLLOUS SCRUB														
Valletta	(1)	3·3	2·3	1·5	0·8	0·4	0·1	<0·1	0·2	1·3	2·7	3·6	3·9	20·3
(36° N. 15° E.)	(2)	59	59	62	66	71	79	84	85	81	76	68	62	
	(3)	51	51	52	56	61	67	72	73	71	66	59	54	
Los Angeles	(1)	3·1	3·0	2·8	1·0	0·4	0·1	<0·1	<0·1	0·2	0·6	1·2	2·6	15·0
(34° N. 118° W.)	(2)	65	66	67	70	72	76	81	82	81	76	73	67	
	(3)	46	47	48	50	53	56	60	60	58	54	50	47	
Eyre	(1)	0·6	0·5	0·9	0·9	1·6	1·6	1·2	1·3	0·9	0·8	0·7	0·6	11·6
(32° S. 126° E.)	(2)	79	79	78	74	70	65	64	66	69	72	75	77	
	(3)	60	61	58	54	49	45	43	44	46	50	54	57	
LONG GRASS STEPPE														
Akmolinsk	(1)	0·7	0·5	0·5	0·6	1·1	1·7	1·6	1·5	1·0	1·1	0·7	0·6	11·7
(54° N. 71° E.)	(2)	4	8	20	40	64	73	78	74	62	42	23	11	
	(3)	-8	-7	1	24	42	51	55	52	41	27	11	-1	
Kansas City	(1)	1·3	1·7	2·6	3·2	4·9	4·8	4·1	4·1	4·6	2·8	1·9	1·3	37·3
(39° N. 95° W.)	(2)	38	41	53	65	74	83	89	87	80	68	53	41	
	(3)	22	24	34	46	56	65	70	68	60	49	36	27	
Buenos Aires	(1)	3·1	2·8	4·3	3·5	3·0	2·4	2·2	2·4	3·1	3·4	3·3	3·9	37·4
(35° S. 59° W.)	(2)	85	83	79	72	64	57	57	60	64	69	76	82	
	(3)	63	63	60	53	47	41	42	43	46	50	56	61	

		Jan.	Feb.	Mar.	Apr.	May	Jun.	Jul.	Aug.	Sep.	Oct.	Nov.	Dec.	Ann.
SHORT GRASS STEPPE														
Odessa	(1)	1·0	0·7	0·7	1·1	1·1	1·9	1·6	1·4	1·1	1·4	1·1	1·1	14·3
(46° N. 31° E.	(2)	28	31	39	52	67	74	79	78	68	57	43	33	
	(3)	22	26	32	41	55	62	65	65	56	47	35	27	
Pueblo	(1)	0·3	0·6	0·7	1·6	1·6	1·3	2·0	1·8	0·9	0·6	0·4	0·5	12·3
(38° N. 105° W.)	(2)	45	48	56	64	73	84	89	87	80	68	55	46	
	(3)	16	19	27	36	46	54	60	59	50	38	25	18	
SEMI-DESERT SCRUB														
Tyub-Karagan	(1)	0·4	0·1	0·4	0·5	0·7	0·3	0·2	0·6	0·6	0·3	0·3	0·6	5·1
(45° N. 50° E.)	(2)	25	29	38	53	67	77	82	79	69	56	43	33	
	(3)	21	21	27	43	55	65	71	67	57	45	36	26	
Salt Lake City	(1)	1·3	1·5	2·0	2·0	2·0	0·8	0·6	0·8	1·0	1·5	1·4	1·4	16·3
(41° N. 112° W.)	(2)	35	41	51	62	73	82	92	90	79	66	49	40	
	(3)	17	24	31	38	45	52	61	60	49	40	28	22	
Sarmiento	(1)	0·2	0·3	0·3	0·4	0·8	0·8	0·6	0·5	0·4	0·3	0·2	0·3	5·1
(46° S. 69° W.)	(2)	78	77	70	62	54	46	45	51	57	66	70	74	
	(3)	52	51	47	42	36	31	29	33	36	41	46	49	

B. TROPICAL AND SUB-TROPICAL RAINFALL RÉGIMES (INCHES)

	Jan.	Feb.	Mar.	Apr.	May	June	Jul.	Aug.	Sep.	Oct.	Nov.	Dec.	Ann.
TROPICAL RAIN FOREST													
Uaupes (0° S. 67° W.)	10·3	7·7	10·0	10·6	12·0	9·2	8·8	7·2	5·1	6·9	7·2	10·4	105·4
Rio de Janeiro (23° S. 43° W.)	4·9	4·8	5·1	4·2	3·1	2·1	1·6	1·7	2·6	3·1	4·1	5·4	42·6
Cooktown (15° S. 145° E.)	14·4	13·7	15·3	8·8	2·8	2·0	0·9	1·2	0·6	1·0	2·5	6·6	69·8
SEMI-EVERGREEN SEASONAL FOREST													
Caceres (16° S. 58° W.)	10·2	7·4	6·8	3·6	1·7	1·0	0·4	0·5	1·9	4·2	7·2	8·1	53·0
Darwin (12° S. 131° E.)	15·2	12·3	10·0	3·8	0·6	0·1	<0·1	0·1	0·5	2·0	4·7	9·4	58·7
Chiang Mai (19° N. 99° E.)	<0·1	0·4	0·3	1·4	4·8	4·4	8·4	7·6	9·8	3·7	1·2	0·5	42·5
DECIDUOUS SEASONAL FOREST													
Raipur (21° N. 82° E.)	0·4	0·9	0·7	0·6	0·9	9·1	15·0	14·3	7·7	2·2	0·5	0·2	52·5
Lusaka (15° S. 28° E.)	9·1	7·5	5·6	0·7	0·1	<0·1	<0·1	0·0	<0·1	0·4	3·6	5·9	32·9
Kingston (18° N. 77° W.)	0·9	0·6	0·9	1·2	4·0	3·5	1·5	3·6	3·9	7·1	2·9	1·4	31·5
AFRICAN SAVANNA													
Mali (12° N. 12° W.)	<0·1	<0·1	0·3	0·8	4·8	9·3	15·2	18·7	12·9	7·5	0·9	0·1	70·5
Kaduna (11° N. 7° E.)	<0·1	0·1	0·5	2·5	5·9	7·1	8·5	11·9	10·6	2·9	0·1	<0·1	50·1
Juba (5° N. 32° E.)	0·2	0·6	1·3	4·8	5·9	5·3	4·8	5·2	4·2	3·7	1·4	0·7	38·1

	Jan.	Feb.	Mar.	Apr.	May	Jun.	Jul.	Aug.	Sep.	Oct.	Nov.	Dec.	Ann.
MICROPHYLLOUS FOREST													
Caetite (14° S. 43° W.)	4·8	2·9	3·0	3·5	0·6	0·4	0·5	0·3	0·9	2·5	4·6	5·6	29·6
Iringa (8° S. 36° E.)	6·8	5·1	7·1	3·5	0·5	<0·1	<0·1	<0·1	0·1	0·2	1·5	3·5	29·3
Mandalay (22° N. 96° E.)	0·1	0·1	0·2	1·2	5·8	6·3	2·7	4·1	5·4	4·3	2·0	0·4	32·6
SEMI-DESERT SCRUB													
Guaymas (28° N. 111° W.)	0·2	<0·1	0·3	0·2	0·2	0·1	1·7	3·6	2·4	0·4	0·6	1·5	11·1
Mendoza (33° S. 69° W.)	0·9	1·2	1·1	0·5	0·4	0·3	0·2	0·3	0·5	0·7	0·7	0·7	7·5
Mogadishu (2° N. 45° E.)	<0·1	<0·1	<0·1	2·3	2·3	3·8	2·5	1·9	1·0	0·9	1·6	0·5	16·9
PLANTLESS DESERT													
Antofagasta (24° S. 70° W.)	0·0	0·0	0·0	<0·1	<0·1	0·1	0·2	0·1	<0·1	0·1	<0·1	0·0	0·5
Walvis Bay (23° S. 14° E.)	<0·1	0·2	0·3	0·1	0·1	<0·1	<0·1	0·1	<0·1	<0·1	<0·1	<0·1	0·9

Glossary of Technical Terms

Acidophyllous, a term applied by the author to plants which make few demands upon mineral nutrients in the soil and thus possess foliage which is poor in mineral matter.

Anaerobic, appertaining or adapted to environmental conditions devoid of free oxygen.

Arenaceous, composed predominantly of grains of sand.

Argillaceous, composed predominantly of clay particles.

Arresting factor, any factor which, if persistent, maintains vegetation at a subclimax stage.

Association, a distinct part of a plant formation dominated throughout by the same characteristic species.

Azonal soils, soils maintained in a permanently immature state by persistent deposition or truncation.

Base exchange capacity, the total capacity possessed by a soil for holding replaceable cations (usually expressed in milli-equivalents).

Base status, the amount of replaceable basic ions present in a soil.

Biotic complex, the interacting complex of soils, plants and animals which, in response to climatic and other environmental conditions, forms a varied covering over much of the earth.

Bog, a wet area covered by acid peat.

Calcicolous, able to thrive only in a lime-rich soil.

Cambium, a tissue which maintains its powers of cell-division.

Capillarity, the spontaneous process of upward movement of liquids in a porous medium.

Carr, subclimax vegetation, dominated by alder and willow, occupying eutrophic peat.

Clay-humus complex, the linked clay and humus which make up the chemically-active fraction of the soil.

Climatic climax vegetation, the plant communities which, given undisturbed conditions and free drainage, ultimately come into stable equilibrium with climate and soil.

Colloids, substances which can exist in either a flocculated or deflocculated state, dependent upon variations in the chemical or physical state of the environment.

Consociation, a climatic climax community dominated by a single species.

Disclimax, the equivalent term for *plagioclimax* (*vide infra*) in American literature.

Dominant, a species of a plant which is one of the tallest in a plant community.

Ecological status, the position occupied by a plant community within the hierarchy of seral and climax communities.

Ecosystem, see *biotic complex*.

Ecotone, the zone of competition between two distinct plant communities.

Eluviated horizons, horizons from which certain fractions of the clay-humus complex have been removed by either the podzolisation or laterisation processes.

Endemic, a plant (or animal) confined to a particular area either because it evolved there and has not been able to spread more widely, or because it has survived there and nowhere else.

Epiphyte, a plant which passes its life lodged upon the aerial structure of another, without abstracting any water or nutrients from it.

Eutrophic, accumulating in ground-water containing appreciable amounts of mineral bases.

Family, a group of plant (or animal) genera which are thought to have evolved from a common ancestor.

Flora, the sum total of plant species which make up the vegetation of an area.

Formation, a geographically-distinct part of a formation-type.

Formation-type, a world vegetation type dominated throughout by plants of the same life-form.

Genus, a group of species which are thought to have evolved from a common ancestor.

Habitat, the environment of an organism.

Halosere, a series of communities which, in turn, occupy an originally-bare salt marsh.

Humus, organic material derived, by partial decay, from the organs of dead plants.

Hydrosere, a series of communities which, in turn, occupy the bottom of a silting water body.

Hygroscopic nucleus, a tiny particle, suspended in the atmosphere, around which water can condense to form a fog or cloud droplet.

Illuviated horizons, horizons in which re-deposition of leached colloidal materials takes place.

Intrazonal soils, soils which occur anomalously due to peculiarities of drainage or parent-material.

Ion, an electrically-charged particle which arises in solution by the dissociation of a dissolved chemical substance.

Ionization, the splitting of a dissolved chemical substance into positive and negative ions.

Leaching, the removal of substances in solution or suspension by downward percolation.

Liane, a woody climbing plant.

Life-form, a distinctive set of morphological and physiological characteristics possessed by a group of related or unrelated species.

Lithological, appertaining to the physical and chemical properties of rocks.

Lithosere, a series of communities which, in turn, occupy an originally bare rock surface.

Marsh, an area where the water-table is maintained at, or only just beneath, the surface of the ground.

Mesophyte, a plant adapted to a moderately moist habitat.

Micelle, a large 'molecule' or chemical aggregate, often of indeterminate size and composition.

Monopodial, the mode of growth adopted by plants such as conifers whose upward growth is maintained by the terminal bud, flowers being produced on lateral shoots.

Mor, humus which, because of its poverty in bases, has a very acid reaction.

Mull, humus which contains appreciable amounts of mineral bases.

Muskegs, peat bogs within the general area of the North American boreal forest.

Mutation, a genetic change in a species which may cause noticeable changes in its morphology.

Mycorrhiza, a mutually advantageous and intimate association between the root of a flowering plant and a parasitic fungus.

Nutrient cycle, the circulation of mineral nutrients between soil and vegetation.

Oligotrophic, developed entirely in downward-percolating rain-water.

Pannage, the practice of turning swine into the forest to forage for acorns or beech-mast.

Pedalfer, an acid soil which contains no calcium carbonate because of leaching.

Pedocal, a soil which has an alkaline reaction because of the presence of free calcium carbonate.

Pedogenesis, soil development.

Permafrost, permanently frozen sub-soil.

Photosynthesis, the synthesis of carbohydrates from carbon dioxide and water in the green organs of a plant.

pH value, the logarithm of the reciprocal of the hydrogen ion concentration in a solution.

Phylogenetic, concerning the ancestral origins of plants and animals.

Phylum, a major division of the plant (or animal) kingdom all of whose members are thought to have arisen from a common ancestor at some time in the remote past.

Pioneer community, the first community to occupy a fresh, untenanted surface.

Plagioclimax, a climax community which is maintained by continuous human activity of a specific nature.

Plagiosere, a series of communities which, in turn, occupy an area as the climatic climax vegetation is removed by continuous and constant human interference.

Plumule, the first shoot to emerge from a germinating seed.

Prisere, a series of plant communities which, in turn, occupy an originally fresh, untenanted surface.

Psammosere, a series of communities which, in turn, occupy an area of originally unconsolidated sand.

Radicle, the first root to emerge from a germinating seed.

Rhyzome, a laterally-growing, underground stem by which vegetative reproduction is achieved.

Sclerophyllous, possessing leaves which are heavily cutinised.

Seral community, a community which, by its very existence, creates conditions more and more favourable for a more demanding one.

Sere, any series of plant communities which, successively, occupy the same area of ground.

Sesquioxide, an oxide which contains two metallic atoms to every three of oxygen.

Soil-parent-material, the rock material from which the inorganic fraction of the soil has been derived.

Soil skeleton, the chemically-inert fraction of the soil composed predominantly of quartz grains.

Species, a discrete type of plant or animal which 'breeds true' from generation to generation.

Subclimax, a community of less than climatic climax status which is permitted to persist because of the continuous intervention of an arresting factor.

Subsere, a sere which is initiated when an arresting factor is withdrawn.

Sucker, a laterally-growing, subterranean offshoot from the base of the main stem of a plant, by which vegetative reproduction is effected.

Swamp, a vegetated or partially-vegetated area subject to permanent inundation in stagnant or only slowly-flowing water.

Symbiotic, adapted to living in intimate physiological association with another species to the mutual benefit of both.

Taxonomy, plant and animal classification according to evolutionary origin.

Transpiration, the process of water vapour release to the atmosphere from the aerial organs of plants.

Truncated soil, soil which has lost material from the upper parts of its profile by the processes of erosion.

Valence, the unit of chemical affinity which binds atoms together in molecules.

Weathering, the comminution of rock materials by atmospheric agency.

Weathering complex, the veneer of weathering fragments overlying the rock from which it is derived.

Xeromorphic, possessing structural features which provide resistance to drought.

Xerophilous, adapted to withstand drought.

Xerophyte, a plant which is adapted to withstand drought.

Xerosere, a prisere which develops on a surface originally prone to drought,

Zonal soil, the mature soil which develops in conditions of free drainage. given sufficient time and a stable climate.

Bibliography

[1] Adamović, L., 1929, *Die Pflanzenwelt der Adrialänder*.
[2] Adamson, R. S., 1938, *The vegetation of South Africa*.
[3] Aikman, J., 1926, 'Distribution and structure of the forests of eastern Nebraska', *Univ. of Nebraska Studies*, 26.
[4] Beard, J. S., 1946, 'The natural vegetation of Trinidad', *Oxford Forestry Memoirs*, No. 20.
[5] Bews, J. W., 1918, *The grasses and grasslands of South Africa*.
[6] Brass, L. J., 1941, 'The 1938–39 Expedition to the Snow Mountains, Netherlands New Guinea', *Journal of the Arnold Arboretum*, Vol. 22.
[7] Brown, R. H., 1948, *Historical Geography of the United States*.
[8] Bunting, B. T., 1965, *The geography of soil*.
[9] Butland, G. J., 1957, 'The human geography of southern Chile', *Institute of British Geographers*, No. 24.
[10] Clark, J. G. D., 1952, *Prehistoric Europe—the economic basis*.
[11] Cline, M. G., 1949, 'Profile studies of normal soils in New York', *Soil Science*, Vol. 68.
[12] Cockayne, L., 1926, 'Monograph on the New Zealand beech forests', *New Zealand State Forest Service Bulletin*, 4.
[13] ——, 1928, *Vegetation of New Zealand* (2nd Edition).
[14] Cole, Monica M., 1960, 'Cerrado, Caatinga and Pantanal; the distribution and origin of the savanna vegetation of Brazil', *The Geographical Journal*, Vol. 126, Part 2.
[15] Comber, N. M., 1960, *An introduction to the scientific study of soil* (4th Edition, revised by W. N. Townsend).
[16] Conway, Verona, 1947, 'Ringinglow bog, near Sheffield', *Journal of Ecology*, 34.
[17] Cooper, W. S., 1922, 'The broad-sclerophyll vegetation of California', *The Carnegie Institution of Washington*, Publication No. 319.
[18] Cumberland, K. B., 1962, 'Climatic change or cultural interference? New Zealand in Moahunter times' in *Land and livelihood: geographical essays in honour of George Jobberns*.
[19] Dallimore, W. and Jackson, A. B., 1923, *A handbook of Coniferae*.
[20] Darling, F. Fraser, 1956, *Pelican in the wilderness—a naturalist's Odyssey in North America*.
[21] Darwin, Charles R., 1881, *The formation of vegetable mould, through the action of worms, with observations on their habits*.
[22] Dawson, J. W., 1962, 'The New Zealand Lowland Podocarp Forest. Is it subtropical?' *Tuatara*, Vol. IX, No. 3.
[23] Denevan, W. M., 1961, The upland pine forests of Nicaragua, *University of California Publications in Geography*, Vol. 12, No. 4.

[24] Ernst, A., 1908, *The new flora of the volcanic island of Krakatau* (translated by A. C. Seward).

[25] Ewing, J., 1924, 'Plant successions of the brush-prairie in north-west Minnesota', *Journal of Ecology*, 12.

[26] Eyre, S. R., 1955, 'Historical implications of the curving plough-strip', *Agricultural History Review*, 4.

[27] ——, 1957, 'The upward limit of enclosure on the East Moor of north Derbyshire', *Institute of British Geographers*, No. 23.

[28] ——, 1966, 'The vegetation of a south Pennine upland' in *Geography as human ecology*, Eyre, S. R. and Jones, G. R. J. (Eds.).

[29] Faegri, K. and Iversen, J., 1964, *Textbook of pollen analysis*.

[30] Fraser, L. and Vickery, J. W., 'The ecology of the Upper Williams River and Barrington Tops District, II. The rain-forest formations', *Proceedings of the Linnaean Society of New South Wales*, 63.

[31] Glinka, K. D., 1931, *Treatise on soil science*; published for National Science Foundation by Israel Programme for Scientific Translations, Jerusalem, 1963.

[32] Godwin, H., 1956, *The history of the British flora*.

[33] Harris, L. E., 1953, *Vermuyden and the Fens*.

[34] Harrison Matthews, L., 1959, 'Man and the world's fauna', *The Advancement of Science*, Vol. XVI, No. 62.

[35] Hastings, J. R. and Turner, R. M., 1965, *The Changing Mile*.

[36] Hochstetter, F. von, 1867, *New Zealand*, (translated by Edward Santer).

[37] Homans, G. C., 1942, *English villagers of the Thirteenth Century*.

[38] Jackson, J. K., 1956, 'The vegetation of the Imatong Mountains, Sudan', *Journal of Ecology*, 44.

[39] James, Preston E., 1941, *Latin America*.

[40] Jones, E. W., 1959, 'Biological flora of the British Isles—*Quercus*', *Journal of Ecology*, 47.

[41] Jones, G. R. J., 1966, 'Rural settlement in Anglesey' in *Geography as human ecology*, Eyre, S. R. and Jones, G. R. J. (Eds.).

[42] Jones, R. B. and G. E., 1966, in *Nottingham and its Region*, British Association for the Advancement of Science, Edwards K. C. (Ed.).

[43] Keay, R. W. J., 1949, 'An example of Sudan zone vegetation in Nigeria', *Journal of Ecology*, 36.

[44] Keller, B., 1927, 'Vegetation of the plains of European Russia', *Journal of Ecology*, 15.

[45] Köstler, J., 1956, *Silviculture*, (Translation by M. L. Anderson).

[46] Kubiena, W. L., 1953, *The soils of Europe*.

[47] Leach, W., 1925, 'Two relict upland oakwoods in Cumberland', *Journal of Ecology*, 13.

[48] —— and Polunin, N., 1932, 'Observations on the vegetation of Finmark', *Journal of Ecology*, 20.

[49] Leet, D. L. and Judson, S., 1954, *Physical Geology*.

[50] Lewis, F. J. and Moss, E. H., 1928, 'The vegetation of Alberta, II. The swamp, moor and bog forest vegetation of central Alberta', *Journal of Ecology*, 16.

[51] Matthews, J. R., 1955, *Origin and distribution of the British flora*.

[52] Milne, G., 1936, 'A provisional soil map of East Africa', *Amani Memoirs*, 28.

[53] Mohr, E. C. J. and Van Baren, F. A., 1954, *Tropical Soils*.

[54] Morison, C. G. T., Hoyle, A. C. and Hope-Simpson, J. F., 1948, 'Tropical soil-vegetation catenas and mosaics; a study in the south-western part of the Anglo-Egyptian Sudan, *Journal of Ecology*, 35.

[55] Moss, E. H., 1932, 'The vegetation of Alberta, IV. The poplar association and related vegetation of central Alberta', *Journal of Ecology*, 20.

[56] Palmer, J., 1966, 'Landforms, drainage and settlement in the Vale of York' in *Geography as human ecology*, Eyre, S. R. and Jones, G. R. J. (eds.).

[57] Pessin, L. J., 1933, 'Forest associations in the uplands of the lower Gulf Coastal Plain (longleaf pine belt)', *Ecology*, 14.

[58] Polunin, N., 1960, *Introduction to plant geography*.

[59] Richards, P. W., 1952, *The Tropical Rain Forest*.

[60] Robbins, R. G., 1961, 'The montane vegetation of New Guinea', *Tuatara*, Vol. VIII, No. 3.

[61] ——, 1964, 'The montane habitat in the tropics', *I.U.C.N. Publications*, New Series, No. 4.

[62] Robinson, G. W., 1934, 'Soils of Wales', *Empire Journal of Experimental Agriculture*, 3.

[63] ——, 1936, *Soils, their origin, constitution and classification*.

[64] ——, 1937, *Mother Earth—letters on soil*.

[65] Salt, G., 1954, 'A contribution to the ecology of upper Kilimanjaro', *Journal of Ecology*, 42.

[66] Sauer, Carl O., 1952, *Agricultural origins and dispersals*.

[67] Schimper, A. F. W., 1903, *Plant geography upon a physiological basis*, (translated by W. R. Fisher).

[68] Schmieder, O., 1927, 'The Pampa—a natural or culturally induced grassland?', *University of California Publications in Geography*, Vol. 2.

[69] Shantz, H. L. and Marbut, C. F., 1923, *The vegetation and soils of Africa*.

[70] Stamp, L. D. and Lord, L. 1923, 'The ecology of part of the riverine tract of Burma', *Journal of Ecology*, 11.

[71] Stamp, L. D., 1925, *The vegetation of Burma*.

[72] Stebbing, E. P., 1937, *The forests of West Africa and the Sahara*.

[73] Suslov, S. P., 1961, *Physical geography of Asiatic Russia* (edited by J.E. Williams).

[74] Tansley, A. G., 1953, *The British Islands and their vegetation*.

[75] Tharp, B. C., 1926, 'Structure of Texas vegetation east of the 98th meridian', *University of Texas Bulletin 2606*.

[76] Thirsk, Joan, 1953, 'The Isle of Axeholme before Vermuyden', *Agricultural History Review*, Vol. I.

[77] Wardle, P., 1961, 'Biological flora of the British Isles—*Fraxinus excelsior*', *Journal of Ecology*, 49.

[78] Watt, A. S., 1934, 'Vegetation of the Chiltern Hills, with special reference to the beechwoods and their seral relationships', *Journal of Ecology*, 22.

[79] Watt, A. S., 1937, 'Studies in the ecology of Breckland. II. On the origin and development of blowouts', *Journal of Ecology*, 25.

[80] ——, 1940, 'Contributions to the ecology of bracken (*Pteridium aquilinum*). I. The rhyzome', *New Phytologist*, 39.

[81] Weaver, J. E. and Clements, F. E., 1938, *Plant ecology.*

[82] Weaver, J. E. and Albertson, F. W., 1956, *Grasslands of the Great Plains.*

[83] Webb, L. J., 1959, 'A physiological classification of Australian rain forests', *Journal of Ecology*, 47.

[84] Whitford, H. N., 1901, 'The genetic development of the forests of northern Michigan; a study in physiographic ecology', *Botanical Gazette*, 31.

[85] Williams, A. B., 1936, 'The composition and dynamics of a beech-maple climax community', *Ecological Monographs*, 6.

[86] Wright, R. L., 1966, 'Landform inheritance and regional contrasts in the Daly River Basin, Northern Australia', in *Geography as human ecology*, Eyre, S. R. and Jones, G. R. J. (Eds.).

Index